Automotive Engine Repair and Rebuilding

By Chek-Chart,
a Division of
The H.M. Gousha Company

Roger Fennema, *Editor*
Miles Schofield, *Contributing Editor*
Leslie Wiseman, *Managing Editor*

HarperCollins*Publishers*

Acknowledgments

In producing this series of manuals for automobile mechanics, Chek-Chart has made extensive use of the help and knowledge of the car manufacturing and aftermarket industry. It is a highly complex, fast-changing field, and we are grateful for the help provided by the following companies and organizations.

American Motors Corporation
Bear, Specialty Automotive Tools
Black and Decker Manufacturing Co.
Central Tool, K-D Manufacturing
Chrysler Corporation
Clayton Test Equipment
Comptech Machine
Cummins Machine Co., Inc.
DeAnza College
ESNA
Ethyl Corporation
Federal-Mogul Corporation
Fel-Pro, Inc.
Ford Motor Company
General Motors Corporation
Goodson
Hand Tool Institute
Iskendarian Racing Cams
K-Line Industries
Lisle Corporation
MAC Tools
Marquette Manufacturing Co., A Division of Applied Power Inc.
Mercedes-Benz
Mitsubishi Corporation
Neway Manufacturing, Inc.
Perfect Circle, a Division of the Dana Corporation
Peterson Machine Tool, Inc.
Sealed Power Corp.
Silver Seal
Sioux Tools, Inc.
Snap-On Tools Corp.
The L.S. Starrett Company
Storm-Vulcan, Inc.
South Bay Suzuki
Sun Electric Corporation
Sunnen Products Company
Super Flow
Tobin-Arp Co.
TRW, Inc.
Toyota Motor Co.
Winona Van Norman Machine Shop
Wright Way Machine Shop

The authors have made every effort to ensure that the material in this book is as accurate and up-to-date as possible. However, neither Chek-Chart nor HarperCollins*Publishers* nor any related companies can be held responsible for mistakes or omissions, or for changes in procedures or specifications by the carmakers or suppliers.

This book was reviewed by a group of men selected because of their vast experience in automotive education and technical writing. These consultants were:

Don Nilson, Chabot College, Hayward, CA.
Gary Lewis, DeAnza College, Cupertino, CA.
Ken Layne, Applied Materials, Santa Clara, CA.

Their suggestions and thoughtful comments are an invaluable part of the book you are reading.

At Chek-Chart, MaryAnn Selkirk, Diane Clare, and Ray Lyons participated in the production of this book, under the direction of Elmer Thompson. Special editorial contributions were given by John D. Decker and Josephine Phelps. Original art was produced by Janet Jamieson, Jim Geddes, F.J. Zienty, Karen von Felten, and Elyse Harada. Art coordinator is Gerald McEwan. The entire project is under the general direction of Robert J. Mahaffay.

AUTOMOTIVE ENGINE REPAIR AND REBUILDING, *Classroom Manual* and *Shop Manual* Copyright © 1982, by Chek-Chart, a Division of The H. M. Gousha Company.

All rights reserved. Printed in the United States of America. No part of this publication may be reproduced, stored in a retrieval system, or duplicated in any manner without the prior written consent of Chek-Chart, a Division of The H. M. Gousha Company, P.O. Box 6227, San Jose, CA 95150.

Library of Congress Cataloging and Publication Data:

Chek-Chart, 1982
   Automotive Engine Repair and Rebuilding
   (HarperCollins*Publishers*, Chek-Chart Automotive Series)
Contents
v. 1. Classroom manual.   v. 2. Shop manual.

ISBN: 0-06-454008-1 (set)
Library of Congress Catalog Card No. 80-17091

Contents

INTRODUCTION IV

HOW TO USE THIS BOOK V

PART ONE — AUTOMOTIVE ENGINE FUNDAMENTALS 1

Chapter 1 — Engine Operation and Construction 2

Engine Operation 2

Major Engine Parts 6

Engine Displacement and Compression Ratio 15

Firing Order 22

Engine — Ignition Synchronization 23

Crankshaft Position 24

Basic Timing 25

Review Question 26

Chapter 2 — Engine Physics and Chemistry — Alternative Engines 27

Engine Physics 27

Chemistry and Combustion 30

Engine Force, Inertia, and Momentum 32

Engine Torque and Horsepower 32

Metric Power and Torque 35

Engine Operation and Pressure 36

Airflow Requirements 39

Air-Fuel Ratios 40

Introduction to Emission Controls 42

Alternative Engines 45

Review Questions 50

Chapter 3 — Basic Metallurgy and Machine Processes 51

Engine Metallurgy 51

Manufacturing Processes 55

Material Treatments 57

Basic Machine Processes 58

Materials and Machining Relationships 63

Review Questions 64

On The Cover:
Front — Sunnen Product Company's CK-10 Automatic Cylinder Resizing Machine.
Back — Hydraulic Surface Grinder from Storm-Vulcan.

Chapter 4 — Cooling and Exhaust Systems 65

Cooling System Function 65

Engine Coolant 67

Cooling System Components 67

Engine Temperature Effects on Performance, Economy, and Emissions 74

Exhaust System Types 75

Exhaust System Components 76

Review Questions 78

PART TWO — ENGINE CONSTRUCTION AND OPERATION 79

Chapter 5 — Cylinder Blocks, Heads, Manifolds, and Miscellaneous Engine Covers 80

Materials and Construction Processes 80

Cylinder Blocks 80

Cylinders and Sleeves 82

Cylinder Heads 85

Manifolds 91

Review Questions 97

Chapter 6 — Valves, Springs, Guides, and Seats 98

Valves 98

Springs 106

Guides 110

Valves Seats 112

Valve, Spring, Guide, and Seat Relationships 113

Review Questions 113

Chapter 7 — Camshafts, Lifters or Followers, Pushrods, and Rocker Arms 114

Camshafts 114

Valve Lifters, Tappets, and Cam Followers 123

Pushrods 126

Rocker Arms 126

Engine Balancing Shafts 128

Review Questions 130

Chapter 8 — Crankshafts, Flywheels, Vibration Dampers, Pistons, and Rods 131

Crankshafts 131

Flywheels and Flexplates 137

Vibration Dampers 137

Connecting Rods 138

Pistons 142

Piston Pins 148

Crankshaft, Rod, Piston, and Pin Relationships 152

Engine Balance 153

Review Questions 155

Chapter 9 — Engine Bearings 156

Bearing Types 156

Bearing Materials 158

Bearing Construction and Installation 161

Oil Clearance 162

Undersize and Oversize Bearings 164

Review Questions 166

Chapter 10 — Engine Lubrication and Ventilation 167

Purposes of Motor Oil 167

Motor Oil Composition and Additives 167

Motor Oil Designations 168

Engine Oiling System and Pressure Requirements 171

Lubrication Effects on Performance, Economy, and Emission Control 176

Review Questions 178

Chapter 11 — Gaskets, Fasteners, Seals, and Sealants 179

Gaskets 179

Seals 182

Sealants and Cements 184

Fasteners 186

Contents

Engine Mounts and Shock Absorbers 191
Review Questions 194

PART THREE — ENGINE SERVICE TOOLS AND EQUIPMENT 195

Chapter 12 — Tools and Precision Measuring Instruments 196
Hand Tools 196
Precison Measuring Instruments 204
Power Tools 210
Review Questions 214

Chapter 13 — Engine Test Equipment 215
The Scope for Engine Condition Diagnosis 215
Power Balance Test 216
Compression Test 217
Vacuum Gauge 219
Stethoscope 220
Tachometer and Dwell Meter 220
Bearing Leakage Tester or Engine Prelubricator 221
Oil Pressure Gauge 222
The Infrared Exhaust Analyzer 222
Engine Temperature Gauges 225
Review Questions 227

Chapter 14 — Cleaning and Inspection Equipment 228
Cleaning Equipment 228
Inspection Equipment 232
Review Questions 234

Chapter 15 — Head and Valve Service Equipment 235
Valve Seat Grinders 235
Valve Seat Cutters 235
Valve Grinding Machines 235
Valve Guide Renewing 237
Seat Insertion Tools or Machines 239

Tools for Positive Valve Seat Installation 241
Rocker Stud Pullers 241
Head Surfacing Machines 242
Spotfacers For Enlarging Spring Seats 243
Spring Tension Testers 243
Review Questions 245

Chapter 16 — Block, Crankshaft, Piston, and Rod Service Equipment 246
Block Service Equipment 246
Crankshaft Service Equipment 249
Piston and Rod Service Equipment 251
Flywheel Service Equipment 255
Review Questions 257

NIASE SAMPLE CERTIFICATION TEST 258

GLOSSARY OF TECHNICAL TERMS 260

INDEX 266

ANSWERS TO REVIEW AND NIASE QUESTIONS 268

Introduction to Automotive Engine Repair and Rebuilding

Automotive Engine Repair and Rebuilding is part of the Harper & Row/Chek-Chart Automotive Series. The package for each course has two volumes, a *Classroom Manual* and a *Shop Manual*.

Other titles in this series include:
- Automotive Electrical Systems
- Automatic Transmissions
- Fuel Systems and Emission Controls
- Engine Performance Diagnosis and Tune-Up

Each book is written to help the instructor teach students to become excellent professional automotive mechanics. The 2-manual texts are the core of a complete learning system that leads a student from basic theories to actual hands-on experience.

The entire series is job-oriented, especially designed for students who intend to work in the car service profession. A student will be able to use the knowledge gained from these books and from the instructor to get and keep a job. Learning the material and techniques in these volumes is a giant leap toward a satisfying, rewarding career.

The books are divided into *Classroom Manuals* and *Shop Manuals* for an improved presentation of the descriptive information and study lessons, along with the practical testing, repair, and rebuilding procedures. The manuals are to be used together; the descriptive chapters in the *Classroom Manual* correspond to the application chapters in the *Shop Manual*.

Each book is divided into several parts, and each of these parts is complete by itself. Instructors will find the chapters to be complete, readable, and well-thought-out. Students will benefit from the many learning aids included, as well as from the thoroughness of the presentation.

The series was researched and written by the editorial staff of Chek-Chart, and was produced by Harper & Row Publishers. For over 50 years, Chek-Chart has provided car and equipment manufacturers' service specifications to the automotive service field. Chek-Chart's complete automotive research facilities were used extensively to prepare this book.

Because of the comprehensive material, the hundreds of high-quality illustrations, and the inclusion of the latest automotive technology, instructors and students alike will find that these books will keep their value over the years. In fact, they will form the core of the master mechanic's professional library.

How To Use This Book

Why Are There Two Manuals?

This 2-volume set—**Automotive Engine Repair and Rebuilding**—is not like any other textbook you've ever used. It's actually two books, the *Classroom Manual* and the *Shop Manual*.

The *Classroom Manual* will teach you what you need to know about how the automotive engine operates and why the relationship between engine parts is so important. The *Shop Manual* will show you how to do the repairs and the major rebuilding procedures.

The *Classroom Manual* will be valuable in class and at home, for study and for reference. It has text and illustrations that you can use for years to refresh your memory about the basics of engine repair.

In the *Shop Manual*, you will learn the basic repair procedures for the engine and when to rebuild and when to overhaul an engine.

What's In These Manuals?

There are several aids in the *Classroom Manual* that will help you learn the material more easily:
1. The text is broken into short bits for easier understanding and review.
2. Each chapter is fully illustrated with drawings and photographs.
3. Key words in the text are printed in **boldface type** and are defined on the same page and in a glossary at the end of the manual.
4. Review questions are included for each chapter.
5. A brief summary of every chapter helps you to review for the exams.
6. Every few pages you will find short blocks of "nice to know" information.
7. At the back of the *Classroom Manual* there is a sample test, similar to those given for National Institute for Service Excellence (NIASE) certification. Use it to help you study and prepare yourself when you are ready to be certified as an expert in one of several areas of automobile mechanics.

The *Shop Manual* has detailed instructions on overhaul, repair, and rebuilding procedures. These are easy to understand, and many have step-by-step, photo-illustrated explanations to help you through the written procedure.

Where Should I Begin?

If you already know something about a car's basic systems and how to keep them operating at peak performance, you will find that parts of this book are a review. If you are just starting in car repair, then the subjects in these manuals will be new to you.

Your instructor will design a course to take advantage of what you already know, and what facilities and equipment are available to work with. You may be asked to take certain chapters of these manuals out of order. That's fine. The important thing is to really understand each subject before you move on to the next.

Study the vocabulary words in boldface type. Use the review questions to help you understand the material. While reading in the *Classroom Manual*, refer to your *Shop Manual* to relate the descriptive text to the service procedures. And when you are working on actual engines, you can look back to the *Classroom Manual* to keep the basic information fresh in your mind. Working on a complicated piece of equipment such as the modern automotive engine isn't easy. Use the information in the *Classroom Manual*, the procedures in the *Shop Manual*, and the guidance of your instructor to help you.

To repair or rebuild a specific automobile engine, you will need the particular specifications for that engine. These are available in carmaker's shop manuals, or an independent guide such as the **Chek-Chart Car Care Guide**. This unique book, which is updated every year, gives you the complete service instructions and troubleshooting tips for most cars.

PART ONE

Automotive Engine Fundamentals

Chapter One
Engine Operation and Construction

Chapter Two
Engine Physics

Chapter Three
Basic Metallurgy and Machine Processes

Chapter Four
Cooling and Exhaust Systems

Chapter 1

Engine Operation and Construction

This is a book about the repair and rebuilding of modern automobile engines. The automobile engine has been evolving continuously for over 100 years, since Dr. Nikolaus Otto invented the 4-cycle engine in 1876. To know enough about an engine to service it, you must understand its design, construction, and operation.

The first four chapters (Part One) of this *Classroom Manual* summarize engine design, operation and construction. Most of this material is a preview of subjects covered in detail in Parts Two and Three of this book. You may already have studied some of the subjects covered in the first four chapters. If so, these chapters will be a quick review to refresh your basic understanding.

If this material is new to you, study it thoroughly. In any case, the information in Part One of this *Classroom Manual* is a necessary foundation for the detailed descriptions of engine components and service equipment in Parts Two and Three.

ENGINE OPERATION

All engines use some kind of fuel to produce mechanical power. The oldest "engine" known to man is the simple lever. Food "fuels" the muscle pushing the lever to move objects that the muscle alone could never budge. In a similar way, the automotive engine uses fuel to perform work.

It is common to think of the automotive engine as a gasoline engine. Most of the automotive engines in the world are fueled by gasoline. But the correct name for the automotive engine is "**internal combustion engine**." It can be designed to run on any fuel that **vaporizes** easily or on any flammable gas. Engines have even been designed that run on the gas from barnyard manure.

The automotive engine is called an internal combustion engine because the fuel it uses is burned inside the engine. **External combustion** engines burn the fuel outside the engine. An example of an external combustion engine is a steam engine. Fuel is burned to produce heat to make steam, but this burning takes place anywhere from a few feet to several miles away from the engine. Figure 1-1 shows the basic differences between internal and external combustion engines.

The internal combustion engine burns its fuel inside a combustion chamber. One side of this chamber is open to a piston. When the fuel burns, it expands very rapidly and pushes the piston away from the combustion chamber. This basic action of burning fuel expanding and

Engine Operation and Construction

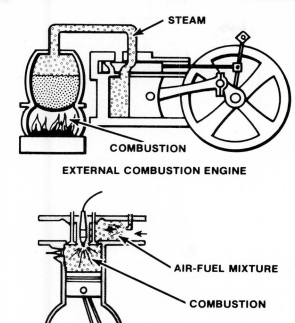

Figure 1-1. The fuel for an internal combustion engine is burned inside the engine. The fuel for an external combustion engine is burned outside the engine.

pushing is the source of power for all internal combustion engines. This includes piston, rotary, and turbine engines.

If fuel burns in the open air, it produces no power. If the same amount of fuel is enclosed and burned, it will expand with some force. To get the most force from the burning of a liquid fuel, it must be vaporized and compressed to a small volume before it is burned. Internal combustion engines are designed to compress the vaporized air-fuel mixture before it is burned.

In a piston engine, the compression is done by a piston in a cylinder. An example of compression can be found in a 2-section mailing tube with metal ends, figure 1-2. Push the inside tube in very quickly, and it will compress the air inside. Release the inside tube quickly and it will fly out. In a similar way, the piston compresses the air-fuel mixture in the cylinder.

When the fuel burns and expands, it produces much more power than was required to compress it. The burning, expanding fuel pushes the piston to the other end of the cylinder.

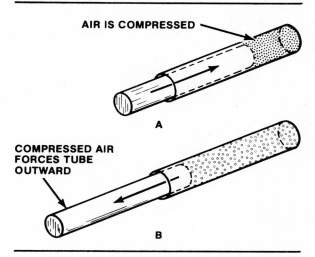

Figure 1-2. Push the inside tube in rapidly, and air is compressed (A). Release the inner tube quickly, and the compressed air forces it out.

■ Air Pressure — High And Low

You can think of an internal combustion engine as a big air pump. As the pistons move up and down in the cylinders, they pump in air and fuel and pump out the exhaust gases. They do this by creating an air pressure differential.

Air has weight and so exerts pressure. As a piston moves down on the intake stroke, it increases the space inside the cylinder that the air must fill. This lowers the air pressure inside the engine. Because the pressure inside the engine is lower than the pressure outside, air flows through the carburetor and into the engine to try to fill the low-pressure area and equalize the pressure.

Although vacuum is technically defined as a space with nothing in it at all, we use the word to describe any low pressure area within the engine. Although you may think of vacuum as sucking air into the engine, it is really the higher pressure of the air outside the engine that forces air into the low-pressure area inside.

Internal Combustion Engine: An engine, such as a gasoline or diesel engine, in which fuel is burned inside the engine.

Vaporize: To change from a solid or liquid into a gaseous state.

External Combustion Engine: An engine, such as a steam engine, in which fuel is burned outside the engine.

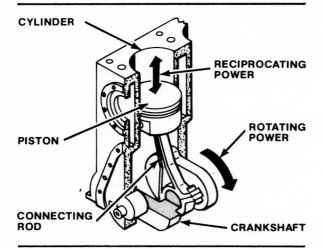

Figure 1-3. Reciprocating motion, developed by combustion force on the piston, is changed to rotary motion by the crankshaft.

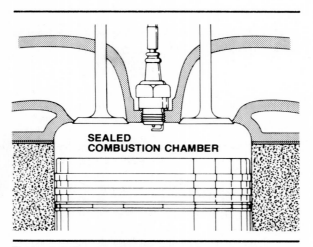

Figure 1-4. For combustion to produce power in an engine, the combustion chamber must be sealed.

Reciprocating Engine

Except for the Wankel rotary engine, all production automotive engines are **reciprocating**, or piston type. Reciprocating means "up and down" or "back and forth." It is the up and down action of a piston in a cylinder that gives the reciprocating engine its name.

Power is produced by the inline motion of a piston in a cylinder. However, this **linear** motion must be changed to rotating motion to turn the wheels of a car or truck. In the following paragraphs we will learn how the pistons, connecting rods, crankshaft, and other parts of an engine produce reciprocating motion and change it to rotating motion, figure 1-3.

Most automobile engines are 4-stroke, spark-ignition, reciprocating engines. Other

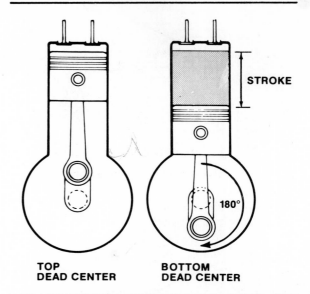

Figure 1-5. One top-to-bottom or bottom-to-top movement of the piston is called a stroke. One piston stroke equals 180 degrees of crankshaft rotation; two strokes equal 360 degrees of crankshaft rotation.

types of engines include 2-stroke and compression-ignition (diesel) engines. Two-stroke engines are used mainly in motorcycles, lawnmowers and other types of equipment. They are seldom used in automobiles. Beyond a basic explanation of their operation, they are not covered in this text. Diesel engines do not use a spark to ignite the air-fuel mixture. The heat from their high compression ignites the fuel. Diesel engine operation is explained later in this chapter. Aside from the differences in ignition and fuel systems, diesel and gasoline engines are physically quite similar. Although this text does not feature diesel engine repair and rebuilding, much of the information on gasoline engines is typical of light-duty diesel engine service.

The Four-Stroke Cycle

Gasoline by itself will not burn. It must be mixed with oxygen (air). In most automotive engines, the gasoline is mixed with air in the carburetor. The air-fuel mixture then goes through passages, or manifolds, to the combustion chamber. To do any useful work, the air-fuel mixture must be compressed and burned in a sealed chamber, figure 1-4. Here, the combustion energy can work on the movable piston to produce mechanical energy.

The air-fuel mixture enters the combustion chamber past an intake valve. The suction or vacuum that pulls the mixture into the cylinder

Engine Operation and Construction

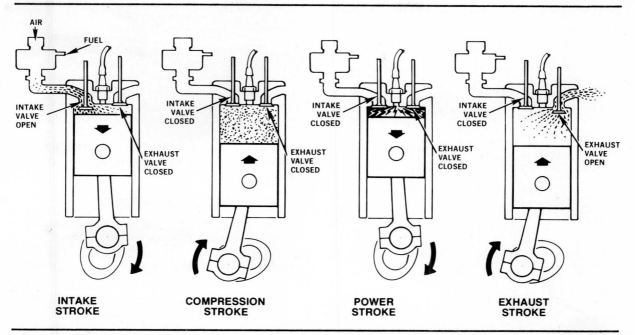

Figure 1-6. The downward movement of the piston draws the air-fuel mixture into the cylinder through the intake valve on the intake stroke. On the compression stroke, the mixture is compressed by the upward movement of the piston with both valves closed. Ignition occurs at the beginning of the power stroke, and combustion drives the piston downward to produce power. On the exhaust stroke, the upward-moving piston forces the burned gases out the open exhaust valve.

is created by the piston. You can create this same suction in the example of the two-piece mailing tube shown in figure 1-2. Draw the assembled tubes apart quickly and let go. The suction tends to pull the tubes back together. If an open intake valve were in the end of the outer tube, pulling the inside tube would draw air past the valve. This suction is known as engine vacuum. Suction exists because of a difference in air pressure between the two areas. We will study vacuum and air pressure in more detail in Chapter 2.

The piston creates a vacuum by moving away from the combustion chamber. The movement of the piston from one end of the cylinder to the other is called a **stroke**, figure 1-5. After the piston reaches the end of the cylinder, it will move back to the other end. As long as the engine is running, the piston continues to move, or stroke, back and forth in the cylinder.

An internal combustion engine must go through four separate actions to complete one operating sequence or cycle.

Depending on the type of reciprocating engine, a complete operating cycle may require either two or four strokes. In the **4-stroke engine**, four strokes of the piston in the cylinder are needed to complete one full operating cycle. Each stroke is named after the action it performs—intake, compression, power, and exhaust—in that order, figure 1-6.

1. Intake stroke: as the piston moves down, the vaporized mixture of fuel and air is drawn into the cylinder past the open intake valve.
2. Compression stroke: the piston returns up, the intake valve closes, the mixture is compressed within the combustion chamber, and ignited by a spark.
3. Power stroke: the expanding gases of combustion force the piston down in the cylinder. The exhaust valve opens near the bottom of the stroke.

Reciprocating Engine: Also called piston engine. An engine in which the pistons move up and down or back and forth as a result of combustion of an air-fuel mixture at one end of the piston cylinder.

Linear: In a straight line.

Stroke: One complete top-to-bottom or bottom-to-top movement of an engine piston.

Four-Stroke Engine: The Otto-cycle engine. An engine in which a piston must complete four strokes to make up one operating cycle. The strokes are: Intake, compression, power and exhaust.

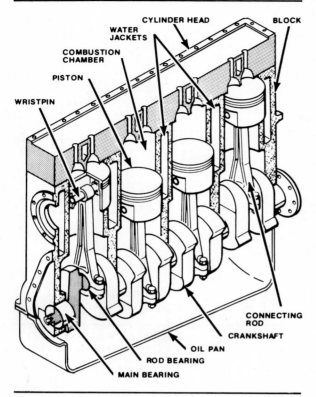

Figure 1-7. Typical engine block and cylinder head assembly.

4. Exhaust stroke: the piston returns up with the exhaust valve open, and the burned gases are pushed out to prepare for the next intake stroke. The intake valve usually opens just before the top of the exhaust stroke.

This 4-stroke cycle is continuously repeated in every cylinder as long as the engine remains running.

Engines that use the 4-stroke sequence are known as 4-stroke cycle engines, or simply 4-stroke engines. This 4-stroke cycle engine is also called the Otto-cycle engine after its inventor, Dr. Nikolaus Otto. A "2-stroke" engine also goes through four actions to complete one operating cycle. However, the intake and compression actions are combined in one stroke, and the power and exhaust actions are combined in the other stroke. The term 2-stroke cycle or 2-stroke is preferred to the term 2-cycle, which is really not accurate.

MAJOR ENGINE PARTS

So far we have discussed pistons and cylinders. It takes many more parts, however, to build a complete engine that will do useful work.

Cylinder Block And Head

Most automobile engines are built upon a cylinder block, or engine block, figure 1-7. The block is usually an iron or aluminum casting that contains the engine cylinders as well as passages for coolant circulation. The top of the block is covered with the cylinder head, which has more coolant passages and which forms most of the combustion chamber. The bottom of the block is covered with an oil pan, or an oil sump.

Connecting Rod And Crankshaft

The piston is attached to one end of a connecting rod by a pin called a piston pin or a wristpin, figure 1-7. The other end of the rod is attached to a crankshaft. As the piston strokes in the cylinder, it rotates the crankshaft. The rotary motion of the crankshaft can be used to turn the wheels of an automobile, rotate the blades of a lawnmower, or turn the propeller of an airplane.

Large bearings, called rod bearings, are used between the connecting rod and the crankshaft. Similar bearings, called main bearings, are used between the crankshaft and the block.

It is important to understand the relationship of the revolving crankshaft to the stroking piston. The piston always makes two strokes for each revolution of the crankshaft. The 4-stroke cycle takes two crankshaft revolutions to complete.

Because only one of the four strokes is a power stroke, the crankshaft must coast for one and one-half revolutions in a single-cylinder engine. It does this because of the inertia of its rotating parts, particularly the flywheel.

Engine Rotation

The front of the engine is always considered to be the end opposite the flywheel. When looking at the front of an engine, the crankshaft and flywheel rotate clockwise. No matter where the engine is situated, front or rear, straight ahead or transversely, the front is always the end opposite the flywheel.

The Flywheel

Because the power in a piston or rotary engine is applied in impulses, the engine tends to jerk or pulse. This tendency is reduced by the flywheel. The flywheel is a large, round, heavy piece of metal attached to the end of the crankshaft, figure 1-8. The flywheel works on

Engine Operation and Construction

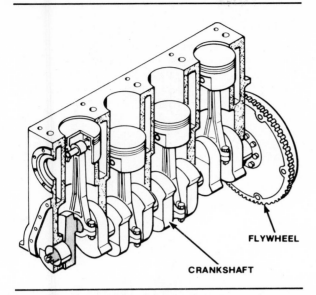

Figure 1-8. The inertia of the flywheel helps to smooth out the impulses of the firing strokes.

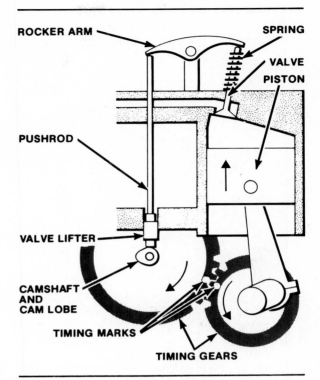

Figure 1-9. Each valve is opened by a lobe on the camshaft and closed by a spring. This simplified cutaway shows the valve lifter, pushrod, and rocker arm arrangement used in an overhead valve engine.

the principle of **inertia**. The inertia of the flywheel resists any change in speed. When there is a power impulse, the heavy flywheel resists a rapid increase in engine speed, and during the coasting period (the exhaust, intake, and compression strokes), the flywheel keeps the engine turning because it also resists a decrease in speed. When an automatic transmission is used, a torque converter is bolted to the flywheel. Because the torque converter is heavy, the flywheel can be much lighter. The flywheel used with a torque converter is often called a "flexplate." The total weight of torque converter and flexplate equals the weight of the flywheel and clutch on a manual transmission engine.

Valve Operation

The opening and closing of the valves in a piston engine is controlled by a camshaft. The camshaft is driven by the engine crankshaft. Lobes on the camshaft push the valves open as the shaft rotates, figure 1-9. A spring closes each valve when the lobe is not holding it open.

Each valve has its own lobe on the camshaft. Once during each revolution of the camshaft, the lobe will push the valve open. The timing of the valve opening is critical to the operation of the engine. Intake valves must be opened just before the beginning of the intake stroke. Exhaust valves must be opened just before the beginning of the exhaust stroke. Because the intake and exhaust valves open only once during every two revolutions of the crankshaft, the camshaft must run at half the crankshaft speed. Study the illustrations to understand the speed relationship between the camshaft and the crankshaft.

Turning the camshaft at half the crankshaft speed is accomplished by using a gear or sprocket on the camshaft that is twice the diameter of the crankshaft gear or sprocket, figure 1-10. If you count the teeth on each gear or sprocket, you will find exactly twice as many on the camshaft as on the crankshaft.

Cylinder And Piston Arrangement

To this point, we have been talking mostly about a single piston in a single cylinder. Actually, an engine can have any number of cylinders. Most car engines have 4, 6, or 8 cylinders, although engines with 5 and 12 cylinders are being built and we are beginning to see a new generation of 3-cylinder engines in commuter cars. Within the engine block, the cylinders are

Inertia: The tendency of an object at rest to remain at rest, and of an object in motion to remain in motion.

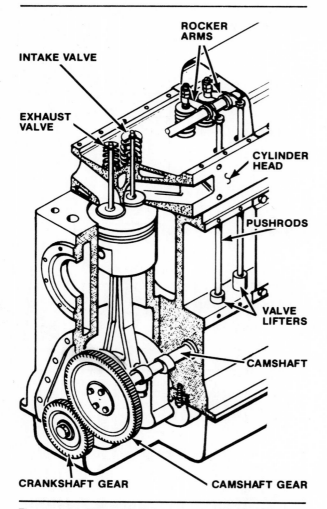

Figure 1-10. The camshaft gear or sprocket has twice as many teeth as the crankshaft gear or sprocket. This causes the camshaft to rotate at one-half crankshaft speed.

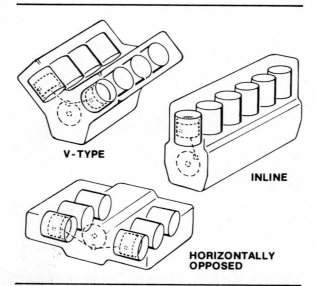

Figure 1-11. Common cylinder arrangements for automotive engines.

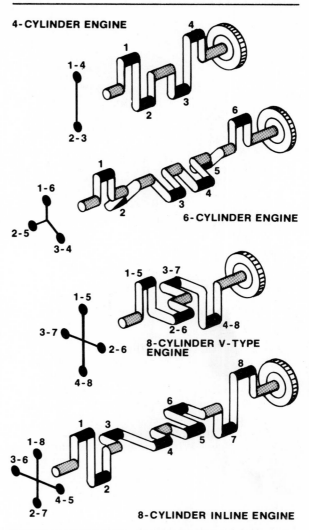

Figure 1-12. The more cylinders an engine has, the closer together the firing impulses are. Here are common crankshaft designs for 4-, 6-, and 8-cylinder engines.

arranged in one of three ways, figure 1-11.
1. Inline engines have a single bank of cylinders arranged in a straight line. Most 4-cylinder and many 6-cylinder engines are of this type. The cylinders do not have to be vertical, as shown in figure 1-11. They can be inclined to either side.
2. V-type engines have two equal banks of cylinders, usually inclined 60 degrees or 90 degrees from each other. Most V-type engines have 6 or 8 cylinders, but V-4 and V-12 engines have been built.
3. Horizontally opposed flat or "pancake" engines have two equal banks of cylinders 180 degrees apart. Air-cooled Volkswagens, Porsches, and Corvairs, and water-cooled Subarus use this design.

Engine Operation and Construction

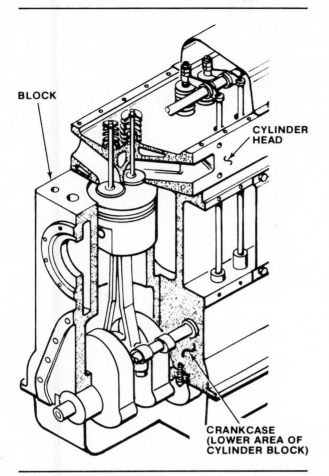

Figure 1-13. The crankcase of most modern engines is the lower portion of the cylinder block casting.

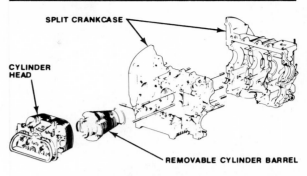

Figure 1-14. The Volkswagen horizontally opposed engine is built with a split crankcase. Individual cylinder castings bolt to the split crankcase, and cylinder heads attach to each pair of cylinders.

When an engine has more than one cylinder, the crankshaft is usually made so that the firing impulses are evenly spaced. Engine speed is measured in revolutions per minute, or in degrees of crankshaft rotation. It takes two crankshaft revolutions or 720 degrees of crankshaft rotation to complete the 4-stroke sequence. If a 4-stroke engine has two cylinders, the firing impulses can be spaced so that there is a power impulse every 360 degrees. On an inline 4-cylinder engine, the crankshaft is designed to provide firing impulses every 180 degrees, figure 1-12. An inline 6-cylinder crankshaft is built to fire every 120 degrees. In an 8-cylinder engine, either inline or V-type, the firing impulses occur every 90 degrees of crankshaft rotation, figure 1-12.

As you can see, the more cylinders an engine has, the closer the firing impulses will be. Therefore, an 8-cylinder engine generally runs smoother than a 4-cylinder engine. Figure 1-12 shows the crankshaft arrangements for some of the most common engine designs. Other arrangements are possible, such as V-4, V-6, opposed 4-cylinder, and several combinations of 2-cylinder layouts. The cylinder arrangement, cylinder numbering order, and crankshaft design all determine the firing order of an engine, which we will study at the end of this chapter.

The Crankcase

All piston engines have a crankcase. It is a housing that supports or encloses the crankshaft, figure 1-13. Early automotive designs used a separate crankcase bolted to the cylinders. Most modern automotive engines use a crankcase that is cast in one piece with the cylinder block. The entire casting of block and crankcase is known as the cylinder block, or simply the engine block. The term crankcase is still sometimes used to describe the open space around the crankshaft.

Some modern automotive engines are still built with separate crankcases and cylinders. An example is the air-cooled Volkswagen engine, figure 1-14. This engine has a 2-piece split crankcase and individual, removable cylinder barrels.

The Valves

All modern automotive piston engines use **poppet valves**, figure 1-15. These are valves that work by linear motion. Most water faucet valves operate with a circular motion. The poppet valve is opened simply by pushing on it. The poppet valve must have a seat on which to rest and from which it closes off a passage-

Poppet Valve: A valve that plugs and unplugs its opening by linear movement.

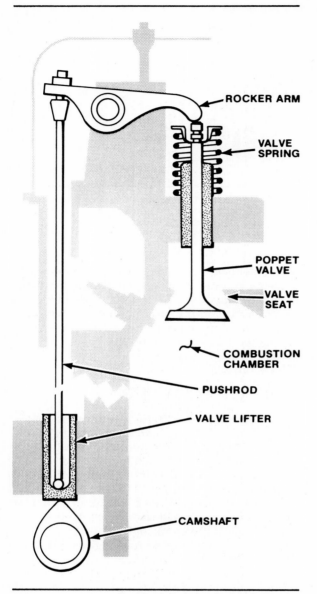

Figure 1-15. Modern automotive engines use poppet valves.

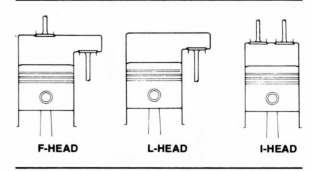

Figure 1-16. Common valve arrangements for 4-stroke automotive engines.

way. There also must be a spring to hold the valve against the seat. In operation, a push on the end of its stem opens the valve, and when the pushing force is removed, the spring closes the valve.

Valve arrangement
Many different locations have been used for the intake and exhaust valves of piston engines. Automotive engines have the valves arranged in one of three ways, figure 1-16:
1. The L-head design had both valves located side-by-side in the engine block. Because the cylinder head is rather flat, containing only the combustion chamber, water jacket, and spark plugs, L-head engines are also called "flatheads." Once very common, this valve arrangement is now obsolete.
2. The F-head design has the intake valve in the cylinder head and the exhaust valve in the engine block, a compromise between the L-head and the overhead valve designs.
3. The I-head or overhead valve design has both the intake and the exhaust valves in the cylinder head. Almost all modern engines are overhead valve design; many also use overhead cams as we will see.

The cylinder head of an overhead-valve engine is heavier and more complicated than that of an L-head engine. It contains not only the valves, but the rocker arms, figure 1-15. Some overhead valve engines also have the camshaft mounted on the head. This is called an overhead-cam engine and will be described later. The camshaft in many overhead-valve engines is in the block. To transfer the push of the cam lobes up to the valve, lifters and long pushrods are used. The pushrods act on one end of the rocker arms. The other end of each rocker arm pushes open the valve. Even though the valve mechanism of an overhead-valve engine is more complicated than that of an L-head, the overhead-valve engine is more efficient because of the compactness of its combustion chamber.

An interesting feature of the overhead-valve engine is that the valves can be placed in several different locations. The head of the valve always forms part of the combustion chamber, but the position and the angle of the valve stem can vary considerably, figure 1-17. Sometimes the stems are parallel to each other. Other designs have the stems separated at various angles. Valves are designed so that their placement creates as little obstruction as possible to the flow of intake and exhaust gases.

Engine Operation and Construction

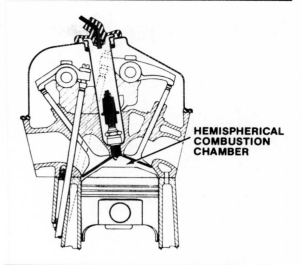

Figure 1-18. The Chrysler Hemi V-8 is one of the best known hemispherical combustion chamber designs.

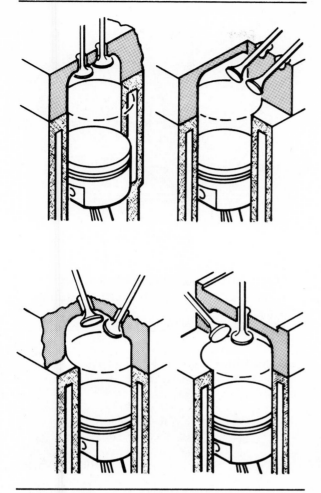

Figure 1-17. The valves can be arranged in the cylinder head at various angles.

Figure 1-19. Ford's Compound Valve Hemispherical engine uses the same elements of the hemi-head engine as the famous Chrysler hemi: the hemispherical shape of the combustion chamber, the compound angled valves and a contoured piston head.

The hemispherical combustion chamber uses very efficient valve locations. Lines drawn through each valve stem intersect beyond the valve heads at about a 60-degree angle, figure 1-18. This design was used by the Chrysler Corporation in its famous "hemi" engine during the 1950's and late 1960's.

The late 1970's and early 1980's saw a resurgence in the hemi-head engine. One example is the Ford Escort 1.6-liter engine, figure 1-19. Ford calls it a compound valve hemispherical engine. Some of the popular imports also introduced hemi-head engines during this time.

Because of the wide angle required between the valve stems on a hemispherical engine, the valve stems are several inches apart. This requires a longer rocker arm and a sturdier mounting. Spark plug and pushrod location complicate the head casting and increase production cost.

Overhead cams
The overhead-valve designs we have talked about so far have had the camshaft in the engine block. When the valves are in the block, as in the L-head engine, it makes sense to have the camshaft in the block close to the valves.

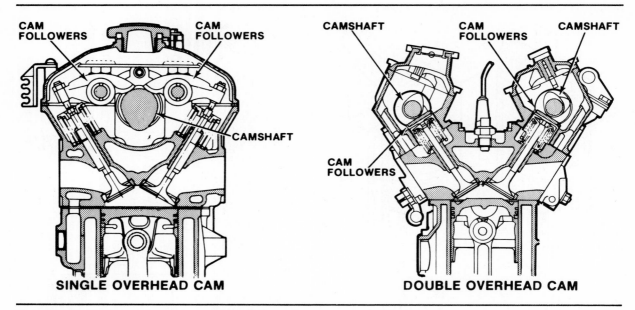

Figure 1-20. Single-overhead-cam and double-overhead-cam comparison.

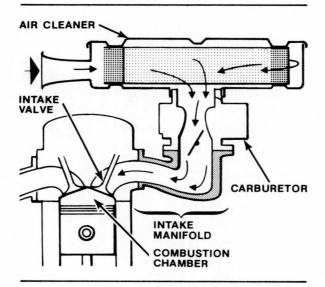

Figure 1-21. The intake manifold has passages that route the air-fuel mixture from the carburetor to the intake valve ports.

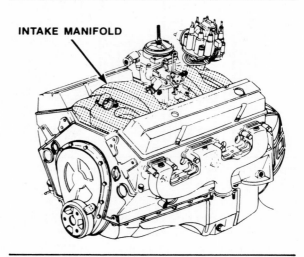

Figure 1-22. V-type engines normally have the intake manifold between the two banks of cylinders. (Chevrolet)

But if the valves are in the cylinder head, having the camshaft in the block is a disadvantage. Pushrods and rocker arms add weight and friction. It is much better to put the camshaft on top of the cylinder head where it can open the valves directly.

Two arrangements are used to do this. If one camshaft is used to operate both the intake and the exhaust valves, it is called a single-overhead-cam (sohc) engine, figure 1-20. If one camshaft is used to open the intake valves and another is used for the exhaust valves, it is called a dual- or double-overhead-cam (dohc) engine, figure 1-20. Some of the highest horsepower engines ever built are the dohc type.

Overhead-cam car engines are usually the sohc type. The 4-cylinder single-overhead-cam engine is widely used. The increased performance of the double-overhead-cam design does not offset the greater complexity and cost of design and manufacture. This means that **the double-overhead-cam design is used only on more expensive cars.**

Engine Operation and Construction

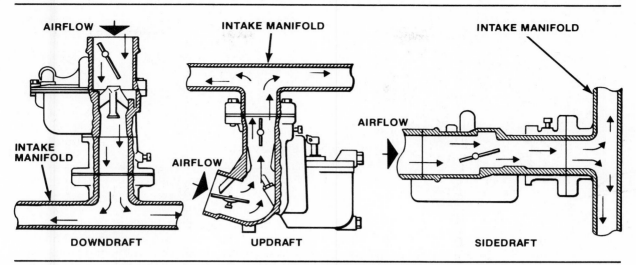

Figure 1-23. Carburetors can be designed as downdraft, updraft, or sidedraft types.

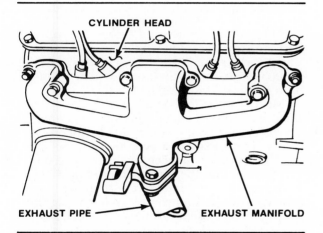

Figure 1-24. The exhaust manifold conducts the exhaust gas from the cylinder head to the exhaust pipe.

Intake And Exhaust Manifolding

Because the piston-engine is an air-breathing engine, there must be passages to route the air-fuel mixture to the intake valve and the exhaust from the exhaust valve. On a carbureted engine, air enters the carburetor and is mixed with fuel. The air-fuel mixture leaves the carburetor and goes through passageways in the intake manifold to passages in the head which lead to the intake valves, figure 1-21.

The intake manifold is usually located in the most convenient place. On a V-type engine, the carburetor is between the two cylinder banks. Underneath the carburetor is the intake manifold, figure 1-22. Overhead-valve inline engines have the intake manifold on the side, but attached to the cylinder head, figure 1-21.

If the carburetor is designed so the air flows through it from top to bottom, it is called a downdraft carburetor, figure 1-23. If the air flows from side to side, it is called a sidedraft carburetor. When the air flows from bottom to top, it is called an updraft carburetor. At one time there were many updraft and sidedraft carburetors, but today almost all carburetors are the downdraft type.

■ Turbocharging Vs. Emissions

Does turbocharging an automotive engine reduce emissions from that engine? Turbocharger manufacturers think so, and point to several closely controlled tests to prove their point. These tests show a 14 percent reduction of HC, 13 percent CO and 8 percent NO_x by simply bolting a turbocharger on a standard production engine. Four reasons explain this effect on exhaust emission:

1. A low-compression engine that runs satisfactorily on low-lead, low-octane gasoline is ideal for turbocharging.
2. The smaller displacement engine is more efficient, burning less fuel and giving off a smaller amount of combustion byproducts under normal use.
3. Turbocharging causes backpressure in the engine's exhaust system during periods of full speed and power. This increases the temperature and pressure of exhaust gases, leading to a more complete burning in the exhaust manifold.
4. Because the intake manifold and carburetor work under pressure instead of vacuum, the turbocharged engine maintains more precise air-fuel mixtures in each engine cylinder.

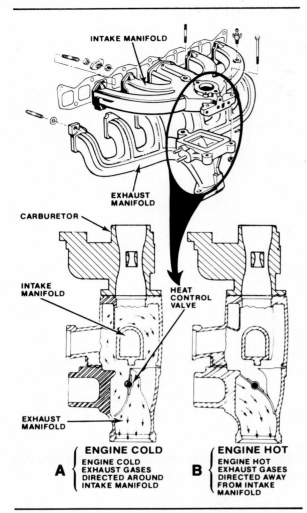

Figure 1-25. The heat control valve forces the exhaust gases to flow either around the intake manifold to heat the carburetor (Position A), or directly out the exhaust system (Position B).

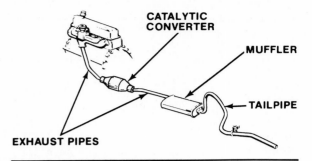

Figure 1-26. Typical exhaust system for a late-model car.

Exhaust passages, also called manifolding, are required to channel the exhaust gases from the engine to the muffler and tailpipe, figure 1-24. The exhaust manifolding is placed wherever it is convenient, with one exception: a portion of the exhaust is used to provide heat to the carburetor, figure 1-25. This helps vaporize the fuel traveling from the carburetor through the intake manifold to the engine.

After leaving the exhaust manifold, the exhaust gases flow into the exhaust pipe, figure 1-26. On late-model cars, the exhaust gases pass through the catalytic converter and then through the muffler and tailpipe.

ENGINE STRUCTURE

We have now covered the basic parts of the engine. At this point, let's review them and examine some of the reasons for the way an engine is put together.

After looking at the illustrations of various engines, it is natural to assume that cylinder heads are always placed on top and crankshafts always on the bottom. This is the conventional way, figure 1-27, but it does not have to be done that way, nor is it always. An engine could be designed with the crankshaft on top, and the cylinder heads and carburetor on the bottom. However, you would have to crawl under the car to service the carburetor, and the air going into the carburetor would be filled with the dust kicked up by the tires.

Most of the components on a modern engine are placed in the same position by every manufacturer. For example, it would be theoretically possible to build an engine with the combustion chamber between the piston and the crankshaft, with the connecting rod passing right through the center of the chamber. Engine design is a compromise. When designing an engine, accessibility and serviceability have to be considered. As a mechanic you can appreciate those two qualities.

Consider the flywheel. Its purpose is to provide inertia. It could, conceivably, be placed at the front of the crankshaft, or even in the middle. But then it would not provide a convenient place for attaching a clutch or torque converter.

Cooling systems usually have the radiator at the front of the car to take advantage of the airflow. But it does not have to be that way either. The radiator could be anywhere there is enough airflow to remove heat from the coolant. The automotive cooling system will be discussed in Chapter 4.

Exhaust systems, like many parts of the car, have also evolved. The best place to get rid of the exhaust from a car is at the rear. But large trucks exhaust straight up in the air and some vehicles exhaust to the side. Exhaust systems will be covered completely in Chapter 4.

At this point, you should have a basic understanding of the internal parts of an engine. It should be clear to you how pistons, cylinders, crankshafts, camshafts, valves, induction

Engine Operation and Construction

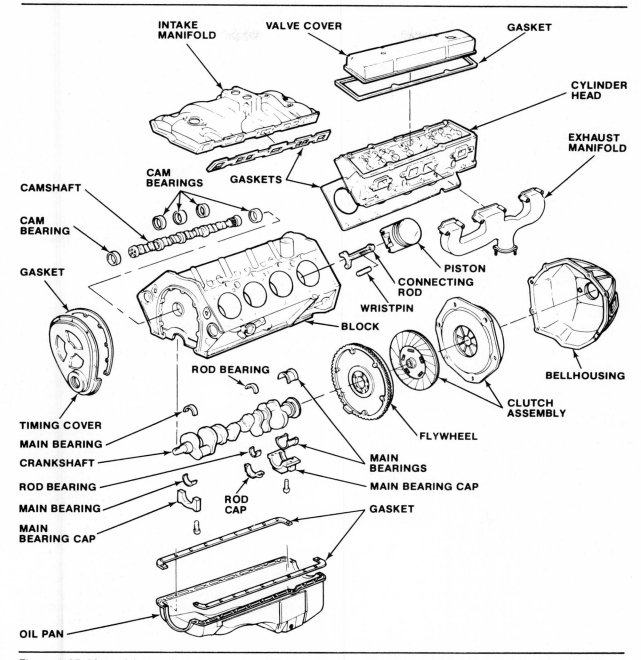

Figure 1-27. Most of the engines you service will probably be built in this conventional manner. (Chevrolet)

and exhaust all work together to form an engine. We cannot overemphasize the importance of knowing thoroughly the basic 4-stroke cycle. It is the foundation upon which the whole engine operates. Knowledge of it is necessary to understand what follows.

ENGINE DISPLACEMENT AND COMPRESSION RATIO

In any discussion of engines, the term "engine size" comes up often. This does not refer to the outside dimension of an engine, but to its **displacement**. As the piston strokes in the cylin-

Displacement: A measurement of the volume occupied by a piston as it moves from the bottom to the top of its stroke. Engine displacement is the piston displacement multiplied by the number of pistons in an engine.

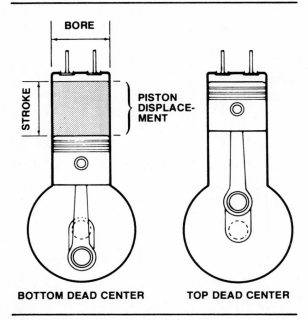

Figure 1-28. Basic engine dimensions.

der, it moves through or displaces a specific volume. Another important engine measurement term is **compression ratio**. Displacement and compression ratio are related to each other, as we will learn in the following paragraphs.

Engine Displacement

This specification is a measurement of engine volume. The number of cylinders is a factor in determining displacement, but the arrangement of cylinders is not. Engine displacement is calculated by multiplying the number of cylinders by the piston displacement of one cylinder. The total engine displacement is the volume displaced by all the pistons.

The displacement of one cylinder is the amount of space through which the piston's top surface moves as it travels from the bottom of its stroke (**bottom dead center**) to the top of its stroke (**top dead center**), figure 1-28. Displacement, then, is the volume displaced in the cylinder by one piston stroke.

We learned earlier that the stroke is one important cylinder dimension. The other important measurement for calculating displacement is the **bore**, figure 1-28. This is the cylinder diameter. Piston displacement is computed as follows:
1. Divide the bore (cylinder diameter) by two. This will give you the radius of the bore.
2. Square the radius (multiply it by itself).
3. Multiply the square of the radius by 3.1416 (pi or π) to find the area of the cylinder cross section.

4. Multiply the area of the cylinder cross section by the length of the stroke.

You now know the piston displacement for one cylinder. Multiply this by the number of cylinders to determine the total engine displacement. The formula for the complete procedure reads: $r^2 \times \pi \times$ stroke $\times$ number of cylinders = displacement. For example, to find the displacement of a 6-cylinder engine with a 3.80-inch bore and a 3.40-inch stroke:

$$\text{radius squared} = \left(\frac{3.80}{2}\right)^2 = 1.9^2 = 3.61 \text{ in.}^2$$

3.61 in.^{2} $\times$ 3.1416 $\times$ 3.40 in. $\times$ 6 = 231 in.^{3}

The displacement is 231 cubic inches. Fractions of an inch are usually not included.

Metric displacement specifications
When stated in English values, displacement is given in cubic inches. The engine's cubic inch displacement is abbreviated as "cu. in." or "cid." When stated in metric values, displacement is given in cubic centimeters (cc) or in liters (one liter equals approximately 1,000 cc). To convert engine displacement specifications from one value to another, use the following formulas:

• To change cubic centimeters to cubic inches, multiply by 0.061 (cc $\times$ 0.061 = cu in.).

• To change cubic inches to cubic centimeters, multiply by 16.39 (cid $\times$ 16.39 = cc).

• To change liters to cubic inches, multiply by 61.02 (liters $\times$ 61.02 = cid).

Our 231-cid engine from the previous example is also a 3,786-cc engine (231 $\times$ 16.39 = 3,786). When expressed in liters, this figure would be rounded to 3.8 liters.

Metric displacement can be calculated directly with the displacement formula, using centimeter measurements instead of inches. Here is how it works for the same engine:

bore = 96.52 mm (9.652 cm)
stroke = 86.36 mm (8.636 cm)

$$\left(\frac{9.652}{2}\right)^2 = 23.29 \text{ cm}^2 \text{ (the radius squared)}$$

23.29 cm^2 $\times$ 3.1416 $\times$ 8.836 cm $\times$ 6 = 3,791 cc

This figure is a few cubic centimeters different from the 3,786-cc displacement we got by converting 231 cubic inches directly to cubic centi-

Engine Operation and Construction

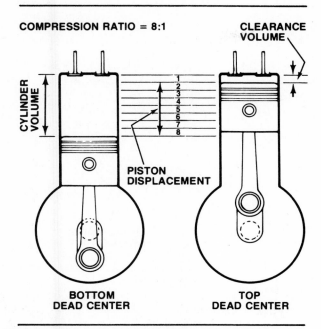

Figure 1-29. Compression ratio is the ratio of the total cylinder volume to the clearance volume.

meters. This is due to rounding. Again, the engine displacement can be rounded to 3.8 liters.

Compression ratio

This specification compares the total cylinder volume when the piston is at bdc to the volume of the combustion chamber when the piston is at tdc, figure 1-29. Total cylinder volume may seem to be the same as piston displacement, but it is not. Total cylinder volume is the piston displacement plus the combustion chamber volume. The combustion chamber volume with the piston at tdc is sometimes called the **clearance volume**.

Compression ratio is the total volume of a cylinder divided by the clearance volume. If the clearance volume is 1/8 of the total cylinder volume, the compression ratio is 8 or 8 to 1. The formula is as follows:

$$\frac{\text{Total volume}}{\text{Clearance volume}} = \text{Compression ratio}$$

To determine the compression ratio of an engine in which each piston displaces 31.12 cu. in. and which has a clearance volume of 4.15 cu. in.:

31.12 + 4.15 = 35.27 (total cylinder volume)

$$\frac{35.27}{4.15} = 8.498$$

The compression ratio is 8.498; this would be rounded and expressed as a compression ratio of 8.5 to 1. This can also be written 8.5:1.

In theory, the higher the compression ratio, the greater the efficiency of the engine, and the more power an engine will develop from a given quantity of fuel. The reason for this is that combustion takes place faster because the fuel molecules are more tightly packed and the flame of combustion travels more rapidly.

■ Nikolaus August Otto (1832 to 1891) — Creator Of The Internal Combustion Engine

In 1860, when a Frenchman, Lenoir, constructed the first gas engine, Nikolaus Otto was working as a traveling salesman. The Lenoir engine was heralded as a great boon to mankind, but could never develop more than 5 horsepower.

Otto made a sketch of a carburetor and had a machinist friend, Michael Zons, build it. Otto attached the carburetor to an engine and while adjusting his "gadget" came upon the idea for the 4-cycle engine. In 1862 he built this feature into a 4-cylinder engine. The bearings and couplings on the engine had to be replaced constantly because of the constant hammering of each explosion.

In 1863, Otto incorporated the idea of atmospheric pressure, borrowed from the steam engine, into his 4-cycle engine. The result was a success and led to the establishment of the first engine-building factory in 1864 in Cologne, Germany.

The engines of Otto and his partner, Eugen Langen sold well, but continued to be noisy and vibrated violently. It was not until 1876 that Otto discovered how to ignite a mixture distributed in layers and to control the previously ungovernable shock against the piston. This was the high compression engine.

The basic engine developed by Otto so many years ago still powers most of the world's automobiles.

Compression Ratio: The total volume of an engine cylinder divided by its clearance volume.

Bottom Dead Center: The exact bottom of a piston stroke. Abbreviated: bdc.

Top Dead Center: The exact top of a piston stroke. Also a specification used when tuning an engine. Abbreviated: tdc.

Bore: The diameter of an engine cylinder; to enlarge or finish the surface of a drilled hole.

Clearance Volume: The volume of a combustion chamber when the piston is at top dead center.

Total Volume: The volume of a combustion chamber when the piston is at bottom dead center.

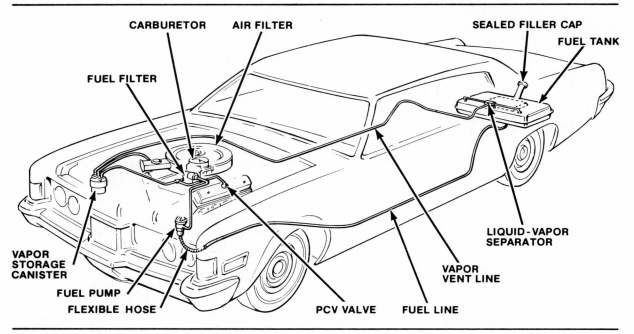

Figure 1-30. Typical automobile fuel system.

But there are practical limits to how high a compression ratio can be. Because of the unavailability of high-octane fuel, most gasoline-burning engines are restricted to a compression ratio no greater than 11.5:1. Ratios this high, however, create high combustion chamber temperatures. This, in turn, creates oxides of nitrogen (NO_x), an air pollutant. Since 1970, compression ratios have been lowered to around 8:1 to reduce NO_x formation.

THE FUEL AND IGNITION SYSTEMS

To understand engine operation completely, you must understand the fuel and ignition systems. The following paragraphs summarize these two major systems and how they relate to overall engine operation. For a complete study of fuel and ignition system operation and service, we recommend the Gousha Chek-Chart book, *Engine Performance Diagnosis and Tune-Up*, also published by Harper & Row.

The Fuel System

Fuel and air must be delivered to the carburetor before the air-fuel mixture can be created. Fuel storage and delivery are the jobs of the fuel system, figure 1-30. The fuel system contains:

- The fuel storage tank
- Fuel delivery lines
- Evaporative emission controls
- The fuel pump
- The fuel filters
- Air cleaners and filters.

Fuel tanks are usually at the rear of the car, underneath the trunk or near the rear axle. On rear engine cars, they are usually toward the front or middle. On pickup trucks, the tank is usually in the cab behind the seat. Fuel is delivered to the engine by a mechanical or electric pump, through hoses and metal lines. Filters are installed in the lines to remove dirt and water.

Fuel pumps can be mounted on the engine, between the fuel tank and the engine, or inside the tank itself. When a pump is mounted outside the tank, it is designed to create a vacuum. Atmospheric pressure then pushes the fuel out of the tank and through the line to the pump. When the fuel reaches the pump, it is forced under pressure to the carburetor. Whenever a fuel pump is mounted outside the tank, it is designed to create vacuum on the suction side and pressure on the pumping side. Most mechanical pumps are mounted on the engine, figure 1-31, and work this way. Pumps that are mounted inside the tank do not need to create vacuum because they are submerged in the fuel. The entire fuel line from the tank to the carburetor is under pressure.

The fuel under pressure is delivered to the carburetor fuel bowl, or float bowl. The fuel enters through a seat and passes around a needle. The needle is pushed into the seat by leverage from a float. When the fuel reaches a preset level, the float rises, pushes the needle into the seat, and shuts off the fuel, figure 1-32. When

Engine Operation and Construction

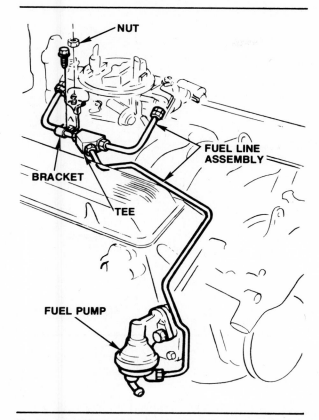

Figure 1-31. Typical fuel pump and line installation on a V-8 engine. (Chevrolet)

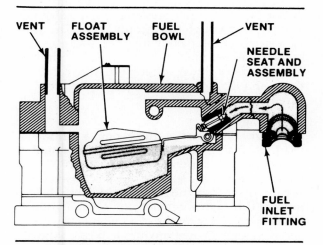

Figure 1-32. Fuel level in the fuel bowl is controlled by the float and needle valve acting against fuel pump pressure.

fuel is consumed by the engine, the float drops and the needle opens the seat to admit more fuel. While the engine is running, the float does not open and close. Instead, it stays in a position that balances the fuel used by the engine with the flow past the needle and seat.

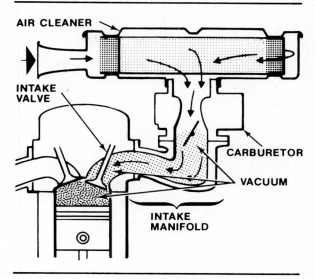

Figure 1-33. Vacuum, created by the piston moving downward, draws the air-fuel mixture from the manifold, through the intake valve, and into the cylinder.

Air and fuel mixing

The carburetor is a device that mixes gasoline with air. Engine vacuum then pulls the air-fuel mixture into the combustion chamber so it can be burned. The mixing must be done to fairly close limits. If there is too much fuel relative to the amount of air, the mixture is said to be rich. If there is too much air, the mixture is lean. The best overall mixture is about 14.7 pounds of air for each pound of fuel. This is a ratio of 14.7:1. For maximum power, a ratio closer to 12:1 works better. To prevent gasoline from passing through the engine unburned and adding hydrocarbons to the atmosphere, a ratio of 15:1 or higher is best.

Outside of these limits, an ordinary engine will not run well. The term "flooded" refers to an engine that will not run because it has too much gas for the amount of air. When an engine runs out of gas, it does not stop because there is no gas in the carburetor, but because there is not enough to maintain the correct air-fuel ratio.

Air-fuel distribution

After the fuel is mixed with air in the carburetor, the mixture passes through the intake manifold to the intake valve of each cylinder. Vacuum is created by the piston on the intake stroke. When the intake valve opens, this vacuum draws in the air-fuel mixture from the intake manifold, figure 1-33.

The scientific explanation of vacuum does not recognize suction. To the scientist or engineer there is no such thing as suction, only

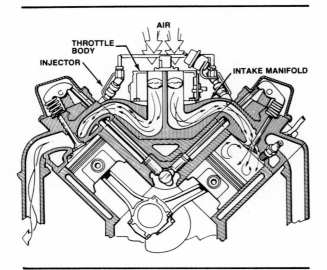

Figure 1-34. With direct-port fuel injection, individual injectors for each cylinder are mounted in the intake manifold. Only air flows through the throttle body. (Cadillac)

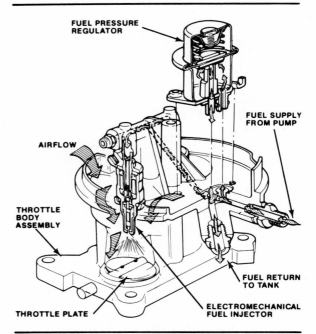

Figure 1-35. With throttle body fuel injection, injectors are mounted on the throttle body. Both air and fuel pass through the throttle plates.

differences in air pressure. When a partial vacuum is created by the piston, this lowers the pressure in the combustion chamber. The outside air pressure, being higher, pushes the air through the carburetor and intake manifold. Air pressure and vacuum are covered more fully in Chapter 2.

Carburetors and intake manifolds do not do a perfect job of distributing the air-fuel mixture. If there is only one carburetor for several cylinders, it is difficult to have the carburetor an equal distance from all cylinders. In the longer intake manifold passages, the fuel has a tendency to fall out of the airstream. Bends in the passageways tend to slow down the flow or cause pockets of fuel to collect.

Multiple carburetors or fuel injection helps to eliminate some of the compromises in carburetor and manifold design. Modern gasoline fuel injection systems are often electronically controlled and are usually one of two types:

- Direct port injection
- Throttle body injection.

With direct port injection, individual injection nozzles for each cylinder are located in the ends of the intake manifold passages, near the intake valves, figure 1-34. Only air flows through the manifold up to the location of the injectors. At that point a precise amount of gasoline is injected into the airflow and vaporizes immediately before it passes through the valve into the cylinder.

With throttle body fuel injection, one or two injection nozzles are located in a throttle body on top of the intake manifold, figure 1-35. The throttle body looks like the lower part of a carburetor, but does not have a fuel bowl or any of the fuel metering circuits of a carburetor. Fuel is injected under pressure into the airflow as it passes through the throttle body. The advantages of throttle body injection over a carburetor are that fuel amounts can be controlled more precisely under different driving conditions, and that injecting the fuel under pressure improves vaporization.

Throttling

To control the speed of the engine, a **throttle** is used. It consists of a butterfly valve mounted in the base of the carburetor. On most carburetors, when the throttle valve is closed tightly, no air-fuel mixture can pass into the cylinders. The engine will not run. Opening the throttle just a few tenths of a millimeter will allow enough air-fuel mixture to pass so the engine can idle slowly, figure 1-36. If the throttle is opened farther, more mixture will pass, and the engine will run faster. The throttle is connected by linkage to the accelerator, or gas pedal. When the driver takes his foot off the pedal, a spring pulls the throttle back to the idle position.

Because of the throttle, vacuum exists in the intake manifold. The highest vacuum is during deceleration when the weight of the car is turn-

Engine Operation and Construction

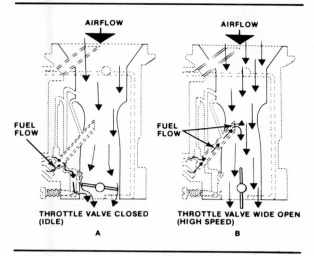

Figure 1-36. The carburetor throttle valve controls engine speed and power by regulating the amount of air and fuel entering the engine.

ing the engine fast with a closed throttle. During idle, the vacuum is lower than deceleration. The lowest vacuum occurs when the throttle is opened to make the car accelerate. At wide-open throttle there is little or no vacuum in the manifold.

The Ignition System

After the air-fuel mixture is drawn into the cylinder and compressed, it must be ignited. The ignition system creates a high electrical potential or voltage. This voltage jumps a gap between two electrodes in the combustion chamber. The arc (spark) between the electrodes ignites the compressed mixture. The two electrodes are part of the spark plug, which is a major part of the ignition system.

The timing of the spark is critical to proper engine operation. It must occur near the start of the power stroke. If the spark occurs too early or too late, full power will not be obtained from the burning air-fuel mixture.

As we have seen, every two strokes of a piston rotate the crankshaft 360 degrees, and there are 720 degrees of rotation in a complete 4-stroke cycle. During the four strokes of the cycle, the spark plug for each cylinder fires only once. In a single-cylinder engine, there would be only one spark every 720 degrees. These 720 degrees are called the **ignition interval**, or **firing interval**. It is the number of degrees of crankshaft rotation that occur between ignition sparks.

Common Ignition Intervals

If four power strokes occur during 720 degrees of crankshaft rotation, then one power stroke must occur every 180 degrees (720 ÷ 4 = 180). The ignition system must produce a spark for every power stroke, so it produces a spark every 180 degrees of crankshaft rotation. This means that a 4-cylinder engine has an ignition interval of 180 degrees.

An inline 6-cylinder engine has six power strokes during every 720 degrees of crankshaft rotation, for an ignition interval of 120 degrees (720 ÷ 6 = 120). An 8-cylinder engine has an ignition interval of 90 degrees (720 ÷ 8 = 90).

Unusual Ignition Intervals

Most automotive engines have 4, 6, or 8 cylinders, but other engines are in use today. Some companies produce 12-cylinder engines, with a 60-degree firing interval. Others make 5-cylinder engines with a 144-degree firing interval.

Other firing intervals result from unusual engine designs. General Motors has produced two different V-6 engines from V-8 engine blocks. The Buick engine, developed in the 1960's, has alternating 90- and 150-degree firing intervals. The uneven firing intervals resulted from building a V-6 with a 90-degree crankshaft and block. This engine was modified in mid-1977 by redesigning the crankshaft to provide uniform 120-degree firing intervals, as in an inline six. In 1978, Chevrolet introduced a V-6 engine that fires at alternating 108- and 132-degree intervals.

Spark frequency

In a spark-ignition engine, each power stroke is caused by a spark igniting the air-fuel mixture. Each power stroke needs an individual spark. An 8-cylinder engine, for example, requires four sparks per engine revolution (remember that there are two 360-degree engine revolutions in each 720-degree operating cycle). When the engine is running at 1,000 rpm, the ignition system must deliver 4,000 sparks per minute. At high speed, about 4,000 rpm, the ignition system must deliver 16,000 sparks per minute. Precise ignition system performance is needed to meet these demands.

Throttle: A valve that regulates the flow of the air-fuel mixture entering the engine.

Ignition Interval: (Firing Interval): The number of degrees of crankshaft rotation between ignition sparks.

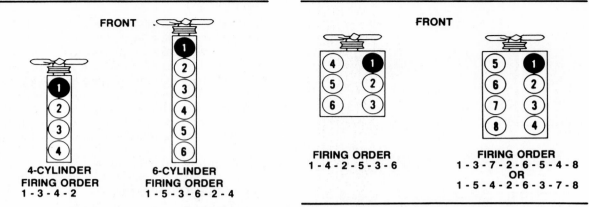

Figure 1-37. Cylinder numbering of an inline engine.

Figure 1-38. American Motors, Chrysler, and most General Motors V-type engines are numbered in this way.

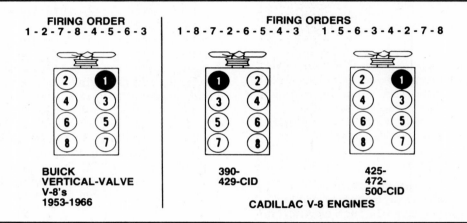

Figure 1-39. Cadillac V-8's and older Buick V-8's have unusual firing orders.

FIRING ORDER

The order in which the air-fuel mixture is ignited within the cylinders is called the **firing order**, and it varies with different engine designs. Firing orders are designed to reduce the vibration and imbalance created by the power strokes of the pistons.

Engine cylinders are numbered for identification. However, the cylinders do not usually fire in the order in which they are numbered. Straight, or inline, engines are numbered from front to rear, figure 1-37. A typical 4-cylinder engine firing order is 1-3-4-2. That is, the number one cylinder power stroke is followed by the number three cylinder power stroke, then the number four power stroke and, finally, the number two cylinder power stroke. Then the next number one power stroke occurs. An inline 4-cylinder engine may also fire 1-2-4-3.

The cylinders of an inline 6-cylinder engine are also numbered from front to rear, but do not fire in that order. The firing order for all inline 6-cylinder engines is 1-5-3-6-2-4, figure 1-37.

Except for Ford products and a few GM engines, the cylinders of domestic V-type engines are numbered the same, figure 1-38. The front cylinder on the left (driver's) side is number one. The front cylinder on the right (passenger's) side is number two. Behind number one is number three; behind number two is four, and so on. The firing order for a V-6 numbered this way is 1-6-5-4-3-2, except for the 2.8 liter (173 cu. in.) V-6 which General Motors introduced in 1980. The firing order for that engine is 1-2-3-4-5-6. For a V-8, the firing order is 1-8-4-3-6-5-7-2.

Besides Ford products, the exceptions to this rule for V-type engine cylinder numbering are:
1. Vertical-valve Buick V-8's built before 1967.
2. Cadillac V-8's.

The older Buick V-8's have the number one cylinder at the front of the right side, figure 1-39, rather than the left side. The firing order is 1-2-7-8-4-5-6-3. Cadillac 425-, 472-, and 500-cid

Engine Operation and Construction

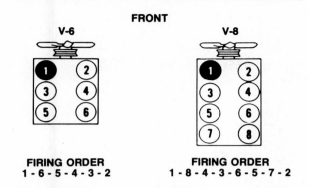

Figure 1-40. Ford numbers its V-type engines in this way.

engines are numbered like the older Buicks, figure 1-39, but the firing order is 1-5-6-3-4-2-7-8. Cadillac 390- and 429-cid engines are numbered in the more conventional manner, with the number one cylinder at the front left, figure 1-39. However, the firing order for these engines is 1-8-7-2-6-5-4-3.

■ How New Cars Are Emission-Certified

The availability of new cars each year depends on whether the carmakers successfully complete the emission certification process. The constant volume sampling test is performed by the Federal Environmental Protection Agency (EPA), although the California Air Resources Board tests some vehicles.

As each new model year approaches, automakers build prototype or emission data cars for EPA use. The automakers are responsible for conducting a 50,000-mile durability test of their emission control systems. The test cars are driven for 4,000 miles to stabilize the emission systems before testing.

The first step is to precondition a car. Then the car stands for 12 hours at an air temperature of 73°F (23°C) to simulate a cold-start. The actual test is done on a chassis dynamometer, using a driving cycle which represents urban driving conditions. The car's exhaust is mixed with air to a constant volume and analyzed for harmful pollutants.

The entire test takes about 41 minutes. The first 23 minutes are a cold-start driving test. The next ten minutes are a waiting, or hot-soak, period. The final eight minutes are a hot-start test, representing a short trip in which the car is stopped and started several times while hot. If the emissions test results for all data cars are equal to or lower than all of the HC, CO, and NO_x standards, the manufacturer receives certification of the engine "family" and the car can be offered for sale to the public.

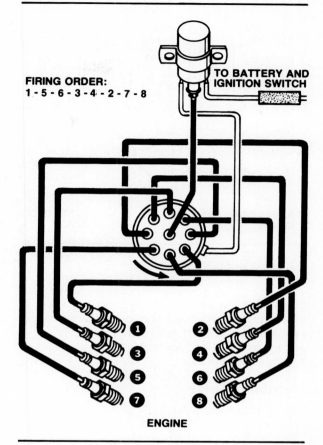

Figure 1-41. The spark plug cables must be connected to the distributor in firing order sequence.

Ford V-type engines have the number one cylinder at the right front, figure 1-40. The numbering continues down the right side, then goes to the left side from front to rear. A Ford V-6 firing order is 1-4-2-5-3-6. Ford V-8 firing orders are 1-3-7-2-6-5-4-8 or 1-5-4-2-6-3-7-8, depending on the engine.

The ignition system must deliver ignition voltage to the correct cylinder at the correct time. To get the correct firing order, the spark plug cables must be attached to the distributor cap in the proper sequence, figure 1-41.

ENGINE—IGNITION SYNCHRONIZATION

During the engine operating cycle, the intake and exhaust valves open and close at specific

Firing Order: The sequence by cylinder number in which combustion occurs in the cylinders of an engine.

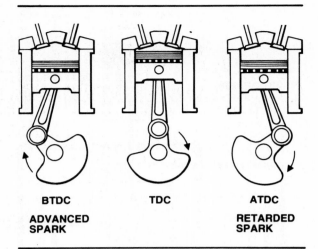

Figure 1-42. Piston position is identified in terms of crankshaft position.

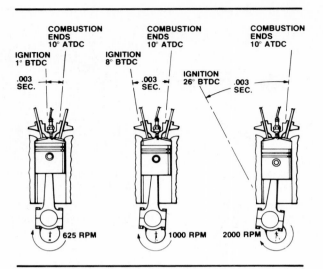

Figure 1-43. As engine speed increases, ignition timing must be advanced.

times. The ignition system delivers a spark when the piston is near the top of the compression stroke and both valves are closed. These actions must all be coordinated, or engine damage could occur.

Distributor Drive

The distributor must supply one spark to each cylinder during each cylinder's operating cycle. The distributor cam has as many lobes as the engine has cylinders, or in an electronic ignition system, the trigger wheel has as many teeth as the engine has cylinders. One revolution of the distributor shaft will deliver one spark to each cylinder. Since each cylinder needs only one spark for each *two* crankshaft revolutions, the distributor shaft need turn at only one-half engine crankshaft speed. Therefore, the distributor is driven by the camshaft, which also turns at one-half crankshaft speed.

CRANKSHAFT POSITION

The bottom of the piston stroke is called bottom dead center (bdc). The top of the piston stroke is called top dead center (tdc), figure 1-42. The ignition spark occurs near tdc, as the compression stroke is ending. As the piston approaches the top of its stroke, it is said to be **before top dead center** (btdc). A spark that occurs btdc is called an advanced spark. As the piston passes tdc and starts down, it is said to be **after top dead center** (atdc). A spark that occurs atdc is called a retarded spark.

Burn Time

Approximately 3 milliseconds (0.003 second) elapse from the instant the air-fuel mixture ignites until its combustion is complete. Remember that this burn time is a function of time and not of piston travel or crankshaft degrees. The ignition spark must occur early enough so that the combustion pressure reaches its maximum just when the piston is beginning its downward power stroke. Combustion should be completed by about 10 degrees atdc. If the spark occurs too soon btdc, the rising piston will be opposed by combustion pressure. If the spark occurs too late, the force on the piston will be reduced. In either case, power will be lost. In extreme cases, the engine could be damaged. Ignition must start at the proper instant for maximum power and efficiency.

Engine Speed

As engine speed increases, piston speed increases. If the air-fuel ratio remains relatively constant, the fuel burning time will remain constant. However, at greater engine speed, the piston will travel farther during this burning time. Ignition timing must be changed to ensure that maximum combustion pressure occurs at the proper piston position.

For example, consider an engine, figure 1-43, that requires 0.003 second for the fuel charge to burn and that achieves maximum power if the burning is completed at 10° atdc.

At an idle speed of 625 rpm, position A, the crankshaft rotates about 11 degrees in 0.003

Engine Operation and Construction

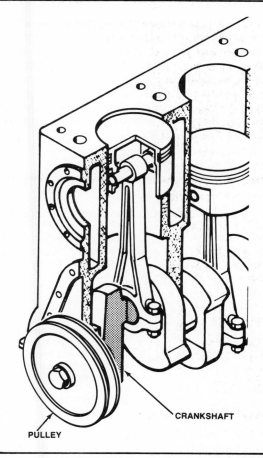

Figure 1-44. Most engines have a pulley bolted to the front end of the crankshaft.

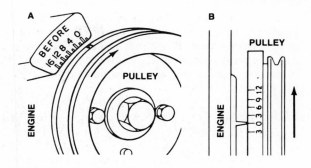

Figure 1-45. Common timing marks.

second. Therefore, timing must be set at 1° btdc to allow ample burning time.

At 1,000 rpm, position B, the crankshaft rotates 18 degrees in 0.003 second. Ignition should begin at 8° btdc.

At 2,000 rpm, position C, the crankshaft rotates 36 degrees in 0.003 second. Spark timing must be advanced to 16° btdc.

Change in timing is called spark advance or ignition advance.

BASIC TIMING

As we have seen, ignition timing must be set correctly for the engine to run at all. This is called the engine's initial, or basic, timing. Basic timing is the correct setting at a specified engine speed. In figure 1-43, basic timing was 1° btdc. Basic timing is usually within a few degrees of top dead center. For many years, most engines were timed at the specified slow-idle speed for the engine. However, engines built since 1974 may require timing at speeds either above or below the slow-idle speed.

Timing Marks

We have seen that basic timing is related to crankshaft position. To properly time the engine, we must be able to determine crankshaft position. The crankshaft is completely enclosed in the engine block, but most cars have a pulley and vibration damper bolted to the front of the crankshaft, figure 1-44. This pulley rotates with the crankshaft and can be considered an extension of the shaft.

Marks on the pulley show crankshaft position. For example, when a mark on the pulley is aligned with a mark on the engine block, the number 1 piston is at tdc.

Timing marks vary widely, even within a manufacturer's product line. There are three common types of timing marks,
- A mark on the crankshaft pulley, and marks representing degrees of crankshaft position, on the engine block, figure 1-45, position A.
- Marks on the pulley, representing degrees of crankshaft position, and a pointer on the engine block, figure 1-45, position B.
- Marks on the engine flywheel and a pointer on the transmission cover or bellhousing.

Some cars may also have a notch on the engine flywheel and a scale on the transmission cover or bellhousing.

Before Top Dead Center: The position of a piston as it approaches top dead center. Abbreviated: btdc. Usually expressed in degrees, such as 5° btdc.

After Top Dead Center: The position of a piston after it has passed top dead center. Abbreviated: atdc. Usually expressed in degrees, such as 5° atdc.

Many late-model cars, in addition to a conventional timing mark, have a special test socket for electromagnetic engine timing. These can be timed with a timing light or, for greater accuracy, with a special test probe that fits into the socket.

SUMMARY

Engine operation and construction can be easily mastered if you study it in small parts as presented in this chapter. As with any new subject it is important to memorize the names of the various parts. A connecting rod or any other part must be called by its proper name.

You are becoming a part of a profession and it is important to understand the language and be able to use it correctly. The ability to communicate makes the study of engines more interesting and understandable.

Review Questions

Choose the single most correct answer.
Compare your answers to the correct answers on page 268.

1. An automotive internal combustion engine:
   a. Burns gasoline
   b. Has pistons which are driven downwards by explosions in the combustion chambers
   c. Is better than an external combustion engine
   d. Uses energy released when a compressed air-fuel mixture is ignited

2. A spark plug fires near the end of the:
   a. Intake stroke
   b. Compression stroke
   c. Power stroke
   d. Exhaust stroke

3. The flywheel:
   a. Keeps the engine running when idling
   b. Provides a mounting plate for the clutch or torque converter
   c. Smoothes out increases and decreases in engine power output, providing continuous thrust
   d. All of the above

4. The camshaft rotates at:
   a. The same speeds as the crankshaft
   b. Half the speed of the crankshaft
   c. Double the speed of the crankshaft
   d. None of the above

5. Which of the following is never in direct contact with the engine valves?
   a. Valve seats
   b. Valve springs
   c. Rocker arms
   d. Pushrods

6. A hemi-head engine has:
   a. An overhead camshaft
   b. Valve stems at a 90-degree angle
   c. Combustion chambers shaped like half of a ball
   d. Hemispherical cam lobes

7. Which parts conduct gases into and away from the cylinder block?
   a. Exhaust pipe
   b. Carburetor
   c. Crankcase
   d. Manifolds

8. The compression ratio of an engine is:
   a. The maximum compression the engine can withstand
   b. The ratio of inlet vacuum to exhaust backpressure
   c. Crankcase capacity divided by cylinder volume
   d. Total cylinder volume divided by clearance volume

9. The displacement of an engine is:
   a. The clearance between the cylinder block and the firewall
   b. The equivalent of the horsepower rating
   c. The number of cylinders multiplied by the square of the bore radius, multiplied by   , multiplied by the length of the stroke

10. Every fuel pump:
    a. Creates a vacuum
    b. Is an electrical pump
    c. Is an immersion pump
    d. Is mechanically driven by the engine

11. The throttle linkage spring:
    a. Returns the butterfly valve to the idle position
    b. Returns the accelerator pedal to its top position
    c. Produces the proper tension in the linkage
    d. All of the above

12. Which of the following shows the ignition interval of an inline 6-cylinder engine?
    a. $600° \div 6 = 100°$
    b. $360° \div 3 = 120°$
    c. $90° \times 6 = 540°$
    d. $720° \div 6 = 120°$

13. How many sparks (spark plug firings) per crankshaft revolution are required by an 8-cylinder engine?
    a. 8
    b. 6
    c. 4
    d. 2

14. The firing order of an engine is:
    a. The same on all V-8's
    b. Is stated in reference books and on the cylinder block
    c. Can be deduced by common sense
    d. None of the above

15. Ignition timing:
    a. Requires a basic setting specified in degrees btdc for a particular engine
    b. Varies while engine is running, synchronizing combustion with proper piston position
    c. Is indicated by the alignment of markings on the crankshaft pulley or flywheel with other markings or a pointer fixed to the cylinder block
    d. All of the above

16. To convert cubic centimeter displacement to cubic inch displacement you:
    a. Multiply by 16.39
    b. Divide by .061
    c. Multiply by .061
    d. Multiply by .155

17. The main function of the carburetor is to:
    a. Increase or decrease engine speed, in response to accelerator pedal pressure
    b. Regulate intake gas pressure
    c. Create the proper mixture of gasoline and air
    d. Control the flow of gasoline supplied by the pump

Chapter 2

Engine Physics and Chemistry— Alternative Engines

In Chapter 1, we examined the basic mechanical parts of an engine. We also studied engine design and 4-stroke-cycle operation. In this chapter, we will take a fast look at some of the principles of chemistry and physics that are applied in an internal combustion engine. We also will examine the basic problems of air pollution and how the automobile engine is involved in the problem. Finally, in the last part of the chapter, we will survey some engines other than the 4-stroke gasoline engine that can be used in automobiles. Some of these designs have been used in past and present cars. Others are still on the drawing boards for future automobiles.

ENGINE PHYSICS

We sometimes talk about engines in terms of their horsepower and torque ratings. We also know that engines burn fuel to release energy and do work. But what do these terms mean? A simple review of these textbook terms will help you to understand why engines of different sizes and designs produce different amounts of power and can do different amounts of work.

Energy

The *ability* to do work by applying force is called **energy**. Heat, light, sound, and electricity are all forms of energy. These things do not have weight or occupy space, but they all have the ability to do work. Energy is contained in the fuel burned in an engine, as well as in the food we eat. The energy of food is commonly measured in **calories**. The energy of fuel is measured in **British thermal units (Btu)**. Both are measurements of heat energy, and one can be converted to the other.

Energy also can be stored in a mechanical device. When you wind a watch spring, you store energy in the spring. However, the energy cannot do any work until the spring is released. Energy, again, is the ability to do work. It can be stored in chemical, mechanical, and other forms.

Work

The application of energy to produce motion is called **work**. Energy cannot be created or destroyed, but it can be changed. When energy is changed from one form to another, work is normally done at the same time. When energy is released to do mechanical work, motion is produced. In order to produce motion and do work the energy must exert force.

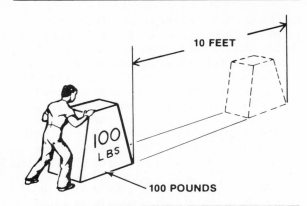

Figure 2-1. Work is calculated by multiplying force times distance. If you push 100 pounds 10 feet, you have done 1,000 foot-pounds of work.

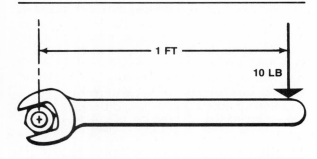

Figure 2-3. Torque is a twisting force equal to the distance to the axis times the force applied.

Force

A push or pull acting on an object is called **force**. Force may start, stop, or change the direction of motion. Without exerting force, energy cannot do any mechanical work. And, as we have said, unless motion is produced by the force, work has not been accomplished. For example, you store energy in your body when you eat food. You can release some of that energy (convert it from chemical to mechanical form) by pushing or applying force on a weight. However, although you may exert force, you can't do any work until you move the weight.

The scientific definition of work is force multiplied by distance. If you push a 100-pound weight 10 feet, you have done 1,000 foot-pounds of work, figure 2-1. To do that work, you "burned" a certain number of calories from the food energy stored in your body. We will see how these same principles work in an automobile engine a little later in this chapter.

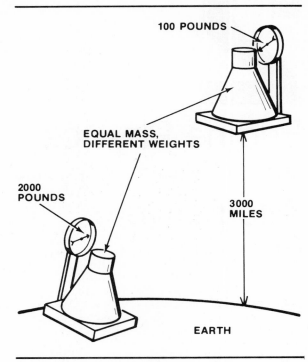

Figure 2-2. Weight will change in relation to gravitational pull, but mass remains constant.

Mass And Weight

Mass and weight are often confused with each other, but they are not the same. **Mass** is the measure of inertia of an object. **Weight** is the measure of the earth's gravitational pull (force) on that object. A 2,000-pound space satellite has the same amount of mass on earth as it has 3,000 miles out in space. But, it weighs 2,000 pounds only on earth. In space, it may weigh just a few pounds, or be virtually weightless, figure 2-2.

In everyday discussions of mass and weight, as well as force, we commonly use pounds and ounces to measure these values. On the surface of the earth, where we operate our automobiles, the mass and weight of an object can be considered to be the same for all practical purposes. It does no harm to think of mass and weight interchangeably in our basic discussion of engine operation. However, when we start comparing torque measurements in the metric and the customary U.S. systems, we will find there is a distinction between mass and weight.

Torque

Torque is related to work and force. When force applied to a body does not act on a single point, the body will tend to rotate. This twist-

Engine Physics

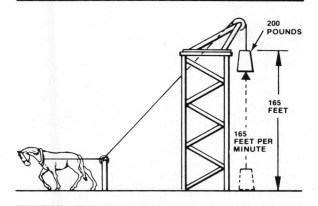

Figure 2-4. One horsepower equals 33,000 foot-pounds (200 x 165) of work per minute.

ing or turning force is called **torque**. Torque, like work, can be measured in foot-pounds or a similar combination unit of measure. When you put a 1-foot wrench on a bolt and apply 10 pounds of force to the end of the wrench to turn the bolt, you exert 10 foot-pounds of torque, figure 2-3.

Torque can be defined by the equation: force × distance = torque. For our purposes, mechanical work can be defined by the same equation. Torque and work are not necessarily the same, however. Mechanical work always includes motion. Torque may or may not include motion. When you pull on the wrench, the bolt may or may not turn. In either case, you are exerting torque.

Power

The rate, or speed, of doing work is called **power**. Engines are commonly rated in terms of **horsepower**. We use horsepower as a unit of measurement because horses were a common form of power used to do work immediately before the industrial revolution. Therefore, horses were something that could easily be compared to machines used to do similar work. In the mid-nineteenth century, James Watt calculated that a horse could move a 200-pound weight 165 feet in 1 minute, figure 2-4. When you multiply 200 pounds by 165 feet, you get 33,000 foot-pounds. The standard definition of one horsepower then is the power needed to do 33,000 foot-pounds of work in 1 minute. Two horsepower can do the same amount of work in half the time; four horsepower in one-quarter the time.

At this point, we can see a relationship between torque, speed, and power. As speed increases, power increases for a constant amount of work. However, as speed and power increase, the amount of torque will decrease.

We won't go into all the complicated scientific relationships of torque, speed, and power. You can find them in a physics text if you wish. But, you should be aware of two other factors that affect the torque and horsepower of an engine. These are inertia and momentum.

Inertia

We learned in Chapter One that inertia is the tendency of an object at rest to remain at rest, and of an object in motion to remain in motion. The greatest amount of inertia in most engines is located in the flywheel. When it is at rest, it is said to have **static inertia**. When it is moving, it is said to have **dynamic inertia**.

Energy: The ability to do work by applying force.

Calorie: The amount of heat energy required to raise the temperature of one gram of water one degree Celsius at a pressure of one atmosphere.

British Thermal Unit (Btu): The amount of heat required to raise the temperature of one pound of water one degree Fahrenheit at a pressure of one atmosphere.

Work: The application of energy through force to produce motion.

Force: A push or pull acting on an object; it may cause motion or produce a change in position. Force is measured in pounds in the U.S. system and in newtons in the metric system.

Mass: The measure of inertia of an object; sometimes defined as the "quantity of matter" in it.

Weight: The measure of the earth's gravitational pull on an object.

Torque: The tendency of a force to produce rotation around an axis; it is equal to the distance to the axis times the amount of force applied expressed in foot-pounds or inch-pounds (customary U.S. system) or in newton-meters (metric system).

Power: The rate or speed of doing work (work × time = power).

Horsepower: 33,000 foot-pounds of work per minute equals one horsepower.

Static Inertia: The tendency of a body at rest to remain at rest.

Dynamic Inertia: The tendency of a body in motion to remain in motion.

Momentum

Momentum is related to inertia. When a body is put into motion by the application of force, it develops **momentum**. This is the product of its mass or weight on earth, times its speed. A body moving in a straight line will keep going in that direction at the same speed forever, if no other forces are exerted upon it. The same is true of a rotating flywheel. However, other forces do act on a body in motion to overcome its momentum. One major opposing force is **friction**. Inertia and momentum are forms of mechanical energy. When motion is opposed by friction, mechanical energy is converted into heat energy. The amount of mechanical energy available to do work is reduced by the amount of energy converted into heat by friction.

At this point, we have come full circle from energy to force, work, power, and back to energy. Let's take a moment to review some of these terms before we go on to see how an internal combustion engine converts the energy stored in fuel into useful power.
Energy—the ability to do work.
Force—the push or pull exerted by energy to do mechanical work (move something).
Work—the application of energy through force to move something a given distance (force × distance = work).
Torque—twisting or turning force.
Power—the rate at which work is done (work × speed = power).
Inertia—the tendency of a body to stay at rest or in motion.
Momentum—the tendency of a moving body to keep moving in the same direction at the same speed.

We have been talking about weight, distance, power, and torque in terms of their customary U.S. units of measure: pounds, feet, horsepower, and foot-pounds, for example. These same things can be measured in other units, however. The international system of measurement is the metric system. Metric units of measurement have been gradually replacing customary U.S. units of measurement in international science and commerce. In the metric system, mass is measured in kilograms, force (weight) in newtons, distance in meters, power in **kilowatts**, and torque in **newton-meters**. We will study these metric measurements more closely later in this chapter.

CHEMISTRY AND COMBUSTION

A gallon of gasoline contains about 123,500 Btu's of energy. This amount of energy represents a lot of explosive power. For that reason, it is easy to suppose that gasoline explodes in the cylinders to make an engine run. At one point in the development of the internal combustion engine, it was called an explosive engine. Now we know that gasoline does not explode, but undergoes controlled burning.

Gasoline is a complex blend of chemical compounds chiefly composed of hydrogen and carbon. The burning of gasoline is a chemical process. When gasoline burns, it combines with oxygen in the air. The result of combining the carbon part of gasoline with oxygen is either carbon monoxide (CO) or carbon dioxide (CO_2). When the hydrogen combines with oxygen, water (H_2O) is formed.

The combustion process is called an **oxidation reaction**. That means that oxygen is combined with gasoline to form new chemical compounds. Combustion is not a simple process. All of the carbon does not combine with oxygen. Some of the carbon is released from the gasoline to form soot, or hard carbon, that is deposited in the combustion chamber. Some of the gasoline clings to the chamber walls and does not burn. If gasoline were a simple chemical compound, it would probably burn clean and go out the tailpipe as nothing more than CO_2 and H_2O. But gasoline is very complex. The carbon and hydrogen in gasoline are in the form of paraffins, olefins, napthenes, and aromatics, all with different chemical structures. To say that gasoline is nothing more than carbon and hydrogen is an oversimplification.

Combustion And Thermal Energy

The gasoline engine is really a heat engine. The more heat that can be produced inside the combustion chamber, the more power that will be produced by the engine. For this reason, engineers talk about gasoline as having a certain amount of thermal energy—approximately 123,500 Btu's per gallon. This thermal energy is converted to power through combustion.

Under some conditions, part of the thermal energy in gasoline may not be released to produce power. Power depends not only on the energy in the gasoline, but on the timing of combustion and on the amount of pressure in the combustion chamber. To produce the most pressure in the combustion chamber, the spark is timed so that combustion is completed slightly after the piston passes through top dead center. This gives the most push on the piston.

Engine Physics

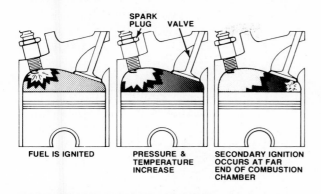

Figure 2-5. Detonation is a secondary ignition of the air-fuel mixture, caused by high cylinder temperatures. It is commonly called "pinging" or "knocking."

As the air-fuel mixture is compressed, a great deal of heat is generated. Unless gasoline is blended to hold up under such high pressures and temperatures, soon after ignition there may be a secondary explosion called **detonation**, figure 2-5. This is popularly called "knocking" or "pinging." Detonation causes a loss of power and overheating of valves, pistons, and spark plugs. The overheating can cause more detonation, and may eventually damage the engine. Detonation also means that part of the thermal energy in the gasoline is being wasted. It is not producing useful power.

If all gasoline contains the same amount of thermal energy, how can gasolines be blended differently to resist detonation and release the greatest amount of this energy? The answer lies in the **octane rating** of a gasoline. In simple terms, the octane rating is a measurement of a gasoline's ability to provide stable combustion at high temperatures and pressures. A higher octane rating permits higher pressures in the combustion chamber without detonation. The higher pressure allows the maximum amount of heat energy to be extracted from the gasoline to produce the most power. At lower pressures, some of the heat energy will go unused.

Because detonation is so quick and powerful, it can damage an engine. To prevent this, either the spark must be retarded, or the octane of the gasoline must be increased. Certain compounds, such as **tetraethyl lead**, can be added to gasoline to increase its octane rating. The high-octane gasoline does not contain any more thermal energy, but it allows more spark advance and higher compression for greater power without engine damage.

With the coming of catalytic converters, lead had to be removed from gasoline. The lead coats the catalyst and "poisons" it, reducing its effectiveness. Without the lead, the octane rat-

■ **"I Wouldn't Touch Metric Conversion With A 3.049-meter Pole"**

While converting the United States system to the metric system will take plenty of time and will cost enormous amounts of money, many measurements are already in metrics.

For instance, cameras and film generally use millimeters (mm) for measuring the size of the lenses or the size of the film. In electronics, we now use the metric system of seconds, volts, watts, amperes and hertz (cycles per second). The drug industry changed to the metric system more than 15 years ago, and the nation's carmakers are not only using liter measurement for displacement, but are building some of their own engines and cars to metric dimensions.

Mechanics need to know the metric system. This nation's tire manufacturers are now using kilopascals as well as pounds per square inch to indicate tire inflation. And the quart of oil may be a thing of the past, as the liter (which is just a little bit larger) takes over.

Momentum: The force of motion. Momentum equals a moving body's mass times its speed.

Friction: The force that resists motion between the surfaces of two bodies when they are in contact. The two basic types of friction are sliding friction and rolling friction.

Kilowatt (kW): The metric unit of power. One horsepower equals 0.746 kilowatt; one kilowatt equals 1.3405 horsepower.

Newton-Meter (Nm): The metric unit of torque. One newton-meter equals 1.356 foot-pounds. One foot-pound equals 0.736 newton-meter.

Oxidation Reaction: A chemical reaction in which oxygen is combined with another element or compound to form a new compound.

Detonation: An unwanted explosion of an air-fuel mixture caused by high heat and compression. Also called knocking, pinging.

Octane Rating: The measurement of the antiknock value of a gasoline.

Tetraethyl Lead: A gasoline additive used to help prevent detonation.

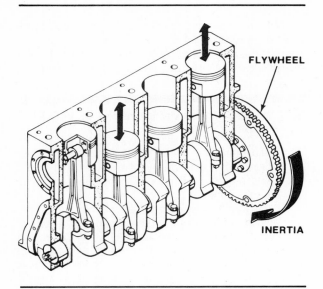

Figure 2-6. The inertia of the flywheel provides the momentum to carry the pistons through the exhaust, intake, and compression strokes.

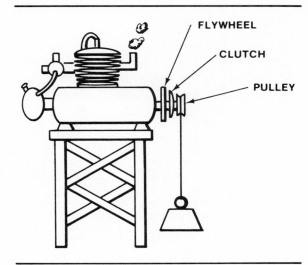

Figure 2-7. Lifting a weight measures the amount of work an engine can do.

ing of the gasoline is reduced. To keep the octane rating up, the gasoline has to have further refining and blending. This results in less gasoline at higher cost.

Another gasoline additive used to reduce knock is MMT, a manganese compound. Manganese was well on its way to becoming a widely used substitute for lead when it was banned by the federal government for causing pollution.

The oil companies are continuing attempts to find a substitute for lead. Compounds made from alcohols and ethers show great promise, and may eventually allow oil companies to produce more gasoline of higher octane at less cost.

ENGINE FORCE, INERTIA, AND MOMENTUM

Combustion releases the energy that exerts force on the piston. The reciprocating force of the piston, through its connecting rod, exerts torque on the crankshaft. This is a form of work, and because it occurs during a period of time, power is produced.

When the piston reaches the bottom of the power stroke, the power has been expended. There is no combustion energy in the cylinder to force the piston to the top of its stroke and rotate the crankshaft along with it. This is where the inertia of the flywheel comes in, figure 2-6. When the piston reaches the bottom of the power stroke, the flywheel is turning. This heavy piece of metal has so much inertia that it keeps turning. In fact, in a 1-cylinder engine, it keeps moving so well that it pushes the piston up on the exhaust stroke, down on the intake stroke, and up on the compression stroke. Remember, there is only power on one of the four strokes. The 1-cylinder engine rotates through the other three strokes only because of the inertia of its rotating parts, principally the flywheel.

When an engine is shut off, the power strokes cease. But the momentum of the engine will continue to turn it until friction has slowed the engine to the point where the compression force stops it. The more cylinders there are, the quicker the engine will stop turning. This is why a 1-cylinder lawnmower engine coasts to a stop through several revolutions, while an automotive V-8 stops almost instantly.

ENGINE TORQUE AND HORSEPOWER

We have already learned that torque is a twisting or turning force. And horsepower, in this case **brake horsepower (bhp)**, is the optimum power the engine has for moving the vehicle. To translate torque into horsepower we assume that the engine is actually moving a weight around a circle. That is, it is doing work.

Engine Torque

The turning or twisting force that is torque is measured at the end of the crankshaft or the flywheel. Through torque, reciprocating motion is changed to rotary motion.

In an earlier example, we measured 10 foot-pounds of torque when you put a 1-foot wrench on a bolt and exerted 10-pounds of

Engine Physics

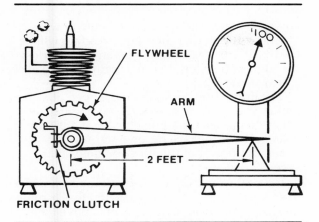

Figure 2-8. An engine exerting 100 pounds of force through a 2-foot arm, delivers 200 foot-pounds of torque.

force on the wrench. It is easy to measure the torque exerted and the work done by a person's muscles because they start and stop almost instantly. An engine, however, must be kept running, or it will not develop any mechanical energy to do the work.

Therefore, we need a device to measure engine torque while the engine is running. This device is called a clutch, or sometimes a friction brake. Imagine a pulley attached to an engine flywheel, with a clutch to connect and disconnect it from the engine. Around the pulley is a rope attached to a weight on the floor, figure 2-7. If we start the engine and slowly engage the clutch, the pulley will turn and lift the weights. If we keep adding weights until the engine cannot lift them, the total weight will stop the engine. In this way, we can measure the amount of work that an engine can do.

We can do this with any engine. Some engines are relatively powerful and can lift very heavy loads. The amount of weight lifted by different engines gives us a standard of comparison. This simple example is a handy way to compare work done by a simple, stationary engine. However, it is not practical for use with an automobile engine that must deliver torque continuously to drive a car.

Now let's attach an arm to the flywheel clutch, or friction brake, on our engine and connect the other end of the arm to a scale. If we engage the clutch, the arm will push on the scale and give a reading in pounds. But to measure torque in foot-pounds, the length of the arm must be considered. If the length of the arm is two feet and the reading on the scale is 100 pounds, we have a torque measurement of 200 foot-pounds, figure 2-8.

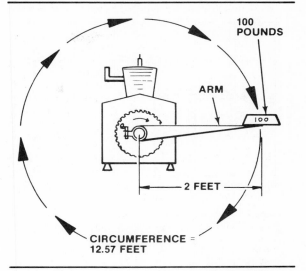

Figure 2-9. The engine now does 1,257 foot-pounds of work.

Torque is a common measurement for comparing engines. It indicates the amount of work an engine can do, or the force it can exert, but it doesn't indicate how fast an engine can do the work or how much power it can deliver.

Engine Horsepower

As we have already stated, when we translate engine torque into engine horsepower, we pretend the engine is actually moving a weight around a circle. The weight is equal to the number of pounds indicated on the scale at the end of the torque arm. If the weight were allowed to rotate, the arm would make a circle with a diameter twice the length of the arm.

In our calculations, we will go from torque to work to power. For example, let's say that the torque scale still reads 100 pounds at the end of a 2-foot arm, so we still have a torque measurement of 200 foot-pounds. Now the engine is rotating, however, so the distance we use to compute work in foot-pounds is no longer just the length of an arm. It is the circumference of the circle described by the end of the rotating arm, figure 2-9. Here is how we calculate the circumference of that circle:

2 × the radius (2 feet) × pi (3.1416) = the circumference

2 × 2 feet × 3.1416 = 12.57 feet.

Now multiply that distance (the circumference)

Brake Horsepower: The power available at the flywheel of an engine for doing useful work.

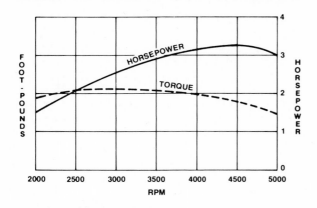

Figure 2-10. Typical engine horsepower and torque curves.

by 100 pounds to get the total work done during one engine revolution:

12.57 feet × 100 pounds = 1,257 foot-pounds.

We know the amount of torque exerted and the amount of work done by the engine, but we still don't know how fast the work is done and how much power is delivered. For that, we must know the speed of the engine in revolutions per minute (rpm). Let's assume the engine speed is 2,500 rpm. If we multiply the foot-pounds of work by the speed, we get the number of foot-pounds per minute:

1,257 foot-pounds × 2,500 rpm = 3,142,500 foot-pounds per minute.

We know that one horsepower equals work done at the rate of 33,000 foot-pounds per minute. Finally, we can compute the horsepower delivered by the engine at 2,500 rpm as follows:

3,142,500 foot-pounds per minute ÷ 33,000 foot-pounds per minute = 95.23 horsepower.

If we know the torque at a given rpm, we can use the formula:

$$\text{Horsepower} = \frac{\text{torque} \times 2\text{ pi} \times \text{rpm}}{33,000}$$

This can be shortened to:

$$\text{Horsepower} = \frac{\text{torque} \times \text{rpm}}{5252}$$

Torque And Horsepower Relationships

The device we used to measure the engine torque reading on a scale is a simple **dynamometer**. In use, the clutch, or friction brake on the dynamometer is tightened to reduce the speed of the engine and produce a reading on the scale. The engine is not stopped, however; it must be kept running. Torque can be measured on an engine dynamometer at any rpm. Readings are usually taken at wide-open throttle so that the torque reading will be the maximum available at a given speed. Most car engines develop their maximum torque at approximately 2,500 rpm.

Maximum horsepower is developed at a higher speed. Most passenger car engines develop their maximum horsepower at 4,000 to 5,000 rpm. Some race car engines develop maximum power at two to three times that speed.

If we take readings of engine torque from idle to several thousand rpm, we would find torque increasing up to about 2,500 rpm. Above that, torque slowly falls off. This is because the faster an engine goes, the more trouble it has getting enough air. But as speed continues to increase, horsepower continues to rise. The speed is going up faster than the torque is falling off.

As engine speed increases, horsepower also increases, as long as the torque does not fall off too rapidly. At the upper end of an engine's speed range, torque drops off more rapidly. Horsepower then begins a similar drop. Figure 2-10 shows typical torque and horsepower curves for a passenger car engine.

Brake horsepower

The horsepower measured on a dynamometer at the engine flywheel is called brake horsepower. This is because the original dynamometers used a friction brake, called a prony brake. Modern dynamometers still act as a brake, but most use an electric generator, a water pump, or a fan blowing air to absorb power. But there are still some that use a friction brake.

The Society of Automotive Engineers (SAE) defines brake horsepower as the power available at the flywheel for doing useful work. However, there are different kinds of brake horsepower. For years, carmakers advertised their engines in terms of **gross brake horsepower**. This is the power output of a basic engine, equipped only with the built-in accessories needed for its operation. These include the fuel, oil, and water pumps and any built-in emission controls. The gross brake horsepower is the maximum power available from an engine. In the days when high horsepower numbers made good advertising material, these were the numbers that the manufacturers emphasized.

A more realistic indication of the power available from an engine to drive a car is **net brake horsepower**. Net brake horsepower is the output of a fully equipped engine with accessories such as the air filter and exhaust systems, the cooling system, the alternator, the

Engine Physics

starter motor, and all bolt-on emission control equipment in place and operating. Since about 1970, carmakers have been publishing the net brake horsepower of their engines, which is considerably lower than the gross power output.

Indicated horsepower and friction horsepower

From this brief comparison between gross and net brake horsepower, we can see that all of the power developed by an engine is not available at the flywheel for useful work. Part of an engine's power is used to drive accessories or is used in other systems. For example, mechanical power is used to develop electrical energy in the alternator. Power also is dissipated as heat energy in the exhaust and cooling systems.

We learned earlier that friction is a form of heat energy. Part of an engine's power is used to overcome the friction between its moving parts. This is called **friction horsepower**, and is lost horsepower. Friction horsepower takes away from the usable brake horsepower available at the flywheel. Through careful engine design and proper lubrication between moving parts, horsepower losses through internal friction are kept to a minimum. These losses still exist, however, and they are an important factor in engine performance. To perform proper engine maintenance and repair, you must be aware of the effects of friction.

The **indicated horsepower** is the theoretical maximum power produced in an engine's cylinders. This indicated power can be determined in a laboratory with an instrument that measures the gas pressure developed in a cylinder during combustion. Indicated horsepower is important to the engineers who design engines, but in the practical use of an engine, we don't have to be concerned with this theoretical value. We should understand, however, that usable brake horsepower is the indicated horsepower minus the losses through friction horsepower.

METRIC POWER AND TORQUE

We learned in Chapter 1 that engine displacement can be calculated in cubic inches, using customary U.S. units. It also can be calculated in cubic centimeters, using metric units. Earlier in this chapter, we said that power and torque, too, could be calculated in the metric system.

Kilowatts

The metric unit for measuring power is the kilowatt (kW). This is simply another way of measuring the amount of energy available to do work. Horsepower is a mechanical measurement—moving weights using a horse. A kilowatt is primarily an electrical measurement, but it can be used equally well to measure mechanical power. Power expressed in kilowatts represents the amount of electricity an engine would produce if it were driving a generator.

There is a direct relationship between kilowatts and horsepower. One horsepower equals 0.746 kilowatt. Therefore, a 100-horsepower engine is also a 74.6-kilowatt engine. To convert from horsepower to kilowatts, you multiply the horsepower rating by 0.746. To convert from kilowatts to horsepower, you multiply the kilowatt rating by 1.3405.

Newton-Meters

In the metric system, torque is expressed in newton-meters (Nm). The newton is the metric unit for measuring force, and the meter is the metric unit for length. Therefore, a newton-meter is a combination measurement, just as a foot-pound is in the customary U.S. system.

You will recall that early in this chapter we mentioned that there was a distinction between weight (force) and mass. In the U.S. customary system the pound is used to measure both of these items. In scientific circles a distinction is made between pounds-force and pounds-mass since only on the surface of the earth are they approximately equal. However, in the metric system the kilogram is the unit for mass and the newton is the unit for force. Although weight is actually a measure of force against a scale, the kilogram is often used as a measure of weight. Therefore torque could be measured in kilogram-meters or kilogram-centimeters, and in fact it used to be. You may see some

---

Dynamometer: A device used to measure the power of an engine or motor.

Gross Brake Horsepower: The flywheel horsepower of a basic engine, without accessories.

Net Brake Horsepower: The flywheel horsepower of a fully equipped engine, with all accessories in operation.

Friction Horsepower: The power consumed to overcome friction within an engine.

Indicated Horsepower: The theoretical maximum power produced in an engine's cylinders.

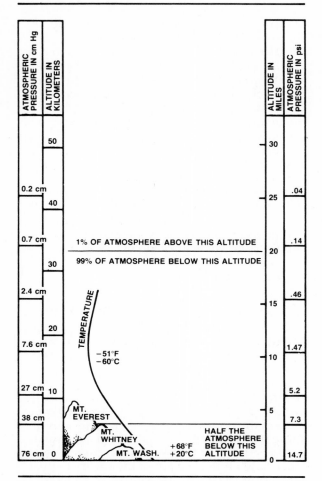

Figure 2-11. The blanket of air surrounding the earth extends many miles above the surface of the earth. Atmospheric pressure decreases at higher altitudes because the blanket is thinner.

older technical books that give torque specifications in these older metric values. Over a decade ago, the international scientific community agreed that torque expressed as a measurement of force times distance was more precise than a measurement of mass times distance. Many modern technical books, such as carmaker's shop manuals, list torque values in newton-meters.

Newton-meters and foot-pounds can be converted back and forth, using the following formulas:

Foot-pounds × 0.7376 = newton-meters
Newton-meters × 1.356 = foot-pounds.

ENGINE OPERATION AND PRESSURE

Up to this point, we have been looking at some of the principles of physics and chemistry that are applied in internal combustion engines. Principally, we have been concerned with energy, work, and mechanical power. To understand engine operation completely, you also should know something of the effects of air pressure and airflow.

We said earlier that you can think of an automobile engine as a heat engine. That is true, and you also can think of it as a large air pump. As the pistons move up and down in the cylinders, they pump in air and fuel for combustion and pump out the burned exhaust. They do this by creating a difference in air pressure.

Air is a substance with weight. The air outside an engine has a specific weight, and so does the air inside the engine. The weight of the air exerts pressure on whatever it touches. The greater the weight, the greater the pressure. When the weight of air outside the engine is greater than the weight of air inside, we say that there is a pressure difference or pressure differential, between the two areas.

Atmospheric Pressure

The weight of air is not always the same. It changes with temperature and height above sea level. Air pressure is measured in pounds per square inch (psi) at sea level at an average temperature of 68° F. In metric units, **atmospheric pressure** at sea level is measured in millimeters of mercury (mm Hg) at 20° C.

At sea level and at an average temperature, one cubic foot of air weighs about 1 1/4 ounce. This seems light enough, but remember that the blanket of air surrounding the earth extends many miles above the earth's surface, figure 2-11. These many cubic feet of air piled one on another all have weight. This increases the weight of air pressing down on an object at sea level to about 14.7 psi or 760 mm Hg.

Effects of temperature

Air expands and becomes lighter as its temperature rises. This reduces the pressure it exerts. As its temperature falls, air contracts. This makes it heavier and increases its pressure. Variations in air temperature account for changing weather conditions. Direct heat from the sun and reflected heat from the earth's surface warm the air. As its temperature increases, air becomes lighter and rises. Cooler air sinks and takes its place, resulting in constant motion. This motion creates wind and weather patterns.

Effect of height

As you climb above sea level, the amount of air pressing down on you becomes less. Since this air weighs less, it exerts less pressure. Air pressure gradually decreases with increased dis-

Engine Physics

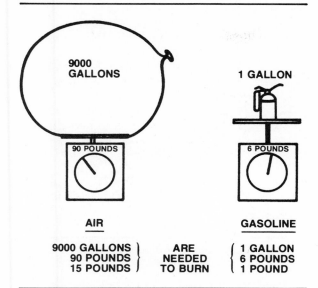

Figure 2-17. The most efficient air-fuel ratio is 15 pounds of air to 1 pound of gasoline (15:1). (Chevrolet)

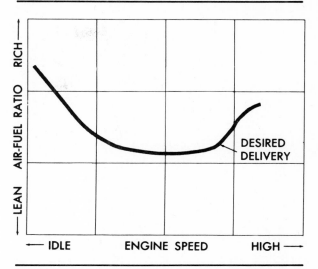

Figure 2-18. The desired air-fuel ratio changes as engine operating conditions change. (Chevrolet)

3. 1 gallon of gasoline = 6 pounds (approximate).

This means that it takes approximately 15 pounds of air to burn one pound of gasoline, so our air-fuel ratio is 15:1, figure 2-17.

Stoichiometric Air-Fuel Ratio

The ideal mixture or ratio at which all the fuel will blend with all of the oxygen in the air and be *completely* burned is called the **stoichiometric ratio**—a chemically perfect combination. In theory, an air-fuel ratio of about 14.7:1 will produce this ratio, but the *exact* ratio at which perfect mixture and complete combustion take place depends on the molecular structure of gasoline, which can vary somewhat.

This relationship between the amounts of air and fuel flow in an engine is sometimes called the fuel-air ratio. Because an engine uses far more air than fuel, the fuel-air ratio is always a number less than one, such as 0.0625. The fuel-air ratio is just a different way of expressing the more familiar air-fuel ratio.

Engine Air-Fuel Requirements

An automobile engine will work with the air-fuel mixture ranging from 8:1 to 18.5:1. But the ideal ratio would be one that provides both the most power *and* the most economy, while producing the least emissions. But such a ratio does not exist because the fuel requirements of an engine vary widely depending upon temperature, load, and speed conditions.

Research has proved that the best fuel economy is obtained with a 15:1 to 16:1 ratio, while maximum power output is achieved with a 12.5:1 to 13.5:1 ratio. A rich mixture is required for idle, heavy load, and high speed conditions; a leaner mixture is required for normal cruising and light load conditions. As you can see, no single air-fuel ratio provides the best fuel economy *and* the maximum power output at the same time, figure 2-18.

Just as outside conditions such as speed, load, temperature, and atmospheric pressure change the engine's fuel requirements, other forces at work inside the engine cause additional variations. Here are three examples:

1. Exhaust gases remain inside the cylinders and dilute the incoming air-fuel mixture, especially during idle.
2. The mixture is imperfect because complete vaporization of the fuel may not take place.
3. Mixture distribution by the intake manifold to each cylinder is not exactly equal; some cylinders may get a richer or leaner mixture than others.

If an engine is to run well under such a wide variety of outside and inside conditions, the

---

Stoichiometric Ratio: An ideal air-fuel mixture for combustion in which all oxygen and all fuel will be completely burned.

Air-Fuel Ratio: The ratio in pounds of air to gasoline in the air-fuel mixture drawn into an engine.

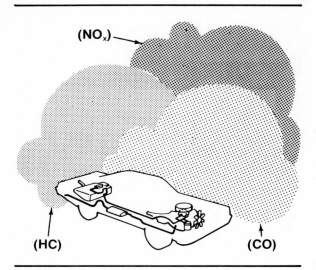

Figure 2-19. Hydrocarbons (HC), carbon monoxide (CO), and oxides of nitrogen (NO_x) are the three major pollutants emitted by an automobile.

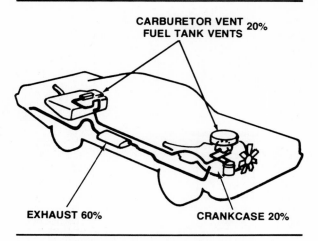

Figure 2-20. Sources of hydrocarbon emissions.

carburetor must be able to vary the air-fuel ratio quickly, and to give the best mixture possible for the engine's requirements at a given moment.

The best air-fuel ratio for one engine may not be the best ratio for another, even when the two engines are of the same size and design. Engines are mass produced, and will have slight variations in manifolding, combustion chambers, valve timing, and ignition timing. To accurately determine the best mixture, the engine should be run on a dynamometer to measure speed, load, and power requirements for all types of driving conditions.

INTRODUCTION TO EMISSION CONTROL

An internal combustion engine emits three major pollutants into the air: **hydrocarbons** (HC), **carbon monoxide** (CO), and **oxides of nitrogen** (NO_x), figure 2-19. An engine also gives off many small liquid or solid particles, such as lead, carbon, sulfur, and other **particulates**, which contribute to pollution. All of these emissions are byproducts of combustion.

Hydrocarbons (HC)

Gasoline is a hydrocarbon material. Unburned hydrocarbons given off by an automobile are largely unburned portions of fuel. For example, every 1,000 gallons (3785 liters) of gasoline used by a car without emission controls produces about 200 pounds (91 kg) of hydrocarbon emissions.

Over 200 different varieties of hydrocarbon pollutants come from automotive sources. While most come from the fuel system and the engine exhaust, others are oil and gasoline fumes from the crankcase. Even a car's tires, paint, and upholstery give off tiny amounts of hydrocarbons. The three major sources of hydrocarbon emissions in an automobile are shown in figure 2-20.
1. Fuel system evaporation—20 percent
2. Crankcase vapors—20 percent
3. Engine exhaust—60 percent

Hydrocarbons are the only major automotive air pollutant that come from sources *other than* the engine's exhaust.

Hydrocarbons of all types are destroyed by combustion. If an automobile engine burned gasoline completely, there would be no hydrocarbons in the exhaust, only water and carbon dioxide. But when the vaporized and compressed air-fuel mixture is ignited, combustion occurs so fast that gasoline near the sides of the combustion chamber does not get burned. This unburned fuel then passes out with the exhaust gases. The problem is worse with engines that misfire or are not properly tuned.

Carbon Monoxide (CO)

The same 1,000 gallons (3785 liters) of gasoline that produce 200 pounds (91 kg) of unburned hydrocarbons when burned without emission controls will also produce 2,300 pounds (1043 kg) of carbon monoxide. Because it is a product of incomplete combustion, the amount of CO produced depends on the way in which hydrocarbons burn. When the air-fuel mixture burns, its hydrocarbons combine with oxygen. If the air-fuel ratio has too much fuel, there is not enough oxygen for this process to happen, and so carbon monoxide is formed. To make combustion more complete, an air-fuel mixture with less fuel is used. This increases the supply of oxygen, which reduces the formation of CO by producing harmless carbon dioxide (CO_2) instead.

Oxides of Nitrogen (NO_x)

Air is about 78 percent nitrogen, 21 percent oxygen, and 1 percent other gases. When the combustion chamber temperature reaches 2,500° F (1371° C) or greater, the nitrogen and oxygen in the air-fuel mixture combine to form large quantities of oxides of nitrogen (NO_x). NO_x is also formed at lower temperatures, but in far smaller amounts. By itself, NO_x is of no particular concern. But when the amount of hydrocarbons in the air reaches a certain level, and the ratio of NO_x to HC is critical, the two pollutants will combine chemically to form smog.

The amount of NO_x formed can be reduced by lowering the temperature of combustion in the engine, but this causes a problem which is difficult to solve. Lowering the combustion chamber temperature to reduce NO_x results in less efficient burning of the air-fuel mixture. This automatically means an increase in hydrocarbons and carbon monoxide, both of which are emitted in large quantities at lower combustion chamber temperatures.

NO_x is created by an oxidation reaction, which means that oxygen is combined with another element (in this case, nitrogen) to form a new compound. If we can prevent high-temperature oxidation, we can prevent NO_x formation in the first place. However, oxidation is necessary to get rid of HC and CO emissions. To get rid of NO_x once it is formed, we must have a **reduction reaction**, which is the opposite of oxidation. In a reduction reaction, oxygen is removed from a chemical compound. It is very difficult to create oxidation and reduction reactions in the same place at the same time, and this has made engineers' jobs very difficult in trying to lower HC, CO, and NO_x emissions at the same time.

Automotive Emission Controls

Early researchers dealing with automotive pollution and smog began work with the idea that all pollutants were carried into the atmosphere by the car's exhaust pipe. But auto manufacturers doing their own research soon discovered that pollutants were also given off from the fuel tank and the engine crankcase. The total automotive emission system, figure 2-21, contains three different types of controls. This picture shows that the emission controls on a modern automobile are not a separate system, but are part of an engine's fuel, ignition, and exhaust systems. Also, the basic mechanical design of a modern engine does a lot to reduce the amount of pollutants it emits.

■ **Touching All The Metric Bases**

Metric measurements of weight (mass) and length are becoming familiar to most of us, yet those are only two of seven base units and two supplementary units. Some of these, too, we already know:

Length = Meter
Weight (Mass) = Kilogram
Time = Second
Electric Current = Ampere
Temperature = Kelvin (Celsius)
Amount of Substance = Mole
Luminous Intensity = Candela
Plane Angle = Radian
Solid Angle = Steradian

We already use the second and the ampere. The base temperature unit, the Kelvin, is commonly used on a Celsius scale, where 0°C equals 32°F and 100°C equals 212°F. The term Celsius is now used in place of the term centigrade. The remaining base units are most often found in scientific or mathematical uses.

Hydrocarbon: A chemical compound made up of hydrogen and carbon. A major pollutant given off by an internal combustion engine. Gasoline, itself, is a hydrocarbon compound.

Carbon Monoxide: An odorless, colorless, tasteless poisonous gas. A major pollutant given off by an internal combustion engine.

Oxides of Nitrogen: Chemical compounds of nitrogen given off by an internal combustion engine. They combine with hydrocarbons to produce smog.

Particulates: Liquid or solid particles such as lead and carbon that are given off by an internal combustion engine as pollution.

Reduction Reaction: A chemical reaction in which oxygen is removed from a compound to form new compounds or to reduce a compound to its basic elements.

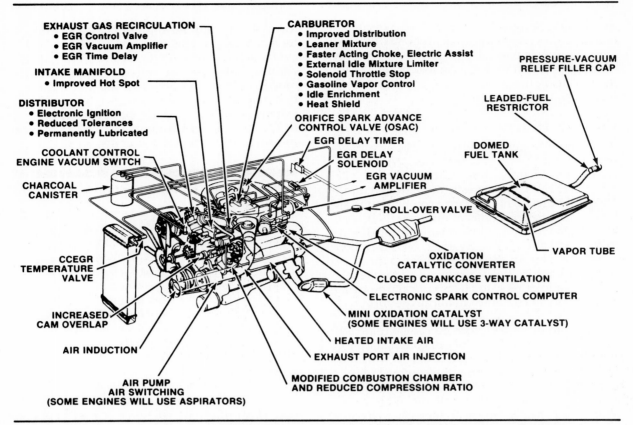

Figure 2-21. The complex emission controls on a modern automobile engine are an integral part of the fuel, ignition, and exhaust systems, as well as the engine itself. (Chrysler)

Automotive emission controls can be grouped into major families, as follows:
1. *Crankcase emission controls*—Positive crankcase ventilation (PCV) systems control HC emissions from the engine crankcase.
2. *Evaporative emission controls*—Evaporative emission control (EEC) systems control the evaporation of HC vapors from the fuel tank, pump, and carburetor.
3. *Exhaust emission controls*—Various systems and devices are used to control HC, CO, and NO_x emissions from the engine exhaust. These controls can be subdivided into the following general groups:
a. *Air injection systems*—These systems inject extra air into the exhaust system to help burn up HC and CO in the exhaust.
b. *Engine modifications*—Various changes have been made in the designs of engines to help eliminate all three major pollutants.
c. *Spark timing controls*—Carmakers have used various systems to delay or retard ignition spark timing to control HC and NO_x emissions. Most of these systems modify the distributor vacuum advance, but most late-model cars have electronic spark controls.

d. *Exhaust gas recirculation*—The most effective way to control NO_x emissions is to recirculate a small amount of exhaust gas back to the intake manifold to lower combustion chamber temperatures.
e. *Catalytic converters*—The first catalytic converters installed in the exhaust systems of 1975 and '76 cars helped the chemical oxidation, or burning, of HC and CO in the exhaust. Later catalytic converters, which began to appear on 1977 and '78 cars, also promoted the chemical reduction of NO_x emissions.

How Basic Engine Design Affects Emissions

Of the three major pollutants, CO is affected the least by internal engine design. CO comes from rich mixtures, and the richness or leanness of a mixture is controlled by the carburetor and air intake. But the distribution of the mixture depends on the intake manifold. If the manifold and the passages through the cylinder head vary greatly from cylinder to cylinder, poor distribution is the result. With poor distribution, some cylinders get rich mixtures, while

Engine Physics

others get lean mixtures. Before the days of emission controls, engineers simply could let the overall mixture at the carburetor be a little rich so that no cylinder would run too lean. They did not have to worry about the amount of CO left over in the exhaust. If engineers simply tried to lean out the mixture at the carburetor to reduce CO emissions, some cylinders would run too lean, which would cause stalling and poor performance.

Another solution to the problem of uneven fuel distribution would be to redesign the intake manifold. In fact, changes have been made in manifold design to improve fuel distribution and vaporization. These changes have been slight, however, when viewed as part of overall engine design. A perfect manifold cannot be designed unless all of the cylinders are an equal distance from the carburetor. Short of arranging the cylinders in a circle, this would be hard to do. This is why most devices to control CO emissions relate to the incoming air-fuel mixture and the outgoing exhaust, rather than to internal engine design.

In contrast to CO emissions, HC emissions are greatly affected by internal engine design. Hydrocarbons will cling to the surface of the combustion chamber, which is cooler than the center of combustion in the chamber. These hydrocarbons that are cooled on the chamber surfaces may go unburned and be emitted in the exhaust. Therefore, modern engines are designed to keep the chamber surface as small as practical.

A sphere has the smallest surface area for its volume. A cube of the same volume will have a greater surface area. The more angles and curves there are in a combustion chamber wall, the greater the surface area will be. Engineers call this the **surface-to-volume ratio**, and they want to keep it as small as possible to reduce HC emissions.

Engineers also discovered that the small area above the top piston ring, between the piston and the cylinder wall, is a gathering place for unburned gasoline. Therefore, some years ago, changes were made in many piston and piston ring designs to reduce this area.

The compression ratio affects HC and NO_x emissions. With a high compression ratio, an engine can develop more combustion heat and produce more power. However, the surface-to-volume ratio also goes up, which can produce more HC emissions.

For some years, engineers were puzzled about how to reduce NO_x emissions. Everything they did to reduce HC and CO either had little effect on NO_x or increased it. Remember that large amounts of NO_x are produced when combustion temperatures go over 2,500° F (1371° C). Design features that get the most power from an engine, such as high compression ratios and advanced ignition timing, also increase NO_x. When engine compression ratios were lowered in the early 1970's, it helped to reduce NO_x formation. However, the most effective way to reduce NO_x is through exhaust gas recirculation (EGR). Exhaust is an inert gas. It will not support combustion because it contains little oxygen. When exhaust is recirculated to the intake mixture, it displaces oxygen from the air-fuel mixture. This, in turn, lowers combustion temperatures and cuts NO_x formation.

ALTERNATIVE ENGINES

Other engine types besides the 4-stroke engine have been installed in automobiles over the years, but only four have been used with any real success—the 2-stroke, the diesel, the rotary, and the stratified charge engines.

The Two-Stroke Engine

While 4-stroke engines develop one power stroke for every two crankshaft revolutions, **2-stroke engines** produce a power stroke for each revolution. Poppet valves are not used in most automobile and motorcycle 2-stroke engines. The valve work is done instead by the piston, which uncovers intake and exhaust ports in the cylinder as it nears the bottom of its stroke.

Two-stroke gasoline engines used in motorcycles and some small cars suck the air-fuel mixture into the crankcase where it is partially compressed for delivery to the cylinders. It is easy to see that, as the piston moves up, pressure increases in the cylinder above it while pressure decreases in the crankcase below it. By using the crankcase to pull in the air-fuel mixture, the 2-stroke engine combines the intake and compression strokes. Also, by using exhaust ports in the cylinder wall, the power and exhaust strokes are combined. The operat-

Surface-To-Volume Ratio: The ratio of the surface area of a three dimensional space to its volume.

Two-Stroke Engine: An engine in which a piston makes two strokes to complete one operating cycle.

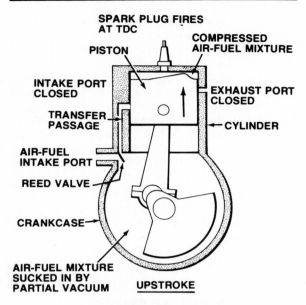

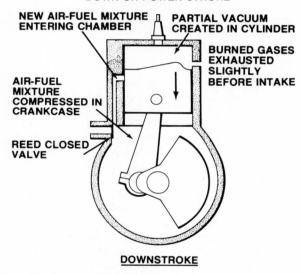

Figure 2-22. Two-stroke cycle engine operation.

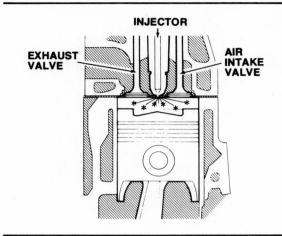

Figure 2-23. Diesel combustion occurs when fuel is injected into the hot, highly compressed air in the cylinder. (Cummins)

ing cycle of a 2-stroke engine shown in figure 2-22 works like this:

1. Intake and Compression: As the piston moves up, a low-pressure zone is created in the crankcase. A **reed valve** opens, and the air-fuel mixture is sucked into the crankcase. At the same time, the piston compresses a previous air-fuel charge in the cylinder. Ignition occurs in the combustion chamber when the piston is near top dead center.

2. Power and Exhaust: Ignited by the spark, the expanding air-fuel mixture forces the piston downward in the cylinder. As the piston travels down, it uncovers exhaust ports in one side of the cylinder. The exhaust flows through the ports and out of the engine. As the piston continues down, it compresses the air-fuel charge in the crankcase. The intake reed valve or rotary valve closes to hold the charge in the crankcase. After uncovering the exhaust ports, the piston uncovers similar intake ports nearer the bottom of the stroke. The compressed mixture in the crankcase moves through a transfer passage to the intake ports and flows into the cylinder. The force of the incoming mixture also helps to drive remaining exhaust gases from the cylinder. A ridge on the top of the piston deflects the intake mixture upward in the cylinder so it will not flow directly out the exhaust ports.

The piston then begins another upward compression stroke, closing off the intake and the exhaust ports, and the process begins again. Because the crankcase of a 2-stroke gasoline engine is used for air-fuel intake, it cannot be used as a lubricating oil reservoir. Therefore, engine oil for a 2-stroke gasoline engine must be mixed with the fuel or injected directly into the cylinder.

In theory, the 2-stroke engine should develop twice the power of the 4-stroke engine of the same size, but the 2-stroke design also has its practical limitations. With intake and exhaust occurring at almost the same time, it does not breathe or take in air efficiently. Mixing fresh fuel with unburned fuel preheats the mixture. Because this increases volume, it reduces efficiency.

Engine Physics

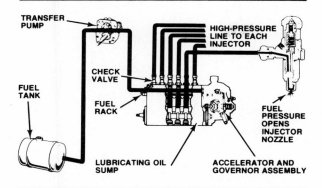

Figure 2-24. Typical automotive diesel fuel injection system.

The Diesel Engine

In 1892, a German engineer named Rudolf Diesel perfected the compression ignition engine named after him. The diesel engine uses heat created by compression to ignite the fuel, so it requires no spark ignition system.

The diesel engine requires compression ratios of 16:1 and higher. Incoming air is compressed until its temperature reaches about 1,000° F (538° C). As the piston reaches the top of its compression stroke, fuel is injected into the cylinder, where it is ignited by the hot air, figure 2-23. As the fuel burns, it expands and produces power.

Diesel engines differ from gasoline-burning engines in other ways. They do not have carburetors, but have a precision fuel injection pump. This device sprays a precise amount of fuel into the combustion chamber at the required time for efficient combustion, figure 2-24. The diesel fuel injection system performs the fuel-delivery job of the carburetor and the ignition-timing job of the distributor in a gasoline engine.

The air-fuel mixture of a gasoline engine remains nearly constant—changing only within a narrow range—regardless of engine load or speed. But in a diesel engine air remains constant and the amount of fuel injected is varied to control power and speed. The air-fuel mixture of a diesel can vary from as lean as 85:1 at idle, to as rich as 20:1 at full load. This higher air-fuel ratio and the increased compression pressures contribute to the diesel's greater efficiency in terms of fuel consumption than a gasoline engine.

Like gasoline engines, diesel engines are built in both 2-stroke and 4-stroke versions. The most common 2-stroke diesels are the truck and industrial engines made by the Detroit Diesel Allison division of General Motors.

In these engines, air intake is through ports in the cylinder walls, and exhaust is through valves in the head. Crankcase fuel induction is not used in 2-stroke diesels, and air intake is aided by supercharging.

Although diesel engines are primarily used in heavy equipment, they have been used in Mercedes-Benz cars for over 40 years. Now, with the appearance of diesel-powered cars from Volkswagen, General Motors and other manufacturers, the trend appears to be toward greater use of diesel engines in automobiles.

The Rotary (Wankel) Engine

The reciprocating motion of a piston engine is both complicated and inefficient. For these reasons, engine designers for years have at-

---

■ Rudolf Diesel

The theory of a diesel engine was first set down on paper in 1893 when Rudolf Diesel wrote a technical paper, "Theory and Construction of a Rational Heat Engine." Diesel was born in Paris in 1858. After graduating from a German technical school, he went to work for the refrigeration pioneer, Carl von Linde. Diesel was a success in the refrigeration business, but at the same time he was developing his theory on the compression ignition engine. The theory is really very simple. Air, when it is compressed, gets very hot. If you compress it enough, say on the order of 20:1, and squirt fuel into the compressed air, it will ignite.

In its first uses, the diesel engine was used to power machinery in shops and plants. By 1910, it was used in ships and locomotives and a 4-cylinder engine was even used in a delivery van. The engine proved too heavy at that time to be practical for automobile use. It was not until 1927, when Robert Bosch invented a small fuel injection mechanism, that the use of the diesel engine in trucks and cars became practial.

Unlike many of the early inventors, Diesel did make a great deal of money from his invention, but in 1913 he disappeared from a ferry crossing the English Channel and was presumed to have committed suicide.

Reed Valve: A one-way check valve. A reed, or flap, opens to admit a fluid or gas under pressure from one direction, while closing to prevent movement from the opposite direction.

Eccentric: Off center. A shaft lobe which has a center different from that of the shaft.

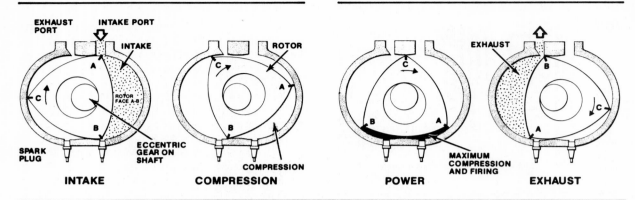

Figure 2-26. This illustration shows the four stages of rotary engine operation. They correspond to the intake, compression, power, and exhaust strokes of a four-stroke reciprocating engine. The sequence is shown for only one rotor face, but each face of the rotor goes through all four stages during each rotor revolution.

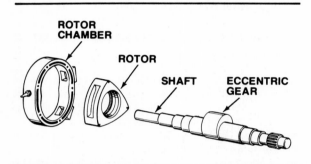

Figure 2-25. The main parts of a Wankel rotary engine are the rotor chamber, the three-sided rotor, and the shaft with an eccentric gear.

tempted to devise engines in which the working parts would all rotate. The major problem with this rotary concept has been the sealing of the combustion chamber. Of the various solutions proposed, only the rotary design of Felix Wankel (as later adapted by NSU, Curtiss-Wright, and Toyo Kogyo-Mazda), has proven practical.

Although the same sequence of events takes place in both a rotary and a reciprocating engine, the rotary is quite different in design and operation. A curved triangular rotor moves on an **eccentric**, or off-center, geared portion of a shaft within a long chamber, figure 2-25. As it turns, the rotor's corners follow the housing shape. The rotor thus forms separate chambers whose size and shape change constantly during rotation. The intake, compression, power, and exhaust functions occur within these chambers as shown in figure 2-26. Wankel engines can be built with more than one rotor. Mazda engines, for example are 2-rotor engines.

About equivalent in power output to that of a 6-cylinder piston engine, a 2-rotor engine is only one-third to one-half the size and weight. With no pistons, rods, valves, lifters, and other reciprocating parts, the rotary engine has 40 percent fewer parts than a piston engine.

One revolution of the rotor produces three power strokes or pulses, one for each face of the rotor. In fact, each rotor face can be considered the same as one piston. Each pulse lasts for about three-quarters of a rotor revolution. The combination of rotary motion and longer power pulses which overlap results in a smooth-running engine. While the rotary overcomes many of the disadvantages of the piston engine, it has its own disadvantages.

The Stratified Charge Engine

The ideal air-fuel mixture for a 4-stroke gasoline engine is about 14.7:1. If the mixture could be made leaner (closer to 20:1), mileage and emissions could be improved, with just a slight loss of power. However, a 20:1 mixture is so lean that it cannot always be ignited by the spark.

The problem then is how to ignite a 20:1 mixture? One answer is to use a **stratified charge**. This means that the mixture is composed of layers, rich in some areas, lean in others. The rich area should be near the spark plug for easy ignition. Once combustion is started, it will spread to the leaner areas for complete burning.

The concept of the stratified charge engine has been around in many forms for many years. Honda, however, was the first carmaker to produce and use one successfully. Honda's Compound Vortex Controlled Combustion (CVCC) design was the first stratified charge gasoline engine used in a mass-produced car.

The CVCC engine has a separate small precombustion chamber located above the main

Engine Physics

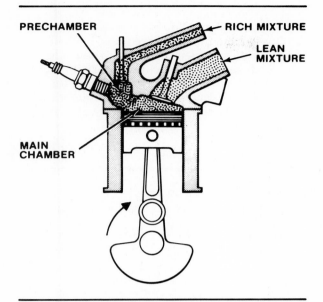

Figure 2-27. Honda CVCC cylinder head, showing main combustion chamber and precombustion chamber.

combustion chamber and containing a tiny extra valve, figure 2-27. Except for this feature, the CVCC is a conventional 4-stroke piston engine. However, it uses a 2-stage combustion process. In the first stage, the rich air-fuel mixture is ignited in the precombustion chamber. In the second stage, the flame front moves down into the main chamber to ignite a mixture with less fuel in it. A close look at the carburetor for the CVCC engine shows that it is really two carburetors in one. The main section supplies the lean mixture to the main combustion chamber. The auxiliary section supplies the rich mixture to the precombustion chamber. Separate passages in the intake manifold carry the lean and rich mixtures.

Another way to provide a stratified charge is with fuel injection. Ford Motor Company has developed its Programmed Combustion (PROCO) engine using this method. In Ford's PROCO engine, the combustion chamber is formed by a bowl in the top of the piston. A fuel injection nozzle is aimed directly at the bowl. The flat top of the piston outside the bowl gives a squishing action to the air as the piston reaches the top of the cylinder. Two spark plugs are located in the top of the cylinder so that they are in the richest part of the mixture.

Another stratified charge engine that uses fuel injection is the Texaco Controlled Combustion System (TCCS) engine. In this engine, fuel is sprayed directly on the spark plug electrodes to ensure ignition and to get charge stratifica-

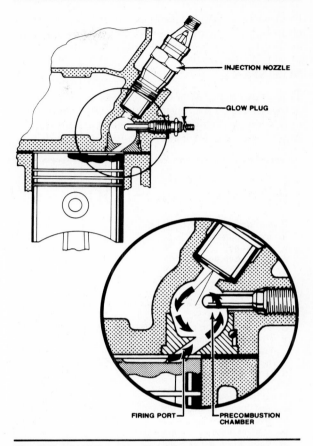

Figure 2-28. Volkswagen's diesel car engine uses a precombustion swirl chamber.

tion. The TCCS engine can be run on many different fuels, including gasoline, diesel fuel, kerosene, and alcohol. This engine is being produced by White Engines, Inc. for use in portable powerplants, forklifts, and utility vehicles.

The stratified charge principle is a method of controlling the combustion process. It does not represent a type of engine construction, such as the reciprocating or rotary engine types. In fact, charge stratification has been applied both to reciprocating diesel engines and to rotary gasoline engines. Most passenger car diesel engines have a precombustion chamber into which the fuel is injected, figure 2-28. This allows the combustion to occur in two stages: in

---

Stratified Charge: A nonuniform air-fuel mixture, arranged more or less in layers. The charge is rich in fuel in some areas, lean in others.

the precombustion chamber and in the main chamber. This improves combustion efficiency and reduces engine noise and vibration.

SUMMARY

This chapter reduces the processes accomplished by an engine to understandable and specific terms. It also gives an understanding of how these terms can be applied to different engine designs. You have seen that internal combustion engines all share the same scientific principles in their operation. Their purpose is to produce work. Some designs do it better than others, with better gas mileage, or longer life. In the last analysis, the engine that is inexpensive to manufacture, and highly efficient, will be the engine of the future.

Review Questions

Choose the single most correct answer.
Compare your answers to the correct answers on page 268.

1. Power is measured in:
   a. Foot-pounds
   b. Kilowatts
   c. Newton-meters
   d. Horses

2. Torque is:
   a. The power required to tighten a nut
   b. A kind of wrench
   c. The force applied to a lever being turned, multiplied by the length of the lever
   d. The rate of doing work

3. In a gasoline engine:
   a. A detonation occurs once in every cycle
   b. Engine output depends on the octane rating of the fuel
   c. All the fuel is burned and produces useful power
   d. None of the above

4. The most useful specification of engine output is:
   a. The indicated horsepower
   b. Stated in gross brake horsepower
   c. Measured at the flywheel
   d. Stated in net brake horsepower

5. One horsepower is equivalent to:
   a. The output of a horse pulling a 100-lb weight a distance of 100 feet in 1 minute
   b. 0.746 kilowatt
   c. 3,300 foot-pounds per minute
   d. None of the above

6. How is engine output in horsepower calculated?
   a. Torque multiplied by rpm, divided by 5252
   b. Foot-pounds multiplied by rpm, divided by the diameter of the flywheel
   c. Crankshaft rotation speed multiplied by average piston thrust
   d. None of the above

7. Air pressure is specified in which of the following units?
   a. Millimeters of mercury
   b. Pounds per cubic inch
   c. Newton-meters
   d. Liters per square meter

8. The octane rating of a fuel indicates:
   a. Which grade of fuel is most powerful
   b. The proportion of pure gasoline to additives
   c. The speed of combustion of a fuel
   d. The ability of a fuel to burn under high pressure without detonation

9. What is the main function of a venturi?
   a. To act as a nozzle through which air and gas are fed into the cylinders
   b. To restrict air supply to guarantee sufficient fuel flow
   c. To create air turbulence, providing a better air-fuel mixture
   d. To increase the air intake speed, drawing additional fuel into the air flow

10. What is the stoichiometric ratio?
    a. The ratio of gallons of fuel to pounds of air
    b. A scientific term for volumetric efficiency
    c. The air-fuel ratio at which all of the liquid fuel present will blend with all the oxygen available and be completely burned
    d. The ratio of exhaust gas volume to inlet mixture volume

11. The purpose of emission control is to minimize the release of:
    a. Hydrocarbons into the atmosphere
    b. Nitrogen oxides into the atmosphere
    c. Carbon monoxide into the atmosphere
    d. All of the above

12. The smallest possible surface-to-volume ratio of a combustion chamber produces:
    a. Minimum fuel consumption
    b. Maximum net brake horsepower
    c. Maximum combustion efficiency
    d. Minimum unburned hydrocarbons

13. Which of the following is not an automotive emission control system?
    a. Exhaust gas recirculation
    b. Air injection
    c. Fuel injection
    d. Catalytic converter

14. A diesel engine requires:
    a. The same number of spark plugs as any other internal combustion engine
    b. Air-fuel mixture kept constant
    c. Compression ratio 16:1 or higher
    d. A modified carburetor

15. In a stratified charge engine:
    a. The air-fuel mixture near the spark plugs is richer than elsewhere
    b. There are two carburetors
    c. The resulting mixture is richer and easier to burn
    d. Each cyclinder has several combustion chambers

16. The momentum of an object is:
    a. The same as inertia, but only applies to rotating objects
    b. The object's weight multiplied by its speed
    c. The opposite of friction
    d. Its tendency to accelerate in a straight line, unless it meets resistance

Chapter 3

Basic Metallurgy and Machine Processes

ENGINE METALLURGY

Metallurgy is the study and science of the properties of metals. Many different metals, and combinations of metals called alloys, are used in an automobile engine. In order to service an engine properly, you do not need to know how to select the raw material to make an engine part. However, you should know something about why different metals are used for different parts of an engine and what the comparative strengths, weaknesses, and compatibilities of these metals are. This knowledge will help you do a better job in the various machining and assembly processes that you will use in the repairing or rebuilding of an engine.

Iron

Iron is the basic metal that comes from iron ore. It is heated in a blast furnace to approximately 1400° F (760° C). Coal is burned in the furnace, and some of the coal combines with the iron. Coal is really carbon. When the liquid iron is drained from the furnace, it is a mixture of iron and carbon, with several impurities. The iron is allowed to flow into channels dug in sand that look somewhat like a mother pig feeding piglets. This is the cast iron called **pig iron**, that is used to make steel, figure 3-1.

The **slag** is a mixture of all impurities in the ore. Slag is lighter than liquid iron. It floats on the top of the iron and is drawn off and allowed to solidify. Slag is mainly rock and stone. It can be used as building blocks or crushed to make gravel.

Steel

Pig iron contains about 2- to 6-percent carbon. This high percentage of carbon makes the iron brittle and of low strength. Steel, on the other hand, has a carbon content of about 0.1 percent to 1.7 percent. This makes the steel much stronger.

To make steel, pig iron is heated in a furnace to about 1600° F (871° C), so that the carbon burns off. Other impurities such as phosphorus, sulfur, silicon, and manganese also burn off, except for small quantities. When the refining process is finished, high-carbon steel has a carbon content of about 0.7 percent to 1.7 percent. Medium- and low-carbon steels run from about 0.1 percent to 0.6 percent carbon.

The best steels for use in an engine are the high-carbon steels. Other elements are added, such as tungsten, chromium, vanadium, or cobalt, to make stronger steel **alloys**, figure 3-2. It may seem peculiar that the best so-called

Figure 3-1. Pig iron is made by combining iron and coal or carbon, heating it, and pouring the liquid into channels dug into the sand. (American Iron and Steel Institute.)

"high-carbon" steels do not have as much carbon in them as pig iron. This situation occurs because the high-carbon steels are not being compared to iron, but to lower carbon steels.

Aluminum

Aluminum is not as strong as steel, but it is much lighter, about one-third the weight of steel. Engine front covers or water pump housings are frequently made of aluminum. Aluminum is more expensive than steel, so it is not usually used where a thin piece of steel will do the job. An example is an engine rocker arm cover, usually stamped of thin sheet steel.

Aluminum has been used to make engine blocks. However, in most cases, an aluminum block requires either steel cylinder liners or special casting and machining processes to make the cylinder durable enough for long-term use in a car engine. Either of these alternatives increases the cost and has definite disadvantages for high-volume engine production. Aluminum blocks are used, however, in some engines for high-priced limited-production cars.

Whereas there are few aluminum cylinder blocks, there are many aluminum cylinder heads. More aluminum parts may be used on engines in the future because of its weight advantage, which translates into better fuel mileage.

Alloy: A mixture of two or more metals.

Pig Iron: Small solidified blocks of iron poured from the blast furnace into channels dug in sand. Used to make steel.

Slag: Impurities in iron ore. Removed before ore is processed into steel.

Basic Metallurgy and Machine Processes

THE EFFECT OF ALLOYING ELEMENTS ON STEEL

Effect	Carbon	Chromium	Cobalt	Lead	Manganese	Molybdenum	Nickel	Phosphorus	Silicon	Sulphur	Tungsten	Vanadium
Increases tensile strength	X	X			X	X	X					
Increases hardness	X	X										
Increases wear resistance	X	X			X		X				X	
Increases hardenability	X	X			X	X	X					X
Increases ductility					X							
Increases elastic limit		X				X						
Increases rust resistance		X					X					
Increases abrasion resistance		X			X							
Increases toughness		X			X	X	X					X
Increases shock resistance		X					X					X
Increases fatigue resistance												X
Decreases ductility	X	X										
Decreases toughness			X									
Raises critical temperature		X	X								X	
Lowers critical temperature					X		X					
Causes hot shortness										X		
Causes cold shortness								X				
Imparts red hardness			X			X					X	
Imparts fine grain structure					X							X
Reduces deformation					X	X						
Acts as deoxidizer					X				X			
Acts as desulphurizer					X							
Imparts oil hardening properties		X			X	X	X					
Imparts air hardening properties					X	X						
Eliminates blow holes								X				
Creates soundness in casting									X			
Facilitates rolling and forging					X				X			
Improves machinability				X						X		

Figure 3-2. This table shows the effects of alloys on steel.

Bearing Alloys

The best bearings for crankshafts, camshafts, and connecting rods are made from **babbitt**, copper alloy, aluminum alloy, or silver alloy. Babbitt is named for Isaac Babbitt, who discovered the formula with a base of tin and small amounts of copper and antimony. This is called tin-based babbitt. There is also lead-based babbitt, which is mostly lead with small amounts of tin and antimony.

The various bearing alloys are selected for four main properties. First, they must be compatible with the rotating shaft. An example of incompatibility is steel on steel. No matter how well it is lubricated, steel rubbing on steel will wear out quickly. But a bearing alloy with good compatibility will rub on the steel crankshaft and last a long time. The wear that does eventually occur will be mostly on the bearing, and only slightly on the shaft. This is what is desired. Bearings can be replaced more cheaply than crankshafts.

The bearing alloy also must be able to carry heavy loads without crushing or cracking. And it must be able to conform to misalignment. Last, and very important, the alloy must have good **embedability**. This means it must be able to absorb small pieces of dirt to a level even with, or below, the surface of the bearing. With good embedability, the dirt will bury itself in the bearing and not scratch the shaft.

All bearing material selection is a compromise. The engineer must select the bearing material that he feels will hold up best under the type of use that the engine will receive. Bearings will be discussed at length in Chapter 9 of this book.

Exotic Materials

Iron, steel, and aluminum are the principal metals used in engine construction. Some other so-called "exotic" metals are used in expensive race car and limited-production engines. Some successful experiments have even been made with engine parts made of nonmetallic materials.

Titanium

Titanium is one metal with some of the properties of both iron and steel. The main advantages of titanium are its light weight and high strength. It is about the same weight as aluminum, but about one-third stronger. Titanium, however, is expensive and its use is generally limited to custom or racing engines. It is often used as the principal element in an

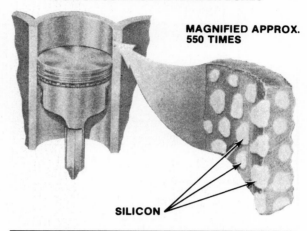

Figure 3-3. Small particles of silicon are a part of the surface of the cylinder in this aluminum silicon engine block. (Chevrolet)

alloy to make valves and valve retainers. Titanium connecting rods also are used in some racing engines.

Aluminum-silicon

Aluminum blocks usually have steel or cast iron sleeves, or liners, in the cylinders. The pistons and rings operate in the sleeves and do not require special manufacturing or maintenance. When the Chevrolet Vega 4-cylinder engine was designed, sleeves were not used.

The Vega engine was die-cast out of aluminum alloyed with 17-percent silicon. The silicon was in the form of small particles, each measuring about 0.001" (0.025 mm). After casting, the surface of the metal was a mix of small particles of silicon and aluminum. The cylinder bores were honed to a 7-microinch finish. Then an electrochemical etching process removed about .00015" (.0038 mm) of aluminum. The resulting finish was pure silicon, with minute gaps between the silicon particles, figure 3-3. However, the silicon particles were so close together that they provided a good wearing surface for the pistons and rings.

To make the aluminum pistons operate in the hard silicon bores without seizing, the pistons were electroplated with a hard iron coating. This gave approximately the same durability as aluminum pistons in a cast iron bore. The Vega engine was used for seven years, and then discontinued. Its main problem was the engine damage that occured when the cooling system was neglected and the engine overheated.

Basic Metallurgy and Machine Processes

Figure 3-4. The 1981 Mercedes-Benz 380 SL and 380 SEL have all-aluminum engines with aluminum cylinders. (Mercedes-Benz)

Figure 3-5. American Motors reduced the weight of the valve covers by the use of composites. (American Motors)

The aluminum-silicon process did not disappear with the death of the Vega though. A similar process is used on the Mercedes-Benz 380 SL and SEL. These cars have all-aluminum engines with aluminum cylinders, figure 3-4. The aluminum has a very high silicon content and before use the inside walls of the cylinders are etched with an acid to remove most of the aluminum, allowing the pistons to ride on silicon particles.

Composites

The general name used by scientists and engineers for a wide variety of manmade nonmetallic materials is "**composites**." This name is usually applied to advanced types of plastic and ceramic materials. A common example of a composite material is fiberglass. The substrate, or ceramic core, of a catalytic converter can be considered a composite material.

It may be hard to picture engine parts made from such materials, but one composite that shows promise for such use is a graphite-reinforced fiber material. Some experimental connecting rods and pushrods have been built successfully from graphite-reinforced fiber materials. Ford Motor Company has built a series of production engines with accessory mounting brackets made from this material.

For 1981, American Motors came out with plastic or glass-filled nylon valve covers that reduced engine weight, figure 3-5.

Weight reduction is the main reason for building engine parts of composite materials. Less engine weight means better fuel mileage. While these materials are not common in production engines at present, engineers will continue to develop them for use in the near future.

MANUFACTURING PROCESSES

All iron parts in an engine are not the same, nor are all steel or aluminum parts. The iron used in a block may differ in its alloy composition from that used in a crankshaft or a camshaft. Equally important, different parts are manufactured by different processes. As a skilled engine serviceman, you should know the differences between **casting**, **forging**, stamping, and other manufacturing processes used to make engine parts.

Casting

A cast metal is poured into a mold. An iron casting usually has a coarse, grainy appearance

---

Casting: The process of making metal parts by pouring the liquid metal into molds where it solidifies.

Forging: Pressing metal in a plastic state into molds under great pressure.

Babbitt: Metal alloy of tin mixed with copper and antimony. Lead-based babbitt is lead with small amounts of tin and antimony. Used to make bearings.

Embedability: The ability to absorb small particles into a bearing metal at a level with, or below, the surface.

Composite: A manmade, nonmetallic material. The term is usually applied to advanced plastic or ceramic materials.

because the mold is made of sand. After the casting is made, the sand mold is broken away from the metal. If the interior of the casting is hollow, the sand that fills the cavities is poured out. Automotive engine blocks are made this way. The round holes in the sides of the block are there for removing the sand. These holes must be closed with expansion plugs when the engine is assembled.

Although sand casting is the oldest and the basic casting method, there are many other ways to cast metals. Aluminum is usually cast in molds that are reused for thousands of castings. These are called permanent molds. In die casting, a permanent mold is also used, but the liquid metal is forced into the mold under pressure. There is also centrifugal casting. As the metal is poured, the mold is spun so the centrifugal action forces the metal into the outer parts of the mold.

Forging

Molds for forging are made in two halves. The metal to be forged is heated to its plastic state, but not enough to be liquid. The two halves of the mold are forced onto the hot metal with tremendous force. The metal is forced into the cavities of the mold and assumes its shape. Considerable metal squeezes out between the halves of the mold and must be trimmed off. Because of the tremendous pressure, forging produces strong parts, without air holes or imperfections.

To understand an important difference between a casting and a forging, you should realize that metal can have a grain just like wood. Metal from raw ore must be formed into different shapes for storage, handling, and use. It would not be practical to keep metal in a molten state from the time it is extracted from the ore until a finished product is made. Therefore, the molten metal is cooled and formed into bars, sheets, **billets**, and other shapes. As this forming is done, the metal develops a grain because of the way the molecules of the material align with each other. You do not need to know all about the chemistry involved, but you should know that this grain exists and that it is an important factor in adding strength to certain metal parts.

When metal is in a molten form, it has no grain. The molecules are arranged in random directions. Because a cast part is made directly from molten metal, it will have no grain. To make a forging, a metal bar or billet is heated until it is plastic (flexible). The grain of the original bar or billet remains intact. The design of the forging is calculated to take advantage of the original grain structure to add strength to the finished product. The high pressure of the forging process actually strengthens the finished part.

There is an old saying in the metal industry that, "a forging is stronger than a casting." This used to be quite true, and still is today for identical parts made from the same material. However, modern alloying processes and casting methods have made it possible to cast some parts that formerly could only be made strong enough by forging. Years ago, crankshafts and camshafts were usually forged because of their strength requirements. Today, cast crankshafts and camshafts are strong enough for hundreds of thousands of miles of use in passenger car engines. Cast parts are cheaper to make because the tooling required is cheaper than forging equipment.

Stamping

If a part is thin and does not have too many complicated angles and surfaces, it can be made by stamping it out of a large sheet of cold metal. If the finished part is to be flat, it is stamped out on a punch press. This works

---

■ The Engine That Couldn't Be Made

Before 1932, there were few V-8 engines in cars. When a V-8 did appear, it usually consisted of two 4-cylinder engines bolted together. The crankcase was separate, and a four cylinder block was bolted on each side to form a V-8.

To mass produce such a V-8 engine was too expensive. The only way that low priced cars would ever get V-8 power was by casting the V-8 block in one piece. But engineers and foundry people said it couldn't be done. There were just too many core pieces that might shift and ruin the casting. Nobody wanted to try it. Nobody but Henry Ford.

Ford had been up against negative thinkers before. Every time he wanted to do something different, everybody told him it couldn't be done. So he just developed his own methods, and went ahead and did it.

The 1932 Ford V-8 was the first mass produced V-8 engine cast in one piece. It was the first time that low priced cars had the advantage of the smoothness and power of a V-8. But it wasn't easy. The cores for the casting totaled over 50 pieces. And all of them had to be held in precise position, or the resulting block had to be scrapped. After the engine came out, everybody hailed it as a tremendous accomplishment. Mr. Ford just repeated his oft-quoted statement, "Anything that can be drawn up can be cast."

Basic Metallurgy and Machine Processes

TEMPERING COLORS AND APPROXIMATE TEMPERATURES FOR CARBON STEEL		
Color	Celsius temperature	Use
Pale yellow	220°	Lathe tools, shaper tools.
Light straw	230°	Milling cutters, drills, reamers.
Dark straw	245°	Taps and dies.
Brown	255°	Scissors, shear blades.
Brownish purple	265°	Axes and wood chisels.
Purple	275°	Cold chisels, center punches.
Bright blue	295°	Screw drivers, wrenches.
Dark blue	315°	Wood saws.

Figure 3-6. Steel turns different colors as it is heated. This chart indicates the colors and uses for carbon steel when heated.

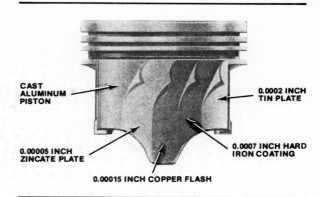

Figure 3-7. When a part is coated with a thin surface of metal it is called plating.

much like a cookie cutter stamping out cookies from a sheet of dough.

If the finished part is not flat, it must be stamped, or drawn, into shape and then trimmed. For a part that is rather shallow, the stamping can be done in one draw, followed by trimming. For deeper parts, two or three draws may be necessary to form the part into its final shape. Engine oil pans, valve covers, and timing covers are usually made by stamping. This is the same basic method used to make most sheet metal car body parts.

Machining From Billets

This is probably the most time-consuming and expensive way to make a part, but it produces some of the strongest and most precise engine parts. A steel billet is placed in a lathe, or similar machine tool, and the part is cut with extreme accuracy to the size and shape wanted. Crankshafts and camshafts for racing engines are often machined from steel billets.

MATERIAL TREATMENTS

After metal parts are formed, machined or made by any process, they usually need further treatment before they are ready for use in an engine. The metal may be treated by applying heat or through any of several chemical processes. Also, the desired surface finish (smoothness or roughness) may be created by machining or by chemical processes.

Heat Treating

Steel is one of the few substances that can be hardened by heating and cooling. When a high-carbon steel is heated to about 1400° F (760° C) and then quickly cooled in water, it hardens. The steel then becomes so hard that it is difficult to cut or work. To make the steel workable, the process of **tempering** is used, figure 3-6. In tempering, the steel is heated to 396° to 567° F (220° to 315° C) and allowed to cool slowly. The result is a steel that is slightly softer and more easily worked.

Annealing is a heating process for completely softening the steel. It is heated to above 1400° F (760° C) and then allowed to cool slowly. After machining, the same steel can be hardened and tempered, if desired.

Plating

When a coating of copper, nickel, chromium, or other metal is applied to the surface of a part, it is called **plating**, figure 3-7. The coating can be painted or sprayed on, but usually it is applied in a process called electroplating. The part is immersed in a tank of chemicals that contain the plating metal in solution. Direct current is then applied to the solution through an electrode. This causes the metal in the solution to be deposited on the part.

Plating can increase the wear resistance of the part and usually improves the appearance of the part.

Billet: A forged piece of steel with a cross sectional area of 4 to 36 square inches (26 to 232 square centimeters).

Temper: The process of reheating steel to temperatures between 396° and 567° F (220° to 315° C), then allowing it to cool slowly to make the steel more workable.

Anneal: To soften steel by heating to 1400° F (760° C) or above and allowing to cool slowly. Removes internal stresses.

Plating: Applying a thin coat of metal onto the surface of another substance.

Anodizing

Electroplating is a process that is performed by running electric current through a solution. The part to be plated is the cathode, or negative electrode. The plating metal is the anode, or positive electrode. When the current is flowing, metal from the anode is deposited on the cathode. In some plating, the anode does not take part in the chemical process. It only supplies the electric current. The metal is part of the solution, and it flows to the cathode when the current is flowing.

In **anodizing**, the plating process is reversed. The metal to be anodized is the anode. When current is flowing, the metal leaves the anode and is deposited at the cathode. A special solution is used that results in oxygen combining with the anode as the metal leaves. The result is a coating of oxide on the anode. In the case of aluminum, this oxide coating is very durable. It does not conduct electricity, and it reduces corrosion of the aluminum. Various processes have been developed to give the anodizing a color tint. Although it is most often used to prevent corrosion, it can also be used for decoration. Magnesium and titanium are some other metals that can be anodized.

Surface Finishes

Every surface has minute grooves or imperfections. A highly polished metal surface may appear to be perfectly smooth, but if it is examined under a microscope, all sorts of hills and valleys will appear. The amount of roughness of a surface can be measured by a surface analyzer. The analyzer moves a diamond stylus over the surface of the metal. As the stylus rises and falls over the uneven surface, its movement is magnified and measured by the machine. The result is a measurement in millionths of an inch of how rough the surface is.

A millionth of an inch is called a **microinch**. It is abbreviated by the lowercase Greek letter Mu, written μ, followed by the word inch or its abbreviation. The surface analyzer usually gives its readings of the roughness of a surface in the form of a graph. The actual hills and valleys, magnified many times, can be seen on the graph. The graph tells the story, but to express in numbers what the graph shows, an average of the height of all the ridges is taken. The surface analyzer does the calculation and gives the reading in microinches.

There are two ways to do the average calculation. If it is done by simple arithmetic, it is called the arithmetic average, or aa method. A more complicated calculation is the root mean square, or rms method. Microinch readings are often listed with aa or rms after the reading, to indicate how the calculation was done. The difference between the two readings is about 10 percent, the aa being smaller. Root mean square measurements are more common.

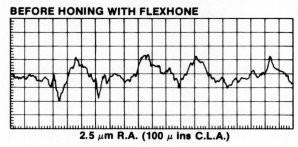

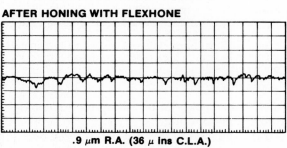

Figure 3-8. Comparing these two charts, you can see the difference in the surface after honing.

A rough cut from a saw gives a finish of about 800 microinches. A casting surface about 700, but there is great variation. The ordinary drill gives a reading of about 70 to 20 microinches. It is only when surfaces are ground, honed, or polished that they become smooth, figure 3-8. Grinding will give a surface finish between 10 and 50 microinches. Honing with fine stones can reduce that to below 5. The smoothest finish is the superfinish. It is a honed finish, but the hones are scrubbed back and forth until they polish the surface. Superfinishing can produce a surface as smooth as 1 microinch.

BASIC MACHINE PROCESSES

In rebuilding an engine, you will use various machine tools and perform many different machining operations. All of these operations can be categorized in groups of basic machine processes. The two primary categories are internal and external machining. Some machine operations actually fall into both groups. Understanding the terms for these basic machine processes, as well as their differences and similarities, will help you do a better job of engine service.

Basic Metallurgy and Machine Processes

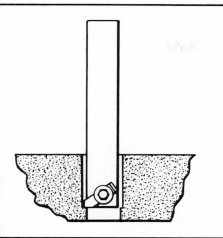

Figure 3-9. Boring, done in a hole that has already been cut, is more precise than drilling.

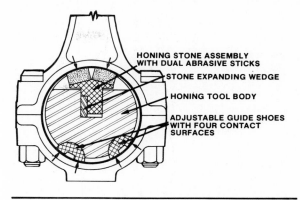

Figure 3-10. Honing finishes a hole to the exact diameter needed.

Internal Machining

Internal machining is working inside the engine cavities. It includes boring, honing, drilling, reaming, tapping, and broaching. In boring, **honing**, **drilling**, and **reaming**, the object is to make a round hole to a particular size. **Tapping** is the process of cutting threads inside a hole. **Broaching** is used to change the shape of an opening or enlarge it.

Boring

The boring process is used in a hole that has already been drilled or cast, figure 3-9. Boring provides a smooth surface that is much more precise than drilling, but not as accurate as honing.

Boring is typically done with a bar that has a small cutting tool on the side. The boring bar is inserted in the hole and rotated to make the cut. As the bar revolves, it is slowly moved through the hole by the feed mechanism. Boring is used most commonly on automotive engines for enlarging the cylinders to the next available piston size.

Another boring process is used to align main bearing seats and camshaft bearings and ensure that they are concentric with or have the same center as each other. This process is called align boring or line boring because a series of holes is bored in a straight line.

The machine used to bore cylinders can only be used for that purpose. Similarly, a line-boring machine can only be used for its specific purpose. Other boring tools are made to bore connecting rods and pistons in order to fit piston pins.

Honing

A hone is a tool with rotating abrasive stones used to finish an opening, such as a cylinder, to an exact dimension and a specific finish, figure 3-10. In cylinder honing the desired finish differs depending on the piston rings used.

Honing can be done by rigid or flexibly mounted stones. The rigid stones are usually several inches long, but not as long as the hole being finished. During the operation, the rigid hone is expanded against the walls of the hole or cylinder. The pressure against the walls can be increased or decreased depending on the

---

Anodize: The process of oxidizing a metal such as aluminum, titanium, or magnesium by running electric current through a special solution containing the metal anode.

Microinch: One millionth of an inch, abbreviated μ inch or μ in.

Hone: To finish the interior surface of a cylinder with an abrasive device. A hone is the device used for finishing the internal surfaces.

Drill: The cutting of round holes. A drill is an end-cutting tool.

Ream: The process of smoothing and sizing round holes in metal. A reamer is the side-cutting tool used for finishing already-drilled holes.

Tap: The process of cutting threads in a hole. A tap is the tool used to form an inside thread.

Broach: To finish or change the shape of a roughly cut hole. A broach is the metal tool for finishing holes, usually used to form an irregular hole from a round hole.

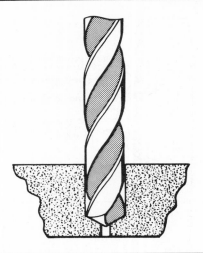

Figure 3-11. The drill uses a point that is forced against the surface to make an opening.

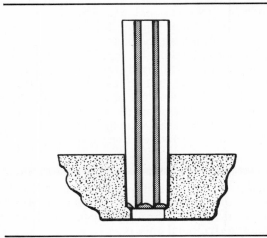

Figure 3-12. The cutting edges of the reamer finish a hole to the exact size.

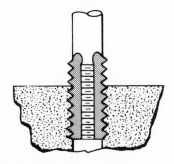

Figure 3-13. The male threads around the piece of metal called a tap are capable of cutting corresponding threads in an opening.

amount of material to be removed and the type of finish desired. As the hone rotates in the hole, it is moved up and down so it covers the entire surface of the opening.

Stones mounted on a flex hone cannot be adjusted to different diameters. They are used only for getting the correct surface finish.

Drilling

A drill is an implement with a cutting edge for making holes in various materials, figure 3-11. A drill has a point that is forced against the material. The point penetrates the material slightly, and the cone-shaped cutting lips of the drill carve out the material. A drill works like a rotating knife blade or chisel. All of the cutting is done with the end of the drill. Drilling produces a relatively accurate hole size with a moderately good surface finish. However, where extremely close tolerances are needed, a drill may not produce a hole diameter that is accurate enough. If the tip of the drill is slightly offcenter, the diameter of the hole will not be the same as the diameter of the drill. Therefore, when a hole of a very precise size is needed, it is usually drilled first and then reamed.

Reaming

After a hole is drilled, reaming will finish it to the desired size. The reamer has cutting edges ground into a rod several inches long, figure 3-12. These cutting edges can be straight, or they can spiral around the rod. Unlike a drill, a reamer cuts with its sides, not its tip. Reaming removes only a few thousandths of an inch of material. The hole must first be drilled accurately, just slightly under the size of the reamer. The reamer is then forced into the hole and turned at the same time. Reaming smooths the hole and straightens the sides, but is not as accurate a method for sizing a large hole as honing.

Tapping

Cutting threads in a hole with a tap is called tapping, figure 3-13. The tap is a solid piece of metal with male threads around it, similar to a bolt. Flutes are ground into the sides of the tap so that there are either three or four cutting edges. The tap is hardened metal and will cut threads in any material that is softer than the tap.

The end of the tap, which may be tapered for easy entry into the hole, is turned until it cuts into the metal. After a tap starts, it is self-feeding while it is being turned.

The hole must be drilled accurately before the tap is used. If the hole is too large, the threads will be too shallow. If the hole is too small, the tap may seize or break. The usual depth of the tapped threads is 75 percent of the threads on the tap. Under 75 percent results in

Basic Metallurgy and Machine Processes

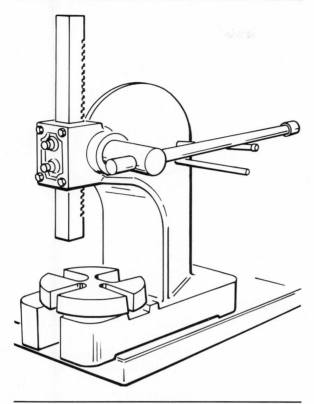

Figure 3-14. To change the shape of an opening, you can use a broaching tool in this press.

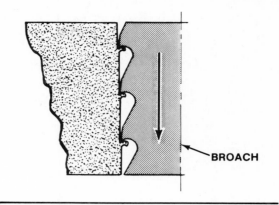

Figure 3-15. The broach cuts only in one direction and removes all the metal required in one pass.

weak threads. Over 75 percent makes it too difficult to turn the tap.

In any specific size of taps there are three types. A starting tap has a long taper so it can be easily started in an untapped opening. A regular tap has a taper, but only for the first few threads on the tap. It is used to clean out holes that have already been threaded. The bottoming tap is used to cut threads all the way to the bottom of the hole. It has a very short taper on the end, just enough to allow the tap to cut without seizing.

Broaching

A broach is a cutting tool most often used to change the shape of openings, figure 3-14. If a round hole is going to be made square, a broach would be forced through to cut out the four corners. Broaches may have cutting teeth all around their sides. In changing the round hole to a square shape, the broach might cut all four corners at once.

Broaches cut in one direction, and all the metal to be removed is taken in one pass through the work. Each tooth on a broach removes about one thousandth (.001) of an inch, or twenty five thousandths (.025) of a millimeter, figure 3-15. The broach is moved through the work with considerable force. The longer the broach, the more material will be taken out, because there are more teeth. Broaches are also designed to do external cutting such as cutting a keyway or a slot in a shaft.

■ The Beginnings of Measurements

The United States now is the last major industrial nation not using the metric system exclusively, and that is changing.

Why the metric system? Because it is far more precise and universal than the presently used system. And where did this present system come from? Mostly, from old England.

The inch, decreed King Edward II, was equal to the length of three barleycorns placed end to end. Unfortunately, that was not too accurate. King Henry I decided that a yard was the distance from the tip of his nose to the tip of his outstretched thumb. Another measurement, the rod (seldom used today), was the distance of sixteen left feet of men who lined up left-foot-to-left-foot. A furlong was 40 rods, or the length of a furrow in a farmer's field. An acre was said to be the amount of land a team of oxen could plow in one day.

When these measurements were first invented, they seemed to work well enough. But obviously, they were terribly inexact. (What happens if this year's barleycorns are bigger than last year's?) The world needed a new system of measurements, one that would not change from country to country or from year to year. The system finally devised to meet this need is called the metric system.

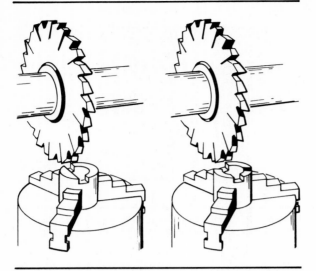

Figure 3-16. Milling is used to cut slots or keyways.

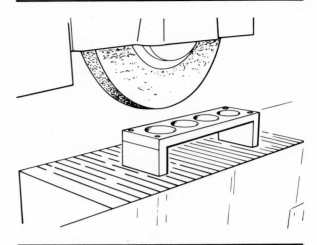

Figure 3-17. When you grind, you remove metal with an abrasive wheel.

External (Surface) Machining

When you work on the outside surface of a part to change its shape or the texture of the surface, you are doing external machining. This includes **milling**, **grinding**, polishing, and **lapping**. Milling and grinding are used to remove metal to a much greater degree than polishing. Polishing is done to achieve a better appearance, or a better surface finish. Lapping, or the fitting together of two parts or surfaces, is only done under special circumstances.

Milling

The first mills used by man were for making flour. A large stone wheel crushed and ground grain into flour. The modern machine mill is also a wheel, usually only a few inches in diameter. Its cutting teeth are usually spaced all around the edge, similar to a circular saw.

The main characteristic of a milling machine is the fixed milling wheel. The work is mounted on a table which can be moved up, down or sideways. Milling can also be used to cut slots or keyways on the outside of parts, figure 3-16.

A mill also can be used with a single cutting edge to remove material from a flat surface. If a surface is warped, grooved, or pitted, it can be mounted on a milling table for resurfacing. A cylinder head is a good example. The surface of the head is mounted vertically, so that the milling tool will move across it when the table is moved from side to side. In one pass a few thousandths of an inch is removed. More passes are made until the entire surface is flat.

Grinding

Grinding is the removal of metal with an abrasive wheel, figure 3-17. The applications of grinding are almost unlimited. There are machines that will grind just about any type surface or shape to a precise dimension. About the only thing grinding is not used for is drilling holes. It could be done, but a drill is much faster. However, after the drilling, a grinder may be used to finish the hole to an exact size.

Grinding is used extensively for resurfacing. When cylinder head and block gasket surfaces are warped or otherwise damaged, they can be resurfaced on a grinding machine. Machine grinding is highly accurate. Hand grinding is less accurate. Hand grinding should be limited to areas where parts do not have to be joined.

Polishing

Polishing makes a surface smooth and usually bright. It does not remove enough metal to change the size of a part, figure 3-18. Polishing is usually done with cloth buffing wheels or a narrow belt sander. Automotive crankshaft journals are polished after they are ground. The object in polishing is to make a surface that rubs against a bearing as smooth as possible to reduce wear.

Lapping

Before we get into the method of lapping, it is important to note that lapping was, at one time, a recommended practice among automotive mechanics. At present it is not recommended as standard practice, but some manufacturers still suggest a final lapping after grinding the valves.

Lapping itself is done by taking a small amount of abrasive paste and putting it on the

Basic Metallurgy and Machine Processes

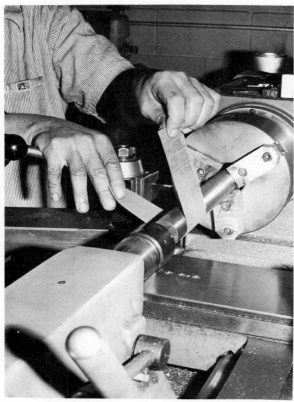

Figure 3-18. Polishing makes a surface smooth, but takes off very little metal.

valve face. Then the valve is rotated on the seat with slight pressure until it makes a perfect fit. All of the lapping compound must be removed after lapping, because it is an abrasive and would damage the engine.

MATERIALS AND MACHINING RELATIONSHIPS

Most flat metal surfaces appear smooth to the naked eye, but as we have already learned, under microscopic examination the true roughness of the surface is revealed.

When using a surface analyzer, the small diamond tip moves over the surface while the machine graphs the peaks and valleys. The amount of movement of the diamond is measured in millionths of an inch, which is expressed in microinches. A perfect surface would measure zero microinches.

From reading about the different machines discussed in this chapter we have learned that it is possible to obtain surfaces ranging from rough to very smooth. Different operating conditions for different parts of the engine require different surface finishes.

Engineers have established that there is an ideal surface finish for every application. For example, if two surfaces are going to be joined together, with a gasket in between, they should not be overly smooth. A smooth or polished surface has a tendency to let the gasket slide, or squeeze out between the pieces. The ideal gasket surface should have what is known as "tooth." It is flat and straight, but there is enough roughness of the surface to make the gasket adhere.

On the other hand, bearing surfaces should be as smooth as possible. If crankshaft journals were rough, they would act as a grinding wheel and remove metal from the bearings. A ball bearing is another good example. The balls and the race must be finely finished so that there is no grinding action between them.

Cylinder walls are a special case. It is impossible to make piston rings that will fit against a cylinder wall perfectly. There are small irregularities in both the rings and the walls that must be worn off in the first few hundred miles. This is known as seating the rings. If the cylinder walls are just the right surface, the rings will wear in quickly and be perfectly seated. Once the rings are seated, they will last many thousands of miles.

If the cylinder wall is too rough, the rings will be ground away. This can happen in only a few hundred miles. The surface produced by a boring bar is about 100 microinches. The boring bar actually plows a furrow in the metal. For best ring seating, the cylinder wall surface should have approximately a 25-microinch finish. The only way to get this finish is with a hone, after boring.

The type of finish is also important. Surfaces can be analyzed to give the measurement of the peaks and valleys in microinches. But this measurement does not give an indication of how sharp the peaks are. A "peaky" finish on a cylinder wall, with lots of sharp points, is not as desirable as a finish with those peaks

Mill: Cutting metals with a round, multiple tooth cutter; also used with a single cutting edge for flat surfaces.

Grind: Removing metal with an abrasive wheel.

Lap: Finishing two metal surfaces to each other by using a fine abrasive compound. Lapping is used to produce a gas-tight or liquid-tight seal.

Plateaued Finish: A finish in which the highest parts of a surface have been honed to flattened peaks.

blunted. The ideal finish for ring seating is one in which the peaks have been cut off, forming small plateaus instead of a point. This is known as a **plateaued finish**.

The plateaued finish is usually achieved on cylinder walls at the end of the honing process. The pressure on the hone is relaxed so that the stones ride over the peaks and cut them into plateaus. A special hone called the flexhone can be used for putting this plateaued finish on a cylinder wall. The flexhone may be used after honing with a standard rigid hone. The rigid hone removes the deep grooves made during the boring operation. The flexhone conditions the surface for the best ring seating.

When using a machine tool on a part, the material of the part affects the machining process. Aluminum is relatively soft, and therefore machines more easily than other metals or alloys. Some machinists sharpen the tool bit differently for each type of material on which they work.

Machining is usually done before heat treating or hardening. Before hardening, the metal is more easily machined. In some situations, such as restoring or rebuilding a part, it is necessary to machine a used part. The part may have been hardened before it was put into service. This makes machining very difficult, if not impossible. A good example is the resurfacing of hydraulic lifters or tappets. If the tappet was surface hardened, and the wear goes through the thin hardening, resurfacing is a waste of time. Resurfacing by grinding will certainly restore the surface and remove the wear, but it does not restore the hardening. For this reason, many engine rebuilders refuse to resurface a hardened part, such as a tappet. They prefer to use new ones.

SUMMARY

Materials and machining are closely related. There are reasons why different kinds of metals are used in specific places in the engine. These reasons include, long life, resistance to wear, or because they are easily machined. Surface treatments given to the finished parts affect not only appearance, but resistance to wear, longevity, and surface texture.

Machining or working of metals is done for the purpose of making the part the correct size, or for getting the correct surface finish.

Review Questions

Choose the single most correct answer. Compare your answers to the correct answers on page 268.

1. Aluminum is often used for:
    a. Cylinder blocks
    b. Rocker arms
    c. Cylinder liners
    d. Cylinder heads

2. Which of the following processes produces the most accurate hole?
    a. Broaching
    b. Drilling
    c. Boring
    d. Milling

3. Bearing alloys must have good:
    a. Embedability
    b. Compatability
    c. Loadability
    d. All of the above

4. Pig iron contains _____ carbon.
    a. 2 to 6%
    b. .7 to 1.7%
    c. Almost no
    d. About 5%

5. High carbon steel has a carbon content of approximately:
    a. 2 to 6%
    b. .7 to 1.7%
    c. .1 to 1.7%
    d. 2%

6. Babbitt is made from:
    a. Tin, copper and antimony
    b. Tin, aluminum and lead
    c. Lead, copper and silver
    d. Lead, tin and aluminum

7. The surface produced by a boring bore is about:
    a. 100 microinches
    b. 50 microinches
    c. 50 millimeters
    d. None of the above

8. In the process of _____ steel is heated to 396° to 567°F (220° to 315°C) and allowed to cool slowly.
    a. Annealing
    b. Forging
    c. Hardening
    d. Tempering

9. Which of the following is NOT a composite material?
    a. Silicon
    b. Fiberglass
    c. Ceramic
    d. Plastic

10. A plateaued finish is produced by:
    a. Etching
    b. A boring bar
    c. A rigid hone
    d. A flexhone

11. Which of the following is used in electroplating?
    a. Direct current
    b. Alternating current
    c. Electromagnetism
    d. High voltage

12. When a precision sized hole is needed, drilling is usually followed by:
    a. Boring
    b. Honing
    c. Reaming
    d. Broaching

13. The symbol for microinch is the Greek letter:
    a. Pi π
    b. Omega ω
    c. Alpha α
    d. Mu μ

14. Lapping is used to resurface:
    a. Valve stems
    b. Cylinder heads
    c. Worn threads
    d. None of the above

15. Surface roughness is measured in:
    a. Millimeters
    b. Angstoms
    c. Multimeters
    d. None of the above

Chapter

Cooling and Exhaust Systems

When the air-fuel mixture in a combustion chamber burns, the cylinder temperature can soar to 6,000°F (3,316°C). During the complete 4-stroke cycle, the average temperature is about 1,500°F (816°C). About one-third of this heat energy is turned into mechanical energy by the engine. The remaining two-thirds must be removed from the engine so that the engine can operate efficiently. If the heat energy remains, several things will happen:
- The engine oil temperature will increase, affecting engine lubrication.
- The incoming air-fuel mixture will become too hot, making it expand and reduce engine efficiency.
- Metal engine parts will expand, and possibly be damaged.

Where does the extra heat energy from combustion go? About half of it is absorbed by the metal in the engine, which is then cooled by the cooling system. The other half remains in the combustion gases, which are removed through the exhaust system. This chapter describes these two systems, and their relationship to performance, economy, and emission control.

COOLING SYSTEM FUNCTION

We have said that about one-third of the total heat energy from combustion is absorbed by the engine's metal. This is both good and bad:
- It is good because an engine that is too cool will have poor fuel vaporization, poor lubrication, excessive acids in the blowby gases, and high HC emissions.
- It is bad because an engine that is too hot will have weakened metal, poor volumetric efficiency, poor lubrication, and high oxides of nitrogen (NO_x) emissions.

Obviously, there must be a "just-right" engine operating temperature that will minimize these problems. This temperature is slightly different from engine to engine, depending on the design. Early engines ran at about 180°F (82°C). Late-model engines run as high as 212°F (100°C). These are average temperatures, and will vary depending on conditions.

Cooling System Operation

Most automotive engines use a liquid cooling system. The liquid, or coolant, is constantly circulated through the engine, absorbing heat from the engine metal. The coolant is then circulated outside of the engine and exposed indirectly to the air. The air absorbs heat from the coolant, so that the coolant can go back into the engine and absorb more heat. The greater the difference in temperature between the coolant

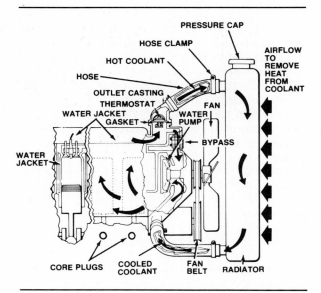

Figure 4-1. A typical liquid cooling system.

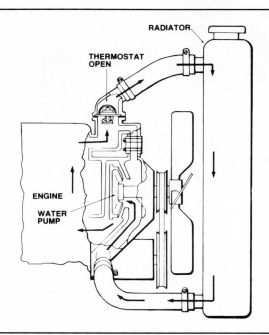

Figure 4-3. As the coolant warms, the thermostat opens and allows it to flow through the radiator.

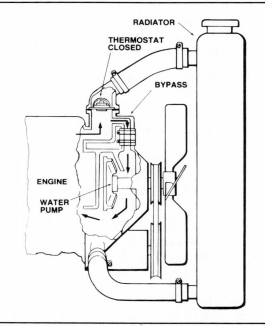

Figure 4-2. The thermostat prevents cold coolant from entering the radiator.

and the air, the more heat will be absorbed by the air.

Circulation Patterns

Within the engine, coolant circulates in passages called water jackets, figure 4-1. Outside the engine, the coolant circulates through hoses and the radiator. To keep the engine running at its ideal temperature, there are two patterns of coolant circulation. The two patterns are controlled by the thermostat which is a temperature-sensitive valve that opens and closes a passage between the engine and the radiator.

Circulation when cold

When the engine is cold, it must be warmed quickly to its ideal temperature. With the cooling system in full operation, it would take a long time for this to happen. To speed the warmup, the thermostat is closed when the coolant is cold, figure 4-2. This keeps the coolant in the radiator from circulating through the engine. Because there is no heat transfer between the coolant in the engine and the air passing the radiator, the heat of combustion quickly warms the engine.

Circulation when warm

As the coolant within the engine gets warm, the thermostat opens, figure 4-3. This allows full coolant circulation between the engine and the radiator. The amount of coolant in the system, the size of the radiator, and the opening temperature of the thermostat help to determine the average operating temperature of the engine.

System Pressure

There are two reasons for pressurizing a cooling system: to increase water pump efficiency, and to raise the boiling point of the coolant.

Pump efficiency is affected by pressure. Without a pressure cap installed in the system, a water pump is only about 85 percent efficient.

Cooling and Exhaust Systems

Ethylene Glycol Proportions By volume	Freezing Point °F	Boiling point With 15-pound Operation Pressure in Cooling system
33 percent	0 degrees	259 degrees
40 percent	−12 degrees	261 degrees
50 percent	−34 degrees	264 degrees
60 percent	−62 degrees	269 degrees
70 percent	−85 degrees	275 degrees

Figure 4-4. If you need a different freezing or boiling point see the table.

With a 14-psi pressure cap, the pump becomes almost 100 percent efficient.

Coolant boiling-point control is equally important. Water boils at 212°F (100°C) under atmospheric pressure (14.7 psi) at sea level. An average solution of engine coolant boils at around 223°F (106°C) under the same conditions. Even this higher figure is dangerously close to the operating temperature of the engine. Pressurizing the system raises the boiling point of the coolant. Each pound increase in pressure raises the boiling point about 3°F (1.7°C). Thus, coolant under 15-psi pressure will boil at approximately 268°F (141°C), or 45°F (7°C) higher than without the cap. Most cooling systems are pressurized at 12 to 17 psi. This is important to remember when removing the radiator pressure cap on an engine that is hot. When the pressure is released, the coolant will boil instantly.

ENGINE COOLANT

Although pure water transfers heat very well, it has disadvantages when used in a cooling system:
• Metal exposed to water will rust, and rust particles will clog water jackets.
• When the car is not running and the system is not pressurized, water will freeze at 32°F (0°C), harming engine parts.

Early cooling systems used water blended with a rust inhibitor and alcohol to lower the freezing point. Alcohol, however, boils at a lower temperature than water and escapes from the system more easily than water.

Modern cooling systems use a mixture of water and antifreeze, an ethylene-glycol based solution. Glycol does not transfer heat as well as water, but its freezing point is much lower and its boiling point is much higher, figure 4-4. Many modern cooling systems need the antifreeze in the coolant to activate the temperature warning lamp, which is set to register high temperatures considerably above the boiling point of water. When there is no antifreeze in the system, the water can boil away and cause severe damage to the engine block; yet the warning lamp will not light.

Antifreeze usually contains corrosion preventives, rust preventives, and water pump lubricants. It may also contain small particles designed to seal minor leaks in the cooling system. A coolant mixture that contains from 35- to 70-percent ethylene-glycol antifreeze will protect the engine and cooling system while improving the system's high- and low-temperature operation.

A 50-percent mixture of ethylene-glycol and water will protect against freezing to approximately −30°F (−34°C). If the car is going to be operated in colder climates, additional antifreeze should be added. A solution of 68-percent antifreeze will give protection down to −90°F (−62°C). If the percentage of ethylene-glycol is any higher than that, the freezing point will actually go back up. One hundred-percent anti-freeze will freeze at slightly below 0° F (−18° C). For this reason, never put more than 68-percent antifreeze in the system.

Although 50-percent antifreeze is the industry standard, some manufacturers recommend only a 33-percent solution or less. Less than a 50-percent solution will protect from freezing but does not give the corrosion protection that a 50-percent solution does. For this reason, the industry recommends filling all cooling systems with about 50-percent ethylene-glycol.

Antifreeze also protects against boiling. A cooling system filled with only water may boil when the engine is turned off. This will open the pressure cap and force water out of the radiator.

This is avoided by using a 50-percent antifreeze mixture. Water boils at about 248°F (120°C) under 14-psi pressure. The 50-percent solution boils at about 263°F (128°C). The 15 degrees of added protection will prevent boiling after shutdown in normal operation.

COOLING SYSTEM COMPONENTS

A typical liquid cooling system, figure 4-1, includes the:
• Water jackets
• Thermostat-controlled bypass
• Core, or freeze, plugs
• Radiator
• Water pump
• Radiator fan
• Thermostat

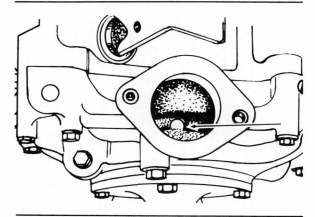

Figure 4-5. This internal passage in the thermostat housing directs cold coolant to the water pump.

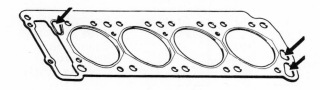

Figure 4-7. These holes in the head gasket carry coolant the full length of the cylinder head.

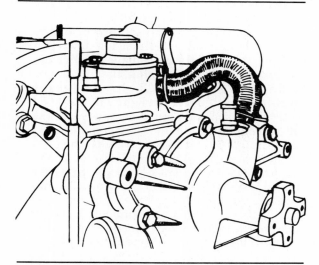

Figure 4-6. This external bypass hose connects the thermostat housing to the water pump.

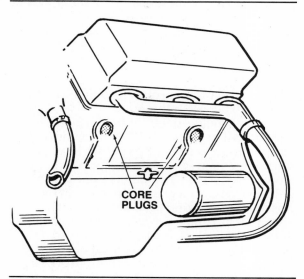

Figure 4-8. Core plugs are pressed into the cylinder block.

- Radiator pressure cap
- Coolant hoses
- Drive belt
- Coolant recovery system.

The following paragraphs describe these parts.

Water Jackets, Bypass, And Core Plugs

Water jackets and coolant passages are cast into the cylinder block and head. They are designed so that the coolant will circulate freely and not remain stationary in pockets.

The thermostat-controlled bypass is a passage that can direct coolant back into the engine before it has circulated through the radiator. The bypass can be an internal passage in the block, figure 4-5, or an external connection, figure 4-6.

Because the coolant enters and leaves at the front of the engine, various means are used to make sure it travels to the back of the cylinder block. On 6-cylinder engines, a water distribution tube often is used. When the coolant first enters the block it runs through the tube and exits near each valve seat. The holes in the tube are graduated so an equal amount of water is distributed the full length of the block.

Head gaskets are also used for water distribution. The front holes in the gasket are frequently smaller than the rear holes. This forces the coolant to flow to the rear of the block, figure 4-7. Because more coolant goes to the back of the block, it means more coolant will flow the full length of the cylinder head. This is because the outlet for the coolant is at the front of the cylinder head.

Core plugs are pressed or clamped into the sides of the cylinder block to close the holes left by the foundry, figure 4-8. They are called core plugs because the holes are used by the foundry to remove core sand from the inside of the block. Core plugs are also called freeze plugs.

Cooling and Exhaust Systems

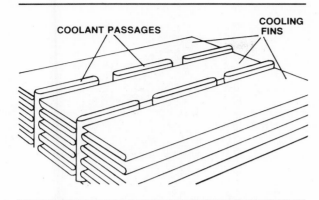

Figure 4-9. The tubes and cooling fins of the radiator core. (Chrysler)

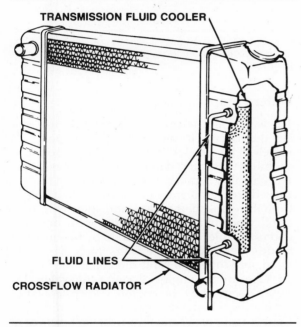

Figure 4-11. The transmission fluid cooler is installed in one of the radiator tanks. (Harrison)

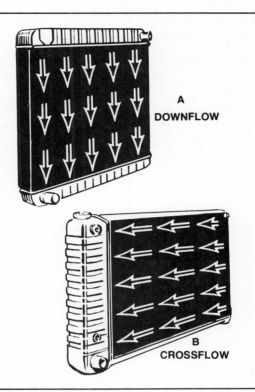

Figure 4-10. A radiator may be either a downflow type or a crossflow type.

Some people mistakenly believe that the plugs are there to relieve pressure in the block if the water should freeze. Core plugs will frequently pop out if the water freezes, but that is not their purpose. If one does pop out, the chances are that the same freeze also cracked the block. Core plugs are definitely not insurance against freeze damage.

Radiators

The purpose of the radiator is to expose a large amount of surface area to the surrounding air. Radiators are made of thin metal tubes, usually copper or aluminum, with cooling fins attached, figure 4-9. Coolant flows through the tubes and transfers its heat to the radiator metal. The air flowing past the radiator tubes and fins absorbs this heat. The number of tubes and fins in a radiator determines the unit's heat-transferring capacity.

The assembly of tubes and fins is called the radiator core. To keep a steady stream of coolant flowing into and out of the core, two larger metal tanks are connected to opposite ends of the tubes. There are two patterns of radiator construction:
- Downflow radiators have vertical tubes and a tank at the top and bottom of the core, figure 4-10, position A.
- Crossflow radiators have horizontal tubes and a tank on both sides of the core, figure 4-10, position B.

Crossflow radiators usually allow a lower hood line on the car. Other than the flow direction, there is no difference in operation between the two types.

One of the radiator tanks, either the top tank or a side tank, has an open neck sealed with a pressure cap. On cars with automatic transmissions, one of the tanks contains a cooler for the transmission fluid, figure 4-11. The fluid and the coolant do not mix, but the extra heat of the transmission fluid adds to the work of the cooling system. For this reason systems on auto-

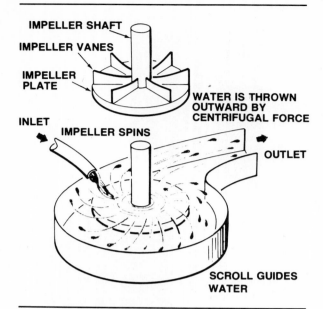

Figure 4-12. The water pump impeller is driven by the engine. Its spinning blades apply centrifugal force to the engine coolant.

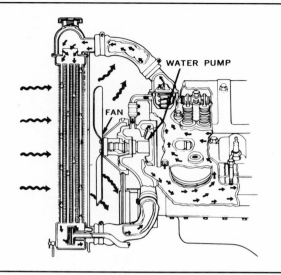

Figure 4-14. The engine-driven fan blows air into the engine compartment through the radiator core. (Harrison)

Water Pumps

The water pump uses **centrifugal force** to circulate the coolant. It consists of a fan-shaped impeller, figures 4-12 and 4-13, set in a round chamber with curved inlet and outlet passages. The chamber is called a scroll because of these curved areas. The impeller is driven by a belt from the crankshaft pulley, and spins within the scroll. Coolant from the radiator or the bypass enters the inlet and is picked up by the impeller blades. Centrifugal force flings the coolant outward, and the scroll walls direct it to the outlet passage and the engine. Because the water pump is driven by the engine, it keeps coolant circulating whenever the engine is running. Centrifugal pumps must turn fast to be efficient. Worn or loose belts will slip, which is not easily detected. Early water pumps had to be externally lubricated, but modern coolant mixtures contain a water pump lubricant. The pump bearings must be well sealed to prevent coolant leaks.

Radiator Fan

When the car is traveling fast, airflow through the radiator core is great enough to absorb all excess heat. At low speed or idle, however, there is not enough natural airflow. To increase the amount of air passing through the radiator, a fan is mounted behind the radiator, figure 4-14. The fan is usually mounted on the same shaft as the water pump impeller, and is driven by the same belt. The fan blades pull extra air into the engine compartment. Many late-model

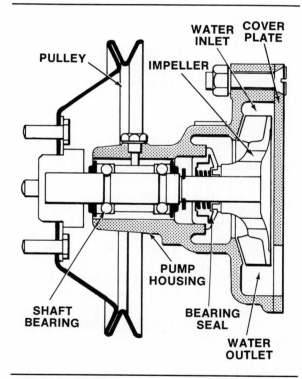

Figure 4-13. Cutaway of a typical water pump.

matic-transmission cars are usually built with a greater capacity than those on manual transmission cars.

Cooling and Exhaust Systems

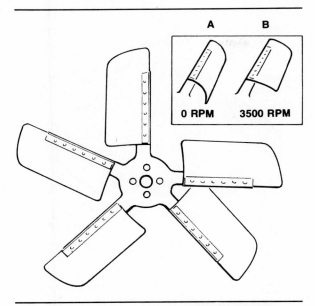

Figure 4-15. The flexible blades of this fan change shape as engine speed changes.

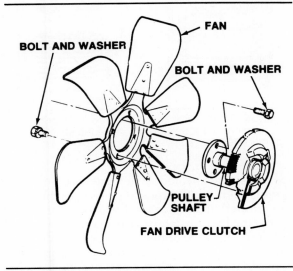

Figure 4-16. This fan has a fluid clutch controlling the speed of rotation. (Ford)

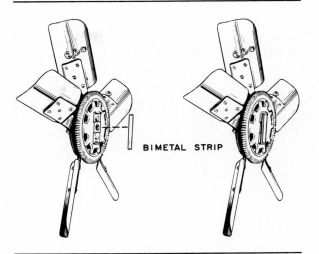

Figure 4-17. The bimetal temperature sensor spring controls the amount of silicone that is allowed into the drive and that, in turn, controls the speed of the fan.

cars have shrouds around the fan to concentrate the airflow where it is needed most.

If engine power drives the fan, this power is wasted at high speeds when natural airflow will do the job. There are three ways to avoid this waste:
- Flexible fan blades
- Clutch fans
- Electric fans.

Flexible fan blades change their angle, or pitch, with changing engine speeds. At low speed, the blades are sharply angled, figure 4-15, position A. This uses a lot of engine power. At high speeds, position B, centrifugal force flattens the fan blades. They pull less air and require less power. This also reduces fan noise at high speeds.

Clutch fans are heavier than conventional fans, but the power saved makes them worthwhile. The drive, figure 4-16, is a clutch assembly that depends on a silicone fluid to transmit motion from the drive belt pulley to the fan itself. These are also called fluid drive fans. When there is a lot of fluid present, the drive pulley and fan rotate at the same speed. When there is little fluid present, the drive slips and the fan turns at less than drive pulley speed. When the drive is slipping, less engine power is used to drive the fan.

The amount of slicone fluid present is controlled by a **bimetal temperature sensor** spring exposed to the airflow from the radiator, figure 4-17. When the airflow is cool, the spring lets very little silicone into the drive, and the fan turns slowly or idles. When the airflow is hot, the spring allows a lot of silicone into the drive, and the fan turns faster.

Centrifugal Force: A force applied to a rotating object, tending to move the object toward the outer edge of the circle in rotation.

Bimetal Temperature Sensor: A sensor or switch that reacts to changes in temperature. It is made of two strips of metal welded together that expand differently when heated or cooled causing the strip to bend.

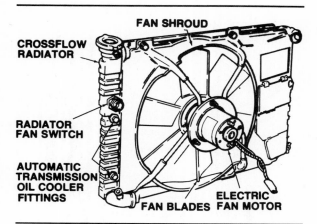

Figure 4-18. In this car, an electric motor drives the radiator fan. (Chrysler)

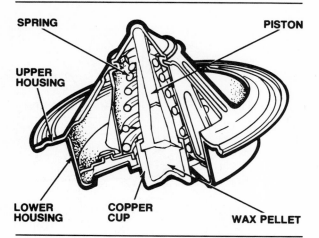

Figure 4-20. A cross section of a wax-actuated thermostat. (Stant)

Figure 4-19. Always disconnect the fan power lead before working near an electric fan.

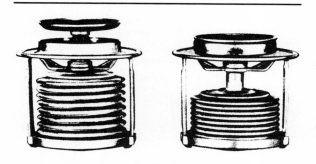

Figure 4-21. These bellows-actuated thermostats are not used in late-model cars because the bellows operation is affected by coolant pressure.

Electric fans are used on some automobiles. The fan is not driven by the engine but by an electric motor, figure 4-18. The motor is switched on and off by a temperature-sensitive switch that senses engine coolant temperature. When you work on a car with an electric fan, disconnect the battery ground cable or the fan power lead before getting near the fan, figure 4-19. The fan can switch on whether or not the engine is running in many cases.

Thermostats

The thermostat regulates coolant flow between the engine and the radiator. When the coolant is cold, the thermostat is completely closed. As the coolant warms, the thermostat begins to open. The thermostat's position varies during normal operation, but it is rarely opened completely, unless the engine is working very hard or the weather is hot.

All late-model thermostats are wax-actuated, figure 4-20. A chamber in the thermostat is filled with wax that is solid when cold, but melts and expands when heated. When the wax is cold, a spring holds the thermostat closed. As the coolant temperature increases, the wax expands and forces the thermostat to open against the spring tension. This type of thermostat will work reliably under normal system pressure. Earlier types that used an **aneroid bellows**, figure 4-21, could be affected by the coolant's pressure and not work reliably.

The thermostat's opening temperature is stamped on the outside. Replacement thermostats must have the correct opening temperature for the system to work properly. The thermostat is mounted in a metal housing at the top front of the engine, figure 4-22.

Radiator Pressure Cap

To improve the performance of the coolant, the cooling system is sealed and pressurized. Part of the pressure comes from the water pump. The rest of the pressure comes from expansion of the coolant as it warms.

The radiator cap must keep the system pressurized, vent any excess pressure, and allow

Cooling and Exhaust Systems

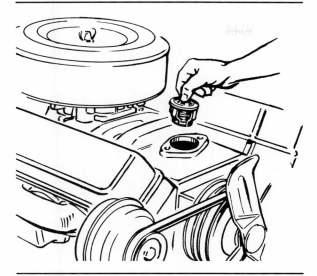

Figure 4-22. A typical thermostat location.

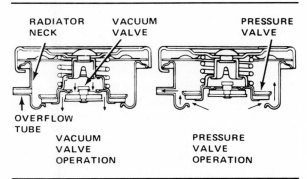

Figure 4-23. A cross-section of a typical radiator pressure cap. (AMC)

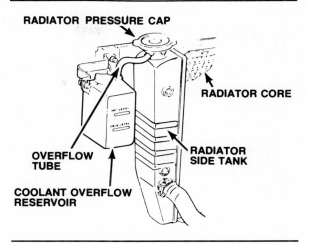

Figure 4-24. Most radiator caps are mounted on one of the radiator tanks. (Ford).

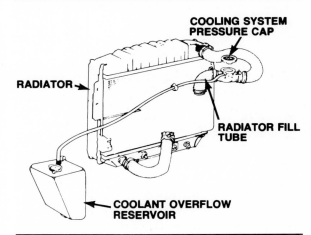

Figure 4-25. Some radiator caps are mounted away from the radiator to shorten the radiator profile. (Ford)

atmospheric pressure to re-enter the system as the coolant cools and contracts after the engine is shut off.

To do this, radiator caps have two valves, figure 4-23. A pressure valve relieves the excess pressure inside the system during engine operation. A vacuum valve allows outside air to enter the system when the engine is off. These pressure caps usually are on the top or side tank, figure 4-24, but occasionally can be found in unusual locations such as on a special hose section, figure 4-25.

Aneroid Bellows: An accordian shaped bellows that responds to changes in coolant pressure or atmospheric pressure by expanding or contracting.

■ **If A Little Is Good, A Lot Can Be Worse**

It doesn't happen too often, but it is possible. Too much antifreeze in an engine's cooling system will *raise* the freezing point. The "right" amount of antifreeze varies from one part of the country to another and from one vehicle to another. However, in almost all cases, maximum freeze protection is provided by a coolant solution with about 67-percent antifreeze by volume. When the antifreeze concentration goes above this level, the coolant may freeze or become slushy at higher temperatures, and coolant heat transfer will be reduced.

Figure 4-26. Various types of hose clamps.

Figure 4-28. When adding coolant to the recovery system reservoir, be sure to take note of the "hot" and "cold" fill marks and fill accordingly

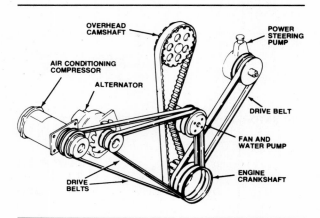

Figure 4-27. The fan and water pump drive belt also drives the alternator in many cases. (Harrison)

Radiator Hoses and Drive Belts

Radiator hoses connect the cooling system passages to the radiator tanks. The hoses are flexible and will absorb the motion between the vibrating engine and the stationary radiator. The hoses are made of rubber, and are often reinforced with a wire coil. The hoses can be molded in a specific shape or they can be ribbed, flexible lengths that adapt to different installations.

The hose ends fit over necks on the engine and radiator and are held in place with hose clamps, figure 4-26. Hose clamps can be held in place by spring tension or by a screw.

The drive belt that operates the radiator fan and water pump is a reinforced rubber V-belt, figure 4-27. It fits tightly over the crankshaft pulley, the fan pulley, and usually, the alternator pulley. Belt tension is usually adjusted by moving the alternator in its mounting.

Coolant Recovery System

Most late-model cars have a coolant recovery system and a closed cooling system. Instead of a pressure cap that vents to the atmosphere, the pressure cap is connected to a coolant tank or overflow reservoir, figure 4-28. When system temperature increases and the coolant expands, the extra coolant flows out of the radiator neck and through an overflow hose. Instead of being lost, as in earlier cooling systems, this overflowing coolant is held in the coolant tank. When the system cools and a vacuum develops in the system, coolant is drawn back into the radiator through the same hose.

This system is necessary in late-model cars that tend to run hotter because of emission control equipment, automatic transmissions, air conditioning, and other power-consuming accessories. The coolant tank ensures that there will always be enough coolant to fill the system because none is lost if the radiator overflows. Coolant recovery systems are said to increase cooling system efficiency about 10 percent.

When more coolant must be added to a recovery system, it is added to the overflow tank, not directly to the radiator. The tank is marked with "hot" and "cold" fill levels, figure 4-28.

ENGINE TEMPERATURE EFFECTS ON PERFORMANCE, ECONOMY, AND EMISSIONS

When the engine is too cold, performance suffers because of poor fuel vaporization. The liquid fuel entering the cylinders also tends to wash oil off the cylinder walls, reducing lubrication. The liquid fuel runs past the piston rings and into the crankcase, where it dilutes the oil and further affects lubrication.

Cooling and Exhaust Systems

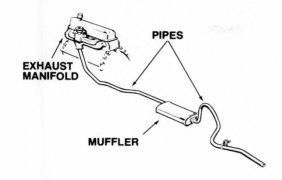

Figure 4-29. The exhaust system used with an inline engine.

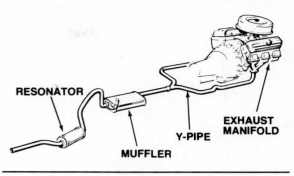

Figure 4-30. This V-type engine uses a Y-pipe to connect the two cylinder banks to a single exhaust pipe.

A cold engine affects economy in several ways. Because the air-fuel mixture is not vaporizing and burning completely, less energy is extracted from the fuel. Cold oil tends to form sludge, which increases engine friction, robs it of power, and increases mechanical wear.

Many emission control systems use coolant temperature to activate switches. If the coolant does not reach the proper temperature, the emission control systems do not work properly. The unburned fuel in the combustion chambers also increases exhaust HC emissions.

An engine that is too hot will have lower volumetric efficiency and decreased performance. Hot oil will form carbon and varnish deposits that affect engine operation.

The increased mechanical wear allowed by hot oil that is too thin will affect economy. The thin oil will also be drawn into the combustion chamber in greater amounts, fouling spark plugs and increasing both HC and CO emissions.

EXHAUST SYSTEM TYPES

The exhaust system routes engine exhaust gases to the rear of the car, quiets the exhaust noise, and, in some cases, reduces the pollutants in the exhaust. The system design varies according to engine design. The three major types are:
- Inline
- Single V-type
- Dual V-type

These designs are described in the following paragraphs.

Inline

When an engine's cylinders are arranged in line, figure 4-29, all of the exhaust valves are on the same side of the engine. An exhaust system pipe connects to this side of the engine at the exhaust manifold. The single pipe connects to one or more units that may include a muffler, a resonator, and a catalytic converter.

Single V-Type

A V-type engine has exhaust valves on both cylinder banks. There are two ways in which these valves can be connected to a single exhaust pipe:
- There can be a Y-pipe behind the engine, figure 4-30.
- There can be a crossover pipe beneath the engine, figure 4-31.

■ **Adding Temporary Antifreeze**

Permanent antifreeze was introduced in the United States by Union Carbide in 1927. Before that, car owners used alcohol, molasses, kerosene, salt, sugar, or even honey to keep their cars running in the winter. There was no problem with freezing as long as the engine was running. But if a car was parked outside, the water might freeze and crack the engine cylinder block.

The most popular antifreeze was alcohol. At the first sign of winter, it was time to drain the water and add alcohol to prevent freezing. But alcohol evaporates quickly. In cold climates, car owners made weekly or even daily hydrometer checks to be sure they still had enough alcohol in the water.

After 1927, alcohol was still sold because it was less expensive than "permanent" antifreeze. Today, alcohol, or "temporary" antifreeze, has completely disappeared from the marketplace.

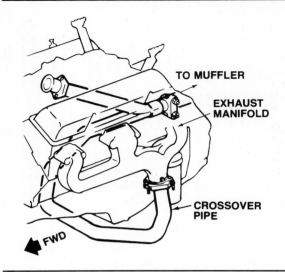

Figure 4-31. This V-type engine has a crossover pipe connecting the two banks to a single exhaust.

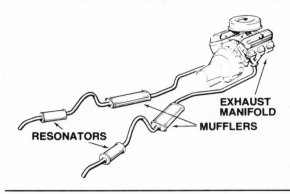

Figure 4-32. This V-type engine has two separate exhaust systems.

Dual V-Type

Some V-type engines have two separate exhaust systems, one for each cylinder bank, figure 4-32. The systems are similar to the single systems used on an inline engine.

EXHAUST SYSTEM COMPONENTS

Major parts of the exhaust system include:
- Manifolds
- Pipes
- Mufflers
- Resonators
- Catalytic converters.

Manifolds and manifold heat control valves were introduced in Chapter 1. We will study them more thoroughly in Chapter 5. The other exhaust system parts are described in the following paragraphs.

Pipes

The flow of exhaust gases from the engine should be as smooth as possible. If there are restrictons in the flow, there will be **backpressure** at the engine. Backpressure is the pressure trying to push the exhaust back into the cylinders. Too much backpressure will not allow all the exhaust to leave the combustion chamber before the next stroke starts. This preheats and leans the incoming air-fuel mixture, hurting efficiency. It can also cause engine mechanical failures such as burned valves. Exhaust system pipes should be as straight as possible, without sharp turns and restrictions. Pipes must also be able to withstand the constant presence of hot, corrosive exhaust gases and undercar hazards such as rocks.

The exhaust pipes on many late-model cars are formed with an inner and an outer skin. Occasionally, the inner skin will collapse and form a restriction in the system. From the outside, the exhaust pipe will look normal even when the inside is partially or almost completely blocked. Close inspection is necessary to locate such a defect.

Mufflers and Resonators

The muffler is an enclosed chamber that contains baffles, small chambers, and pipes to direct exhaust gas flow. The gas route through the muffler is full of twists and turns, figure 4-33. This quiets the exhaust flow but also creates backpressure. For this reason mufflers must be carefully matched to the engine and exhaust system.

Some mufflers consist of a straight-through perforated pipe surrounded by sound-deadening material, usually fiberglass. These "glasspack" mufflers, as they are often called, reduce backpressure but are not nearly as quiet as conventional mufflers.

Some engines also add resonators to the system. These are small mufflers specially designed to "fine tune" the exhaust and give it a pleasant, quiet, "resonant" tone.

Catalytic Converters

In order to meet tightening exhaust emission standards, carmakers turned to the catalytic converter in 1975. These were oxidation converters that changed HC and CO to harmless CO_2 and H_2O. Reduction converters first appeared on some 1978 cars to help eliminate NO_x emissions.

In the simplest arrangement, one catalytic converter is installed in the exhaust system between the manifold and the muffler, figure 4-34.

Cooling and Exhaust Systems

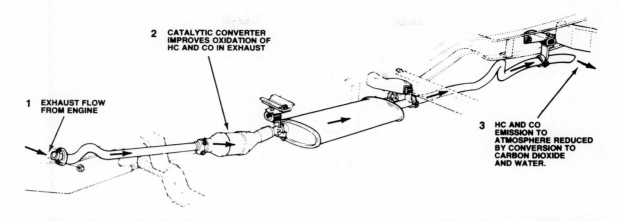

Figure 4-34. Typical catalytic converter installation. (Ford)

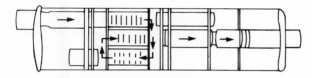

Figure 4-33. Exhaust gases must twist and turn to travel through this muffler.

Many cars, however, use two converters. Some cars have converters bolted directly to the exhaust manifold.

Catalytic converters are simple. They have no moving parts and never need adjusting. Most converters have a guaranteed lifespan of 50,000 miles. The pellets in early General Motors converters can be removed and replaced by removing a plug in the bottom of the unit. Tetraethyl lead, the gasoline octane booster described in Chapter 2, will coat or "poison" the catalyst and reduce its efficiency. This may make converter replacement necessary.

The link between a converter and a well-tuned engine is important. An engine that is misfiring or improperly tuned can also destroy a catalytic converter, because the converter cannot accept exhaust temperatures above 1,500°F (815°C). Neither will it work right if the air-fuel mixture is too rich. Two spark plugs misfiring in succession for a prolonged time will raise the temperature in the converter and shorten its life.

Exhaust System Effects on Performance, Economy, and Emissions

The main effect of the exhaust system on the engine is the backpressure it creates. Too much backpressure will restrict the flow of exhaust gases from the cylinders and hurt engine performance. The catalytic converter affects emissions, as we learned in Chapter 2.

SUMMARY

The cooling and exhaust systems must remove about two-thirds of the heat produced by a car engine. Most cooling systems use coolant to remove this heat. It circulates through the engine block where it picks up the heat, then goes through the radiator where the outside air flowing past removes heat. The coolant is partly

---

Backpressure: A pressure that tends to slow the exit of exhaust gases from the combustion chamber; usually caused by restrictions in the exhaust system.

■ The Engine Without A Water Pump

All modern liquid-cooled engines have a pump to circulate the coolant. But there was a time when the water pump was something that only the more expensive cars had. In the early days of the automotive industry, many automotive engines had cooling systems operated by the thermo-syphon principle. This means, simply, that hot water rises and cold water sinks. As the water in the block was heated, it rose out to the top of the radiator. Once in the radiator, it started to cool, and slowly sank. This heating and cooling created enough circulation to keep the engine from overheating, as long as the weather was cool. In hot weather, or when pulling hard up a hill, the engines would sometimes overheat. Because overheating was common, motorists accepted it as just one more hazard of motoring, like flat tires or getting stuck in the mud.

water and partly a number of additives that allow a greater range of engine operating temperatures.

The coolant is pushed through the system by a water pump. As it heats, it expands, often overflowing the radiator. Newer systems recover this excess coolant in a separate reservoir, where it can be reused as the engine cools. An electric or engine-driven fan cools the radiator by drawing air through it at low speeds and idle. A pressure cap keeps the system pressurized, and a thermostat regulates the flow of coolant by sensing engine temperature. Engine operating temperatures that are either too hot or too cold will produce greater emissions and poor performance, and can damage the engine.

Exhaust systems carry the exhaust away from the combustion chambers through a series of pipes, mufflers, resonators, and catalytic converters. The system must be as straight as possible to reduce backpressure in the cylinders, and generally must dampen the noise from combustion.

Review Questions

Choose the single most correct answer. Compare your answers to the correct answers on page 268.

1. Which of the following terms describes a type of radiator?
   a. Downflow
   b. Backflow
   c. Thruflow
   d. Acrossflow

2. In order to work properly, a catalytic converter must have:
   a. Exhaust gas hotter than 1500°F (815°C)
   b. A rich air-fuel mixture
   c. Gasoline without MMT in it
   d. None of the above

3. The automobile engine converts about _____ of its total heat energy into mechanical energy.
   a. One fourth
   b. One third
   c. One half
   d. Two thirds

4. All of the following conserve power at high engine speeds except:
   a. Flexible blade fans
   b. Multiblade fans
   c. Clutch fans
   d. Electric fans

5. In extremely cold climates the cooling system should be filled with:
   a. 100% ethylene glycol
   b. 73% ethylene glycol and 27% water
   c. 68% ethylene glycol and 32% water
   d. 50% ethylene glycol and 50% water

6. Core plugs are installed in the engine block to:
   a. Relieve pressure if the coolant should freeze
   b. Allow the block to be flushed during engine overhaul
   c. Provide inspection parts for internal block condition
   d. None of the above

7. An engine that is too hot will have:
   a. Poor volumetric efficiency
   b. Poor cooling system circulation
   c. High oxides of carbon
   d. High thermal inefficiency

8. A 50% water and anti-freeze mixture under a pressure of 14 psi will boil at approximately:
   a. 248°F (120°C)
   b. 255°F (124°C)
   c. 263°F (128°C)
   d. 270°F (132°C)

9. The water pump uses _____ to circulate the coolant.
   a. Positive pressure
   b. Gravitational force
   c. Centripetal force
   d. Centrifugal force

10. The greater the difference in temperature between the coolant and the air, the:
    a. Hotter the engine will run
    b. Greater the cooling system pressure will be
    c. More heat will be absorbed by the air
    d. Faster the coolant will circulate through the system

11. A pressurized cooling system will:
    a. Reduce engine temperature
    b. Increase the coolant circulation rate
    c. Reduce the coolant boiling point
    d. Increase water pump efficiency

12. If the incoming fuel mixture becomes too hot, it will:
    a. Cause the engine to backfire
    b. Expand which will reduce engine efficiency
    c. Preignite and cause pinging
    d. Wash the oil from the cylinder walls

13. A disadvantage of ethylene glycol when used as a coolant is that:
    a. Its freezing point is much lower than water
    b. It does not transfer heat as well as water
    c. Its boiling point is much higher than water
    d. None of the above

14. The thermostat regulates coolant flow between the engine and the:
    a. Cylinder head
    b. Heater
    c. Radiator
    d. Transmission cooler

15. Which of the following is NOT a part of the water pump?
    a. Pulley
    b. Impeller
    c. Stator
    d. Scroll

16. For every one psi increase in pressure, the coolant boiling point is raised about:
    a. 1°F (0.6°C)
    b. 2°F (1.1°C)
    c. 3°F (1.7°C)
    d. 4°F (2.2°C)

17. A major type of exhaust system is the:
    a. Inline
    b. Single V
    c. Dual V
    d. All of the above

18. Catalytic converters help change:
    a. HC and CO to H_2O and CO
    b. HC and CO to H_2O and C
    c. HC and CO to HC_2O
    d. HC and CO to CO_2 and H_2O

PART TWO

Engine Construction and Operation

Chapter Five
Cylinder Blocks, Heads, Manifolds, and Miscellaneous Engine Covers

Chapter Six
Valves, Springs, Guides, and Seats

Chapter Seven
Camshafts, Lifters or Followers, Pushrods, and Rocker Arms

Chapter Eight
Crankshafts, Flywheels, Vibration Dampers, Pistons, and Rods

Chapter Nine
Bearings

Chapter Ten
Engine Lubrication and Ventilation

Chapter Eleven
Gaskets, Fasteners, Seals, and Sealants

Chapter 5

Cylinder Blocks, Heads, Manifolds, and Miscellaneous Engine Covers

MATERIALS AND CONSTRUCTION PROCESSES

Automotive cylinder blocks, heads, and manifolds are usually cast from iron or aluminum. The casting is made by filling the interior of the prospective engine with sand. After the casting cools, the sand is poured out through the core holes. The interior space then becomes the water jacket, figure 5-1.

The rough engine castings are drilled, tapped, and machined until they are ready to be assembled, figure 5-2.

Engine covers are cast or made from sheet metal. The front cover of an engine is often cast from aluminum. Oil pans, rocker covers, and valve covers may be stamped out of sheet steel or aluminum, or molded out of plastic, because they do not have to provide any structural strength. Their main purpose is to prevent oil from escaping, but they also aid heat dissipation and act as oil reservoirs.

CYLINDER BLOCKS

The cylinder block, which is the basic framework of the engine, is a very heavy casting. Inline engines usually have four, five, six, or eight cylinders. Inline fours, fives, and sixes are still made today, figure 5-3. The inline eight has been replaced by the V-8, which is shorter and lighter. However, the V-8 is now being replaced by the V-6.

Most engine blocks are cast in one piece. Oil passages are drilled or cast into the block. The cylinders are supported at both top and bottom. In some cases there are internal webs that also connect to the cylinders, figure 5-4.

Horizontally opposed engines are also called pancake or flat engines, figure 5-5. Most are air-cooled and have fins on the cylinders and heads to assist heat dissipation, figure 5-6. Flat engines usually do not have a block, but consist of a crankcase to which the cylinders and heads are attached.

As we have already seen, water jackets are cast into the blocks of most engines. The water jacket surrounds the cylinders to provide heat dissipation. If there were no cooling system in the engine, the pistons would seize and engine parts would melt.

On V-type engines, the left and right bank water jackets do not connect inside the block. Water under pressure from the pump is divided and flows separately into each bank. Coolant flows around the cylinders and through the head gasket into the cylinder heads. The coolant flows out of each head and joins at the thermostat housing, figure 5-7.

Cylinder Blocks, Heads, Manifolds, and Miscellaneous Engine Covers 81

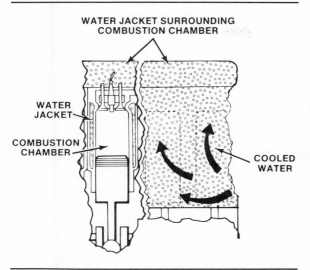

Figure 5-1. The space left in the lining of the block after it is cast becomes the water jacket.

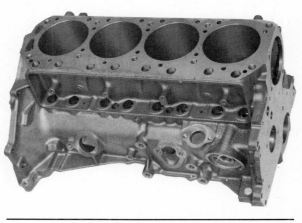

Figure 5-3. This is a Pontiac inline 4-cylinder engine block. (Pontiac)

Figure 5-2. A newly machined cylinder block on the Oldsmobile engine line is being inspected for bore size prior to being assembled. (Oldsmobile).

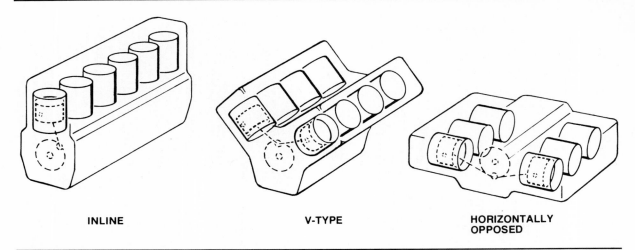

Figure 5-5. These are the common cylinder arrangements for automotive engines. The horizontally opposed or pancake engine is on the right.

Figure 5-4. The web or saddle is cast into the engine block.

Figure 5-6. The Volkswagen engine includes four air-cooled, individual cylinders like these.

CYLINDERS AND SLEEVES

The cylinders are part of the block casting. In most cases, the rough cast cylinders are bored and honed to a smoothly finished size so that the piston will fit with the proper clearance. The rings wear away the cylinder bore. When the cylinder wall becomes tapered from the ring wear, it is bored out to fit the next larger size piston, or bored to accept a sleeve that will accept a standard size piston.

Types Of Sleeves

There are three kinds of **cylinder sleeves**. The steel sleeve is a thin steel shell that is pressed into the cylinder. It provides no strength, and is only a renewable surface for the piston rings. This type of sleeve has not been used for many years.

The other two types of sleeves are cast iron. They are made with a thick wall which ranges from $1/8$ to $1/4$ of an inch (3 to 6mm). Cast iron sleeves are installed in many ways. Some are pressed or screwed into the cylinder. Others are a snug fit, held in place by a locking ring or by the clamping action of the cylinder head.

A **dry sleeve** is installed in a cylinder that has been scored or otherwise damaged, figure 5-8. The dry sleeve does not come in contact with the engine coolant.

The **wet sleeve** is so named because it does come in contact with the coolant.

In some engines, the cylinders are bored originally to accept a sleeve. When the sleeve wears

Cylinder Blocks, Heads, Manifolds, and Miscellaneous Engine Covers

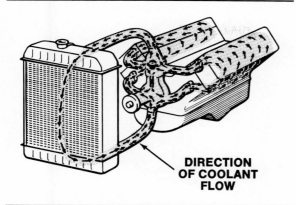

Figure 5-7. This illustration shows the flow direction of the coolant as it comes out of each head and is joined at the thermostat housing.

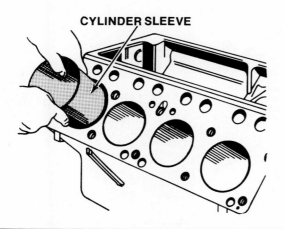

Figure 5-8. This cast iron sleeve is being installed in a cylinder with a complete wall. The sleeve is then known as a dry sleeve.

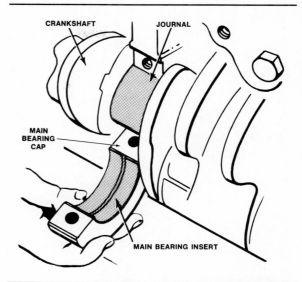

Figure 5-9. The main bearings support the crankshaft.

out, it is replaced. In theory, the block should last forever because installing new sleeves makes it like new. This feature is valuable in high mileage industrial or truck engines. The mileage can be so high in commercial use that the engines would, otherwise, have to be replaced every year.

Main Bearings

Main bearings are the bearings that support the crankshaft, figure 5-9. They must give support without interfering with the rotation of the crankshaft or the action of the connecting rods.

At the bottom of the cylinder block, webs are cast to support the main bearings, figure 5-4. On some engines, there can be as few as two main bearings for the crankshaft. Designers have found that putting a main bearing on both sides of each connecting rod results in the longest engine life. A 6-cylinder inline engine would have seven main bearings in this configuration.

Because of the practice of pairing connecting rods on each crankshaft throw in a V-8 engine, main bearings are usually placed on both sides of every rod pair. This results in a crankshaft with five main bearings for a typical V-8 engine. However, V-type engines with individual crankshaft throws can have bearings on both sides of each throw. A V-6 engine would also have seven main bearings in this configuration.

Although most engines could be designed with the minimum number of main crankshaft bearings, the structural rigidity necessary for long engine life is best supplied by the maximum number of bearings allowed by the engine configuration.

Design And Dimensions

The major axis or datum line for the engine is the centerline of the crankshaft. Cylinder bores must be perpendicular to the crankshaft center-

Cylinder Sleeve: A liner for a cylinder which can be replaced when worn to provide a new cylinder surface.

Dry Sleeve: A sleeve that does not come in direct contact with the engine coolant.

Wet Sleeve: A sleeve that comes in contact with the engine coolant.

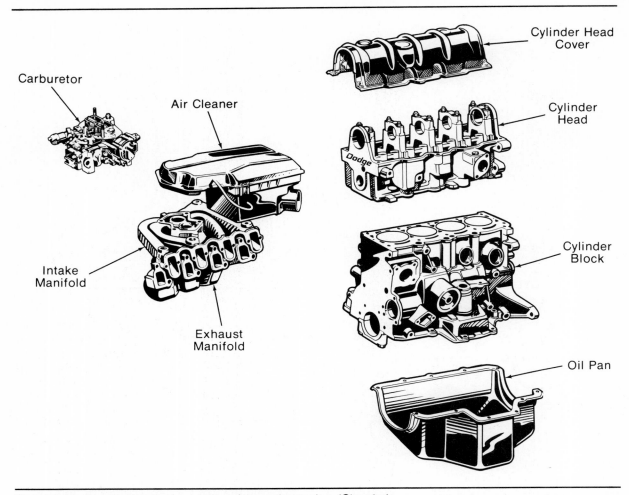

Figure 5-10. The block is the foundation of the entire engine. (Chrysler)

line. The centerline of the camshaft and the gasket surface at the top of the cylinders must be parallel to the crankshaft. Ideally, cylinders should be bored with the equipment located in relation to the crankshaft. However, ordinary boring equipment used in the field locates either on the gasket surface at the top of the cylinders, or on the oil pan rail at the bottom of the block. Because those two surfaces were originally made parallel to the crankshaft, cylinders bored in relation to them will be perpendicular to the crankshaft.

When the gasket surface at the top of the cylinders is resurfaced, it should be made parallel to the crankshaft centerline. Some field equipment will actually locate the block by mounting it on the crankshaft main bearing bores. This ensures that the resurfaced block deck will be parallel to the crankshaft. However, there are many types of equipment used that locate the block in relation to the gasket surface itself. These types of equipment have been used for many years, and are generally accepted as being accurate enough for most engines.

Blocks must be designed as compactly as possible. A large engine is not only heavy, but difficult to fit into an engine compartment. For this reason, only enough space is allowed inside the crankcase for the parts to move without interference. The room needed by the crankshaft is determined by the length of the piston stroke. A long stroke engine has a longer crankshaft throw, and therefore needs a bigger crankcase.

Crankshafts are usually counterweighted to balance the weight of the crankpin and other reciprocating parts. But the **counterweights** are up when the pistons are down. This gives the designer another clearance problem.

Another critical dimension is the thickness of the cylinder walls. If they are too thick, they will retain heat and the pistons will run hot. If too

Cylinder Blocks, Heads, Manifolds, and Miscellaneous Engine Covers

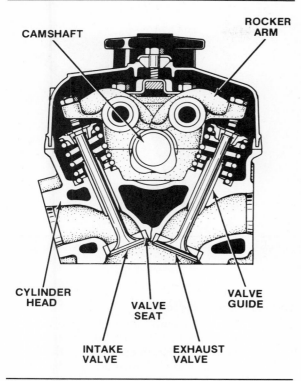

Figure 5-11. The seats and the guides for the valves are in the cylinder head.

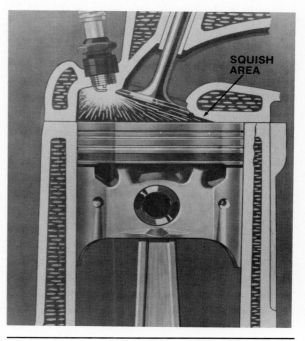

Figure 5-12. The wedge design provides good breathing because it has a quench and a squish area in the same location.

thin, they will be weak and distort or warp when heated and cooled.

The block must be heavy enough to withstand all the clamping and pulling forces that are imposed on it by the engine parts and accessories. Everything that mounts on the engine is bolted to the block or the heads. The heads, of course, are bolted to the block. The block is not only the framework of the engine, but also its foundation, figure 5-10.

CYLINDER HEADS

Cylinder heads for overhead valve and overhead camshaft engines do much more than provide a lid for the combustion chamber. They must also support the rocker arms of the camshaft, and provide guides and seats for the valves, figure 5-11.

Passages for the intake and exhaust gases go through each head to the intake and exhaust valves. The flanges around the exhaust and intake ports must support the weight and clamping action of the intake and exhaust manifolds. In effect, the overhead valve cylinder head is the upper half of the engine

Combustion Chamber Design

Combustion chamber design affects combustion efficiency and emissions. Three factors affecting the efficiency of the combustion are turbulence, squish and quench. These factors are determined, in part, by the shape of the combustion chamber.

Wedge-shaped chambers

The most popular design of the 1970's was the wedge, figure 5-12. The wedge provides good breathing efficiency because it has **quench** and **squish areas** and the air-fuel mixture is swirled about providing turbulence.

The squish area can be seen in figure 5-12. As the piston nears tdc, the air-fuel mixture is

---

Crankshaft Counterweights: Weights cast into, or bolted onto, a crankshaft to balance the weight of the piston, piston pin, connecting rod, bearings, and crankpin.

Quench Area: An area in the combustion chamber that cools the mixture to extinguish combustion and reduce detonation.

Squish Area: A narrow space between the piston and the cylinder head. As the piston approaches top dead center, the mixture squishes or shoots out of the squish area across the combustion chamber.

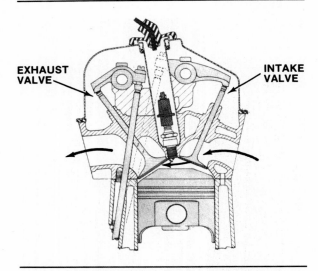

Figure 5-13. Because the intake and exhaust valves are on opposite sides of the combustion chamber, the hemispherical chamber allows the mixture to flow through in almost a straight line.

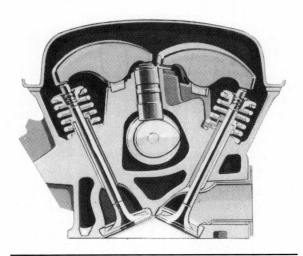

Figure 5-14. The valves on the hemispherical engine are canted at an angle. (Ford)

pushed out of the squish area causing turbulence. Turbulence mixes the air and fuel to assure more uniform combustion. Knocking or detonation can occur if the last part of the mixture reaches a very high temperature and ignites before it reaches the flame. To prevent this, some heat is extracted from the mixture in the quench area. The excess heat is drawn into the relative coolness of the metal.

In the wedge, the squish and quench area are the same. Another advantage of the wedge shape is that the valves are inline. This makes the engine inexpensive to manufacture.

Figure 5-15. The head on this Ford piston has been machined slightly to provide a degree of squish. (Ford)

Hemispherical chamber

The hemispherical combustion chamber provides the greatest breathing efficiency, figure 5-13. The valves are on opposite sides of the chamber, which allows the intake and exhaust to flow through the chamber in an almost straight line. The spark plug is centrally located, so when the mixture is ignited, the flame front has only a short distance to travel.

The hemi chamber has no squish or quench areas so there is little turbulence. On the other hand, the air-fuel mixture has such a short way to go to reach the flame that there is little tendency for detonation.

In 1981 Ford Motor Company came out with a new engine called a Compound Valve Hemispherical. The hemispherical combustion chamber was accomplished through canting the valves at an angle, figure 5-14. The heads of the pistons are machined to provide some amount of squish, figure 5-15.

Other chamber designs

Sometimes combustion chamber design changes are not obvious. Big block Chevrolet engines used what was called the porcupine head. It had the valves tilted slightly out of line for better breathing. It was basically a wedge design combustion chamber, but with some of the advantages of the hemi chamber.

To get maximum horsepower from the engine, the combustion chamber was completely enclosed. The walls of the chamber came straight down from the valves. A large part of the top of the piston was covered by the cylinder head. After several years of running the closed chamber, it was discovered that the engine would produce more horsepower if the

Cylinder Blocks, Heads, Manifolds, and Miscellaneous Engine Covers

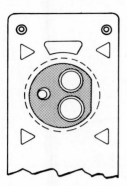 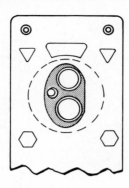

Figure 5-16. Opening the combustion chamber produced more horsepower.

chamber was opened up. The open chamber walls were slanted so that much of the top of the piston was uncovered, figure 5-16. This open design resulted in a lower compression ratio, but the horsepower was raised because of better breathing. This illustrates the importance of breathing efficiency in cylinder head design.

Because the hemi head, with its straight flow path through the chamber, breathes the most efficiently, it follows that the intake valve ought to be on one side of the head and the exhaust valve on the other side. If intake and exhaust are on the same side, the mixture has to make a U-turn to go out the exhaust. When the intake and exhaust ports are on opposite sides of the head, the design is known as a **crossflow head**, figure 5-17. However, an inline engine with a crossflow head needs more engine compartment room, and its greater complexity makes it more expensive to produce.

On the crossflow design, both sides of the cylinder head must be machined, whereas on the design with both intake and exhaust valves on the same side of the head, machining is simpler. In spite of this, new engines are appearing with the crossflow design because it lends itself to transverse positioning of a 4-cylinder engine. With the cylinders running in-line in a transverse position, the intake fits easily at the front and the exhaust at the rear.

Surface-To-Volume Ratio

The surface-to-volume ratio is important in combustion chamber design. A typical surface-to-volume ratio might by 7.5:1. This means that the surface area, divided by the volume, is 7.5. Another way of saying this is that the surface equals 7.5 times the volume.

If the surface-to-volume ratio is too high, there will be a lot of surface to which fuel can cling. The fuel next to the walls of the chamber

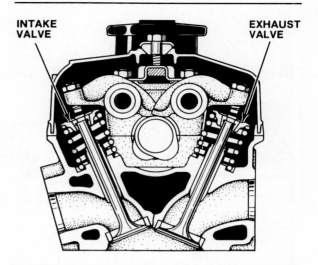

Figure 5-17. The crossflow head design has the intake and exhaust valves on opposite sides of the chamber. (Chrysler)

may not burn completely because the chamber walls cool the mixture below ignition temperature. When the exhaust valve opens, the unburned fuel goes out the exhaust. Unburned fuel includes hydrocarbons, a contributor to smog. So, as you can see, there is a relationship between the surface area of the combustion chamber and the number of unburned hydro-

■ Cylinders In A Circle

Engine cylinders can also be arranged in a circle. This is called a radial engine and is used in aircraft. The connecting rods from the pistons all use the same large end. Radial engines can be designed with any number of cylinders. If there are too many cylinders for one circle, a second circle is added. This is called a double row engine. Many aircraft engines have been built with nine cylinders in each of two rows, for a total of 18 cylinders.

Radial engines are usually air-cooled. The cylinders are not cast together, as in an inline or V-type engine. Air circulates between the cylinders for cooling.

Crossflow Cylinder Head: A head with the intake port and exhaust port exiting from opposite sides of the head. The design provides an almost straight path for the mixture to flow across the top of the piston.

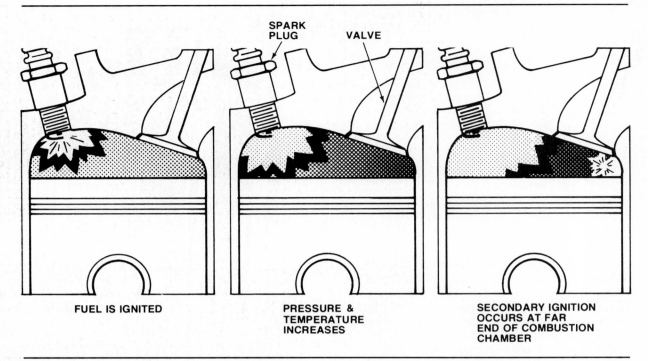

Figure 5-18. Detonation is a secondary ignition of the air-fuel mixture caused by high cylinder temperatures. It is commonly called "pinging" or "knocking."

carbons. To minimize the amount of unburned hydrocarbons, the designer must keep the surface area as small as possible. This is done by eliminating ripples and angles, but at the same time keeping a shape that is easy to manufacture.

Quench Area

We touched briefly on the quench area; let's take a closer look now.

Detonation occurs when the flame front travels across the chamber, pushing unburned fuel ahead of it. The unburned fuel is compressed until it explodes or detonates, figure 5-18. To prevent detonation, an area farthest from the spark plug is designed to quench or cool the air-fuel mixture.

As we saw in the wedge-shaped chamber, the squish area was farthest from the spark plug, so it doubled as a quench area. The water jacket is designed to bring coolant to the outside of the cylinder, reducing the temperature of the quench area.

In the wedge, when the piston rises, air-fuel mixture squirts out of the squish area into the chamber. After ignition, the piston is moving down, which opens up the squish area and allows it to quench the mixture. This known as a squish-quench design.

Valve Location

Valve location in the cylinder head is an important factor in breathing efficiency. Probably the best example of a poor breathing design is the obsolete L-head engine. The incoming mixture had to make a 90-degree turn to get to the piston. After combustion, the exhaust had to make another 90-degree turn to get out. Compare this with the hemispherical combustion chamber. A straight line can actually be drawn from the intake port through the combustion chamber to the exhaust port, figure 5-19.

In the wedge chamber, the mixture enters in a straight line, but the exhaust has to make a turn before it can exit.

Intake and exhaust ports

Ports are the passages that bring the mixture to the intake valve and take the exhaust out past the exhaust valve. Port design has a big effect on how much mixture will pass through the intake valve opening. If the intake port restricts the flow, the chamber will not develop maximum power.

Exhaust ports have the same effect. If the exhaust does not flow freely out of the chamber, some of it will remain and take space that should be filled with fresh mixture. The engine then puts out less power than it could.

Port design is not as simple as it may seem.

Cylinder Blocks, Heads, Manifolds, and Miscellaneous Engine Covers

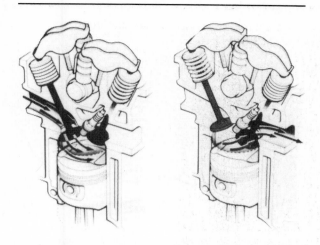

Figure 5-19. In a hemispherical combustion chamber, an almost straight line can be drawn from the intake port to the exhaust port. (Ford)

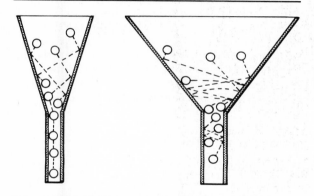

Figure 5-20. This illustration shows that in port design, the larger port or opening does not necessarily make a more efficient system.

The biggest ports do not necessarily do the best job. It is similar to pouring ball bearings through a funnel. A large funnel will allow the bearings to bounce around, and they will not pass through very fast. A small funnel with steeper sides will guide the ball bearings through to the exit more quickly, figure 5-20.

Smaller ports also keep the velocity of intake and exhaust gases high. When the throttle is opened, the response is quicker with smaller ports. However, smaller ports do not pass as much mixture at wide open throttle. The operating range of the engine must be considered when designing ports.

Much of port and valve design is done on a flow bench, figure 5-21. The port shape is made out of plastic or clay and mounted on a bench with a blower. Turning on the blower and measuring how much air comes out the other end of

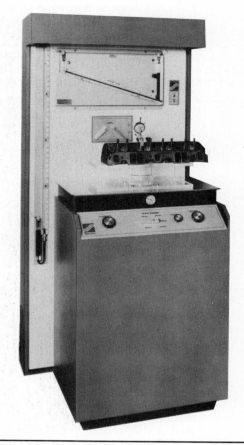

Figure 5-21. Port and valve design is often done on a flow bench. (Super Flow)

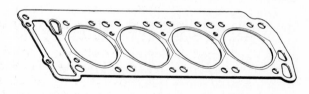

Figure 5-22. Holes in the cylinder head gasket match the holes in the head itself. (American Motors)

the port will tell whether or not the port is efficient.

Sometimes the engineers do not know why a change in port shape produces better flow. But the flow bench allows them to test all kinds of shapes and select the best one.

Cylinder Head Installation

Cylinder heads are designed to fit against the top of the cylinder block with a head gasket in between to prevent leaks. Holes in the gasket match the cylinder bores and coolant passages, figure 5-22.

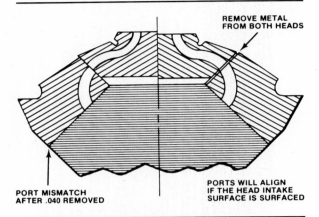

Figure 5-23. When a stud is screwed into a casting it should have coarse threads to make sure it will hold.

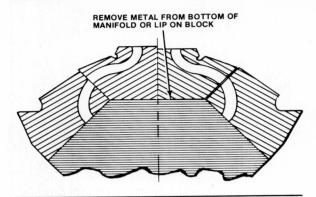

Figure 5-25. When the intake manifold seals against a lip at the front and rear of the block, the lip may be too high. This could prevent the manifold from seating against the cylinder heads. Remove metal from the bottom of the manifold or the lip to correct this.

Figure 5-24. Metal must be removed from the intake side of the heads if the capscrews will not line up with the holes.

The cylinder heads are attached to the block with studs and nuts or with capscrews. These nuts and studs usually have fine threads, which help to distribute tightening pressure evenly, and stay tight. This prevents warping. The end of the stud that is screwed into the casting has coarse threads because a casting is not strong enough to hold fine threads, figure 5-23.

The gasket surfaces of the head and the block must be flat so that the gasket will be compressed equally everywhere. If the head or block are warped so that the gasket is compressed more at the edges than in the middle, then the middle may leak combustion gases or coolant. When designing an engine, the headbolts are positioned as close together as required to equalize the clamping forces over the entire gasket surface. In some places the clamping force may be weak if the design of the engine did not allow the bolts to be close enough together. In that case, the gasket may leak or blow out.

Headbolts must always be torqued starting in the center of the gasket and working out to the edges. As a gasket is compressed, it becomes longer. If the tightening starts at the edges, the center will wrinkle because the edges prevent it from stretching out.

Head surfacing can affect the valve train in one of two ways, depending on whether the camshaft is in the head or the block. If it is in the block, the head moves closer to the cam when material is removed from the head. This causes the pushrods to sit higher in relation to the head, which in turn causes the rocker arms to ride at an improper angle. On an overhead cam engine, the cam will be lowered in relation to the crankshaft, and this will allow the timing chain to lose tension which may retard the cam timing.

Head To Manifold Relationship

Intake manifolds are attached to the cylinder heads with capscrews or with studs and nuts on some inline engines. On an inline engine the position of the cylinder head does not affect the intake manifold. If the head gasket is thick or thin, or if the head has been resurfaced to make it straight, the manifold will be mounted a little higher or a little lower in the engine compartment, but its relationship to the parts will not change.

On a V-6 or V-8 engine, however, the position of the cylinder head can change the fit of the intake manifold. If the heads are resurfaced to correct warping, the distance between them becomes less. The heads move closer to the cylinder block and, therefore, the intake manifold mounting points are closer together.

Any change in the position of the heads requires that the alignment and fit of the intake manifold be checked. If the intake manifold is out of alignment with the head, it is necessary to remove metal from the intake side of the head to restore the original distance between the heads, figure 5-24

The manifold will now line up with the heads, but it is sitting lower on the engine. On an engine where the intake manifold seals

Cylinder Blocks, Heads, Manifolds, and Miscellaneous Engine Covers

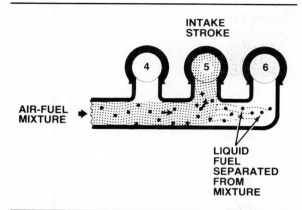

Figure 5-33. An intake manifold with large passages and sharp angles will cause liquid fuel to separate out of the air-fuel mixture. (Chevrolet)

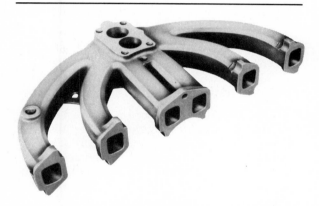

Figure 5-34. This Chrysler intake manifold has curved runners for better mixture distribution. (Chrysler)

Cylinder-to-cylinder mixture distribution is more equal with the single-plane design, but it usually allows less airflow in low-to-intermediate speed ranges. Short runner length causes a drop in mixture velocity. The single-plane manifold design can produce more horsepower at high rpm than can the 2-plane design.

Inline engines use a single-plane manifold. The carburetor usually is placed in the middle of the manifold, leading to the air-fuel flow condition illustrated in figure 5-33. Cylinders closer to the carburetor will receive too much air in the mixture, while those farther away will receive too much fuel. By using siamesed ports, an inline engine needs fewer manifold runners. Chrysler inline 6-cylinder engines do not have siamesed intake ports, but the intake manifolds have long, curved runners with no sharp turn, figure 5-34

V-8 engine intake manifolds may be either a single-plane or a 2-plane design. A 2-plane in-

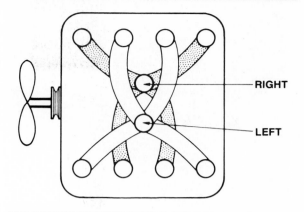

Figure 5-35. The two chambers of a 2-plane manifold are of different heights. (AC Delco)

take manifold has two separate plenum chambers connected to the intake ports of the engine, figure 5-35. Each chamber feeds two central and two end cylinders. Mixture velocity is greater at low-to-intermediate engine speeds than with the single-plane manifold, but mixture distribu-

■ **Were They The Smoothest?**

In the late 1920's and early '30's the automotive industry was maturing to the point where cars were quite dependable. It was time to think about comfort and elegance. And what could be more elegant than increasing the number of cylinders in the engine? If four cylinders were common, and eight were better, then 16 ought to be the ultimate.

In 1930, Cadillac came out with a V-16 engine. In 1931, Marmon also had a V-16. These were probably among the smoothest engines ever built. With a power impulse every 45° of crank rotation, they didn't just run, they hummed. There were also V-12's, made by Cadillac and other companies, but nothing could top the smoothness and novelty of the V-16's.

Two-Plane Intake Manifold: A manifold design used on V-type engines, where one half of the carburetor feeds half the cylinders on one bank and half the cylinders on the other bank.

Single-Plane Intake Manifold: A manifold design used on V-type engines, where each half of the carburetor feeds all the cylinders on one bank of the engine.

Figure 5-36. The job of the exhaust manifold is to carry the exhaust out of the engine as efficiently as possible. (Pontiac)

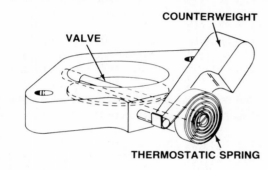

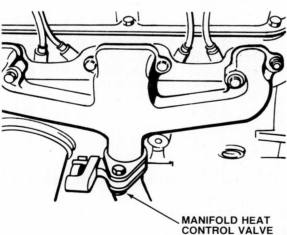

Figure 5-37. Top: The manifold heat control valve, when closed, forces exhaust to the carburetor where it hastens warmup. Bottom: The heat control valve can also be within the manifold itself.

tion is usually more uneven. When a 4-barrel carburetor is used with a 2-plane manifold, each side of the carburetor (one primary and one secondary barrel) feeds one plenum chamber.

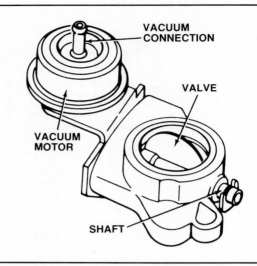

Figure 5-38. A vacuum-operated heat control valve.

When a 2-barrel carburetor is used, each barrel feeds one chamber.

Exhaust Manifolds

The exhaust manifold, figure 5-36, must carry the exhaust out of the engine as efficiently as possible. There is often a clearance problem with exhaust manifolds on V-8 engines. Because the exhaust manifolds are usually low on the engine, they can hit the frame or part of the chassis. Therefore, exhaust manifolds are frequently squeezed into shapes that detract from the performance and efficiency of the engine.

A properly designed exhaust manifold will give good **scavenging** of the combustion chamber. The rush of exhaust gas out of the pipe should create a vacuum at the exhaust valve which will draw the exhaust gases out of the cylinder. This will improve performance and efficiency.

Part of many exhaust manifolds is a **manifold heat control valve**, figure 5-37. This valve, when closed, forces the exhaust to go through a passageway in the intake manifold that leads to the base of the carburetor. This heat hastens warmup and vaporization of the fuel. However, the heat valve reduces engine efficiency when it is closed because it causes backpressure in the exhaust. Most manifold heat valves are spring loaded and counterweighted. This enables them to open more easily when the exhaust manifold gets to normal operating temperature. The flow of exhaust against the valve opens it. On some late-model cars the valve is opened by intake manifold vacuum, figure 5-38

Some late-model cars use an exhaust heat passageway to conduct exhaust gas to the ex-

Cylinder Blocks, Heads, Manifolds, and Miscellaneous Engine Covers

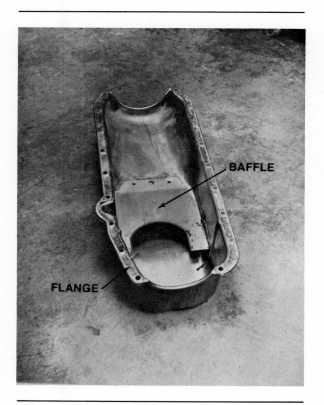

Figure 5-39. One of the most important engine covers is the oil pan.

haust gas recirculation (EGR) valve. This valve allows a small amount of exhaust gas to enter the intake manifold and mix with the air-fuel mixture to help control NO_x emissions.

A considerable amount of design work has gone into accessory **headers** for cars. A header is a very efficient exhaust manifold. It creates a free flow of exhaust gas that does a better scavenging job than a stock exhaust manifold. This is done by having the tubes from each cylinder exactly the same length This "tuning" of the exhaust is designed for a specific engine rpm, but will also increase efficiency when the engine is running close to the optimum speed.

Miscellaneous Engine Covers

The cylinder block and heads are sealed against outside elements by several covers, such as the valve covers, oil pan, and timing chain covers. These covers keep oil in the engine and dirt out of it.

The most important of these covers is the oil pan, figure 5-39. It covers the bottom of the crankcase and is also the oil reservoir for the engine. The part of the pan that contains the oil is known as the sump. Within, or near, the sump there may be baffles to keep oil from splashing around during hard braking, cornering, or acceleration. At the bottom of the sump is the oil drain plug.

The oil pan is bolted to the engine block and sealed with a gasket. It must be removed to inspect bearings or repair the oil pump.

The timing cover protects the timing sprockets and chain or pulleys and belt at the front of the engine. The cover may be a simple piece of sheet metal with nothing attached to it, figure 5-40. Or it can be a large aluminum casting that provides water passages and a mounting for the water pump. On some engines with front-mounted distributors, the front cover also provides a place to mount the distributor and oil pump.

On most overhead valve engines, the front cover holds the front crankshaft oil seal. The oil pan then seals against the bottom of the front cover.

On a V-8 engine, there is a large area at the top of the engine between the cylinder banks that must be covered. The intake manifold acts as a cover for this valley in the engine. On other

■ Is A Distributor Necessary?

For the last 50 years, all production cars have had distributors. An ignition coil generates the high voltage, and the distributor gets it to the right cylinder. But it wasn't always that way. In the early days of the car industry, distributors were unheard of. The Ford Model T is a good example. It had four cylinders, and a coil for each cylinder. A crude timer, called a commutator, started each coil sparking at approximately top dead center on the compression stroke. Because each coil was only connected to its own cylinder, and only sparked when its own cylinder was ready to fire, no distributor was necessary.

Scavenging: The ability of the exhaust system to create vacuum, which draws the exhaust from the cylinder when the exhaust valve opens.

Header: A free-flowing exhaust manifold, designed for a specific engine speed, with tubes of equal length which meet in a collector; it improves scavenging and increases engine efficiency at or close to its design rpm.

Manifold Heat Control Valve: A valve that diverts part of the exhaust through a passageway under the carburetor to provide heat for the mixture that is in the intake manifold.

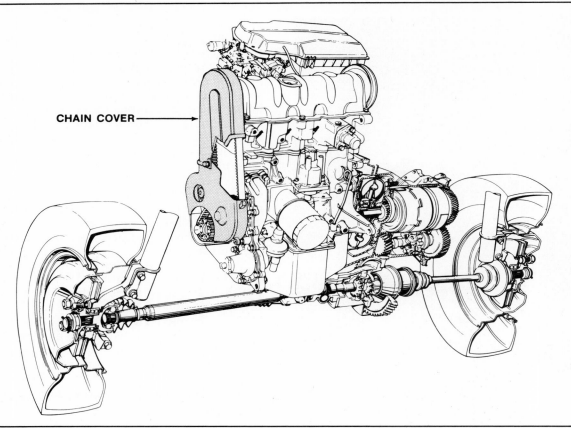

Figure 5-40. The timing chain cover protects the timing sprockets and chain or pulleys and belt at the front of the engine.

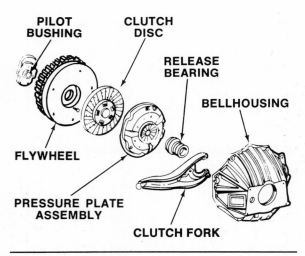

Figure 5-41. The bellhousing is used to cover the flywheel or clutch.

engines a sheet metal valley cover is used. The intake manifold then sits above the valley cover.

On most engines, the valve lifters can be lifted out after removing either the valley cover or the side cover and the pushrod or valve. On overhead valve, V-type engines the intake manifold must be removed before you can remove the valley cover. On some engines, the lifters can be pulled up through the pushrod area in the head without removing the intake manifold or valley cover.

A bellhousing is used to cover the clutch and flywheel, figure 5-41. On an engine with an automatic transmission the bellhousing covers the torque converter, or fluid coupler. On most automatic transmissions, the bellhousing is made in one piece with the transmission, but on manual transmissions the bellhousing is usually separate. The bottom side of many bellhousings is left open for inspection and repair access. Clutches can sometimes be removed through this bottom opening. To prevent stones or other debris from hitting the moving parts inside the bellhousing, the bottom opening is covered with sheet metal or plastic.

Valve covers and overhead camshaft covers are stamped out of sheet steel, molded from plastic, or cast out of aluminum.

Cylinder Blocks, Heads, Manifolds, and Miscellaneous Engine Covers

They are mounted with a gasket to prevent oil leakage. They are usually lightweight and fragile, and require care when installing. Overtightening the bolts or capscrews that hold the covers in place could cause warping, leaking or cracking of the cover.

SUMMARY

When you design an engine, you start with the block, the foundation of the entire engine. You then add the cylinders, cylinder head, intake and exhaust valves, and also add an intake and exhaust manifold.

You have seen that in each engine part, from the block to the various engine covers, design and dimensions are critical to engine efficiency. We have looked at the different types of blocks: inline, V-type, and horizontally opposed. We have seen the overhead valve engine, the overhead cam engine and even the L-head engine. We have looked at the various locations of the intake and exhaust valves and the different manifold designs. In every case, the design, the dimensions and the placement affect how the engine runs.

But don't stop with the design and dimensions. When you are looking at engine efficiency you also have to consider the relationship of one part to another and how they all work as a whole.

Review Questions

Choose the single most correct answer.
Compare your answers to the correct answers on page 268.

1. Another name for a horizontally opposed engine is:
   a. Air-cooled engine
   b. Flat-four
   c. Pancake engine
   d. All of the above

2. In a wedge-shaped combustion chamber, squish and quench areas are:
   a. The same
   b. On opposite sides of the head
   c. Not as important as in the hemi
   d. Not necessary for good combustion

3. Which of the following is a type of production intake manifold?
   a. Staggered runner
   b. Double runner
   c. Bi-plane
   d. Single plane

4. Which of the following affects the efficiency of the combustion chamber?
   a. Turbulence
   b. Squish
   c. Quench
   d. All of the above

5. Which of the following is NOT a type of cylinder sleeve?
   a. Dry sleeve
   b. Reusable sleeve
   c. Wet sleeve
   d. Steel sleeve

6. The main purpose of engine covers is to:
   a. Prevent oil from escaping
   b. Provide additional structural strength
   c. Provide necessary heat dissipation
   d. All of the above

7. The manifold heat control valve is part of the:
   a. Cylinder head
   b. Carburetor
   c. Intake manifold
   d. Exhaust manifold

8. Surface-to-volume ratio should be ideally:
   a. As low as possible
   b. As high as possible
   c. Above 7.5:1
   d. As close to 7.5:1 as possible

9. Head bolts should be torqued:
   a. Starting at the edges, and working in toward the center
   b. Starting at the center, and working out toward the edges
   c. Evenly, to approximately 75 foot-pounds
   d. None of the above

10. Small intake and exhaust ports:
    a. Reduce the velocity of intake and exhaust gases
    b. Reduce throttle response
    c. Increase the flow of gases at wide open throttle
    d. Keep the velocity of intake and exhaust gases high

11. The main bearings in an engine support the:
    a. Crankcase
    b. Crankshaft
    c. Camshaft
    d. Connecting rods

12. A crossflow head:
    a. Requires a wedge-shaped combustion chamber
    b. Requires that the mixture make a U-turn before it can exit
    c. Has the intake and exhaust valves on opposite sides
    d. Needs less room in the engine compartment

13. A V-8 engine with paired rods on each throw usually has _____ main bearings.
    a. 3
    b. 5
    c. 7
    d. 9

14. A manifold area used to store the air-fuel mixture is called a:
    a. Runner
    b. 2-plane chamber
    c. Plenum chamber
    d. Reservoir chamber

15. Head surfacing will often affect the:
    a. Compression ratio
    b. Timing chain tension
    c. Rocker arm angle
    d. All of the above

16. The main purpose of exhaust headers is to:
    a. Reduce exhaust gas for scavenging
    b. Tune the exhaust for a specific engine load
    c. Reduce the emissions at idle speeds
    d. None of the above

Chapter 6

Valves, Springs, Guides, and Seats

VALVES

All 4-stroke piston engines use intake and exhaust valves. The intake valve opens to allow the air-fuel mixture to enter the cylinder. The exhaust valve opens to let the piston push the burned mixture out of the cylinder.

Most engines use one intake and one exhaust valve per cylinder, figure 6-1. The poppet valve is used in almost all 4-stroke engines. It consists of a large head and a long stem. In the closed position the head rests on a seat, which shuts off the passageway or port. To open the port, the stem of the valve is pushed by the rocker. This raises the head of the valve off the seat and allows the mixture to flow into the combustion chamber.

The size of the valve is a factor in how much airflow there will be when the valve head is raised off its seat. To get more airflow, designers sometimes use two small intake and two small exhaust valves in each cylinder instead of one large intake and exhaust valve. You will find this in some motorcycle engines. A notable example in the automotive field is the Cosworth Vega, figure 6-2.

These smaller valves do have the advantage of low mass over the larger, heavier valves. The greater mass of the large valve may cause it to bounce when it hits the seat, and when it is pushed open it does not stop easily when it reaches its full opening. This causes slack in the valve train and increased wear. In extreme cases, the valve may stay open long enough for the piston to hit it, resulting in major engine damage.

On the stem side of the valve head is a ground face, figure 6-3. The face is circular and, on most valves, it is at a 45-degree angle to the top of the head. The valve face seals against a seat in the cylinder head on an overhead valve engine.

Valve Angles

Many engines used to have a 30-degree face and seat angle. This angle was thought to allow a better flow of the intake mixture over the edge of the seat. 30-degree valve faces have been almost universally abandoned in favor of 45-degree valve faces. The 45-degree valve face is better at self-centering the valve as it closes and wedges it tighter in the seat for a better seal.

Actually, the 45-degree figure is a general dimension. The valve face is actually ground at 44 degrees and the seat at 45 degrees, figure 6-4. The difference in these two angles is called the **interference angle**. The purpose of the interference angle is to improve valve sealing at high engine speeds and temperatures, and promote better sealing of the valves during initial engine break-in.

Valves, Springs, Guides and Seats

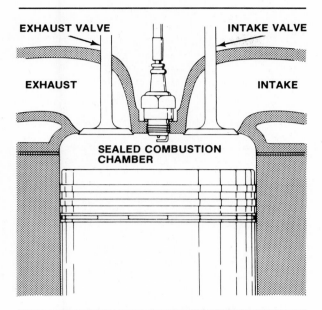

Figure 6-1. Normally, each cylinder has one intake and one exhaust valve.

Figure 6-2. Some motorcycle engines use two intake and two exhaust valves per cylinder. You will also find this in the Cosworth Vega engine pictured here. (General Motors)

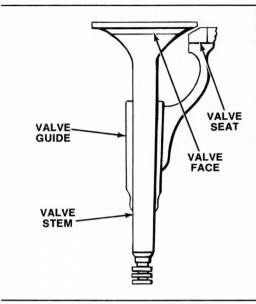

Figure 6-3 This illustration shows the parts of a valve. The face is on the stem side of the valve. (Ethyl)

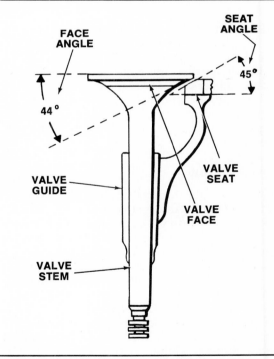

Figure 6-4. The valve face is ground at a 44-degree angle, while the seat is ground at a 45-degree angle. This one degree difference is known as the interference angle.

Interference Angle: The angle or difference between the valve face angle and the valve seat angle.

All valve angles are measured from the horizontal or a line drawn level with the top of the valve head, figure 6-5. This is known as the included angle. The angle itself includes or contains the head of the valve. If the angles are measured from the stem, which is vertical, the slant of the valve face would put the open end in the combustion chamber.

Valves are mounted with a compression spring around the stem. The spring holds the valve closed until the stem is pushed to open it. The stem is pushed by the lobe on the cam-

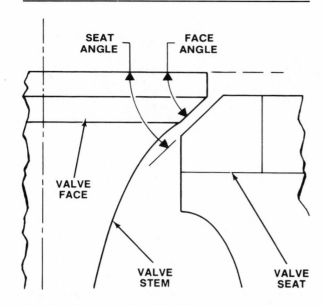

Figure 6-5. Valve angles are measured from the horizontal.

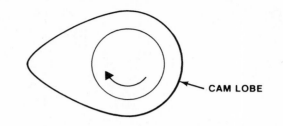

Figure 6-6. The gradual rise from the low point to the high point on the cam lobe allows the valve to be lifted gently from its seat and set back down again.

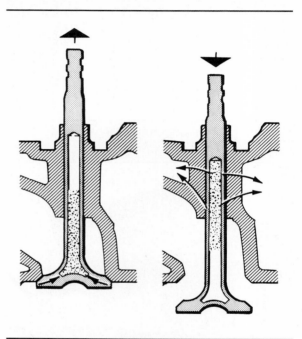

Figure 6-7. The stem of the valve is hollow and partially filled with sodium to conduct heat away from the valve.

shaft operating through the valve train. The lobe is ground so that there is a gradual rise from the low point to the high point, figure 6-6. This allows the valve to be gently lifted from its seat and set back down again.

The length of time the valve stays open is determined in part by the angle from the beginning of one cam lobe ramp to the end of the other, figure 6-6.

Valve Cooling

Because the valve heads form part of the combustion chamber, they get very hot. The intake valve runs cooler than the exhaust valve. The incoming air-fuel mixture holds the intake valve head temperature down to about 800° F (372° C). Exhaust valves run hotter because the hot exhaust gases flow past them to escape from the combustion chamber. The exhaust valve temperature can average 1200° F (558° C).

This heat must be removed or the valve will warp, leak, or burn. Some heat is transferred to the valve stem and then to the guide, but the maximum heat transfer is from the valve to the valve seat. Of course, this can only happen when they are in contact with each other. It is essential that the valve be held tightly against its seat. If the valve springs are weak and not able to do this, the valve will overheat and burn.

A valve will burn if a leak develops at any point around the valve seat. The area around the leak will not cool because it is not in contact with the seat and so cannot transfer heat. As the valve gets hotter, it will start to warp which will, in turn, result in more heat buildup. This chain reaction quickly results in a burned valve and the accompanying loss of engine efficiency and power.

Sodium-cooled valves

Filling valve stems with sodium is a method of cooling the valve to reduce wear. The stem is hollow and about half of the cavity is filled with pure sodium, figure 6-7. Sodium melts at just below the boiling point of water, 207.5° F (97.5° C). During normal engine operation, the valves are much hotter than this, so the sodium stays in liquid form.

Sodium conducts heat well. During engine operation, the sodium moves around in its cavity and transfers the heat from the valve head to

Valves, Springs, Guides and Seats

Valve Train

The **valve train** consists of the components located between the cam lobe and the valve stem. These can include the pushrod, tappet or lifter, and rocker arm.

There are three basic types of valve trains.

L-head valve train

The L-head valve train is the simplest type and is no longer used in modern engines. In it, a barrel-shaped lifter or tappet rests on the cam lobe. The lifter is raised by the cam lobe and, in turn, pushes directly on the valve stem to open it, figure 6-8.

On many L-head engines, the lifter is adjustable and is set with a specific amount of clearance. This clearance is taken up when the engine gets warm.

■ **Why The Poppet Valve?**

Have you ever wondered why the poppet valve is so widely used in automotive engines? In the early days of the industry, there were other types of valve arrangements.

The Stearns-Knight car used the Knight sleeve valve engine. The valves consisted of two concentric sleeves surrounding the piston located between the piston and the cylinder walls. A separate shaft moved the sleeves up or down to open and close the intake and exhaust ports.

There were also rotary valve engines. A long shaft with holes in it allowed air and exhaust to flow when the holes aligned with the intake and exhaust ports. There were even slide valve engines, which had a sliding bar with holes in it that aligned with ports in the cylinders.

But all of these valve designs proved to be inferior to the simple poppet valve, mainly because they all had a tendency to leak. The poppet valve wedges into its seat when it closes, which reduces leakage. And the higher the combustion pressure, the tighter it seals. But it still opens easily with a push. It's no wonder that the poppet valve won out over all the other designs.

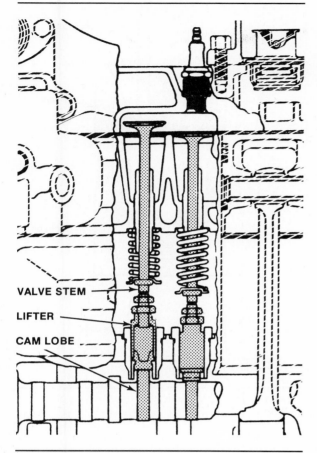

Figure 6-8. In the L-head valve train, the lifter is raised by the cam lobe and pushes on the valve stem.

the stem. From the stem, the heat goes into the guide. On a normal valve, very little heat goes into the guide, so a sodium-filled valve is an excellent way of providing additional cooling for the valve.

Very few passenger cars use sodium-filled valves. The exceptions are some air-cooled Porsches and the Audi Turbo. Both of these vehicles have engines that reach extremely high temperatures. You will also find sodium-cooled valves on turbocharged truck engines.

When you turbocharge an engine, you increase combustion pressure which creates more heat, so there is a greater need for valve cooling. CAUTION: Sodium will burst into flame when it comes into contact with water. If sodium comes in contact with your skin, it will cause a serious burn. If the hollow stem of the valve is cracked or broken, you have a potentially dangerous situation. Be very careful in handling sodium valves. Not all sodium valves are marked as such, but they generally have oversized stems.

Valve Train: The assembly of parts that transmits force from the cam lobe to the valve.

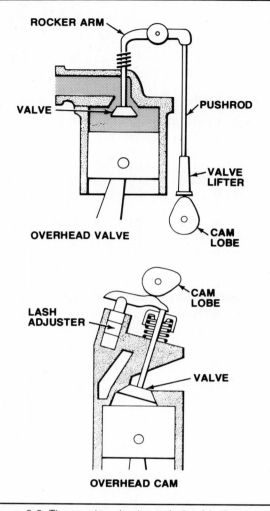

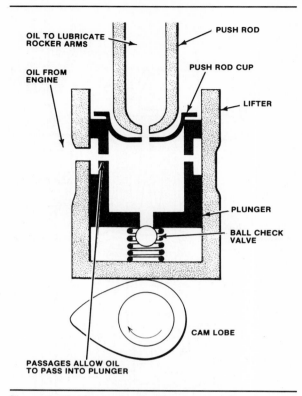

Figure 6-9. The overhead valve train (top) is the most complicated of the three valve trains; the overhead cam, the most efficient.

Figure 6-10. The hydraulic lifter uses engine oil to transmit motion.

Overhead valve train

The overhead valve train, figure 6-9, is the most complicated because it has to transfer the motion of the camshaft all the way to the top of the head where the valves are located. The cam lobe pushes on a lifter which lifts a pushrod. The pushrod moves a rocker arm which pushes on the valve stem causing the valve to be lifted off its seat.

On an overhead valve engine there are two methods of providing for heat expansion. Hydraulic lifters automatically compensate for the changes in the clearance in the valve train, or there may be adjustable pushrods or rocker arms.

Overhead cam valve train

The overhead cam valve train is, by far, the most efficient system. It has the simplicity and light weight of the L-head design while retaining the efficiency of the overhead valve design. In this system, the camshaft sits directly on top of the head and the cam lobe pushes on a tappet which pushes a stem. Or, a cam pushes on a rocker arm which transfers the push to the valve stem, figure 6-9.

On the overhead cam valve train the valves are adjusted one of three ways. If the camshaft sits directly above the valve stems, a shim may be used between the tappet and the valve stem to provide the proper clearances. The overhead cam engine may also have adjustable tappets or adjustable rockers.

Valve lifters

Valve lifters can be mechanical or hydraulic. Mechanical lifters are solid metal as shown in figure 6-9 (top). A hydraulic lifter is a metal cylinder containing a plunger that rides on oil, figure 6-10. A chamber below the plunger fills with engine oil and, as the camshaft lobe lifts the lifter, the chamber is sealed by a check valve. The trapped oil transmits the lifting motion of the camshaft lobe to the valve pushrod. Hydraulic lifters are generally quieter than mechanical lifters and do not need to be adjusted

Valves, Springs, Guides and Seats

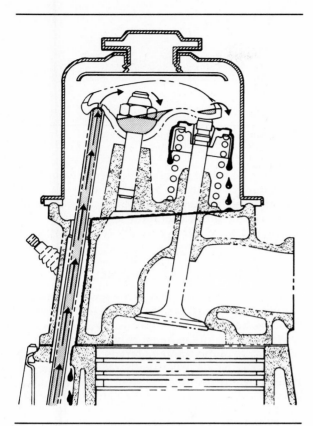

Figure 6-11. On some overhead valve engines, the oil is pumped through the pushrods to the rocker arms. It then runs down the guides and valve stems to lubricate the wear points. (Chevrolet)

as often because the amount of oil in the chamber varies to keep the valve adjustment correct.

Valve Lubrication

In modern engines, valves are lubricated in their guides by oil dripping from above. In an L-head engine, the valve stems are lubricated by oil vapor and splash. This is possible because the valve chamber is close to the spinning crankshaft, which throws off large amounts of oil.

In an overhead valve engine, the valves are a long way from the crankshaft. Oil is pumped through passages to the rocker arm shaft and the rockers. From there the oil runs onto the valve stems and into the guides. In the process, the wear points on the pushrod, rocker arm, valve stem, and guide are lubricated.

A seal is often installed on the stems and guides to prevent too much oil from traveling down the guide and onto the fillet and back of the valve head. When the valve opens, this oil may be drawn into the combustion chamber where it oxidizes into hard carbon that clings to

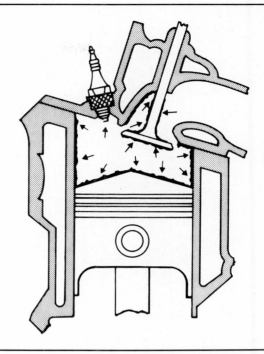

Figure 6-12. If oil is drawn into the combustion chamber, it oxidizes into carbon and clings to the chamber walls. (Black and Decker)

the chamber walls, figure 6-12. Deposits in the chamber raise the compression level and may cause detonation, or become red hot and cause preignition.

Because of the high valve temperature, oil can oxidize on the fillet and the back of the valve

■ Desmodromic Valves

The idea of a desmodromic valve system was first patented in 1910. The system was used in the French Delage GP cars during the 1914 season and in other racing cars in the Twenties and Thirties. Mercedes-Benz had W196 GP cars and 300SLR racers which used desmodromic valve gear.

What is a desmodromic valve system?

It is a system in which both opening and closing of the valves are cam-actuated, rather than using springs for closing. If valve springs are used, they are only for initial sealing. This system has an added cam follower that closes its valve via a projection on the valve stem. The advantage of this valve train is that it allows more aggressive cam profile without valve float. The main drawback is the relatively high reciprocating weight that tends to limit engine speed.

The only production vehicle using desmodromic valves today is the Italian Ducati motorcycle.

Figure 6-13. Oil can also oxidize on the fillet and the back of the head obstructing the flow of the air-fuel mixture. (Ethyl)

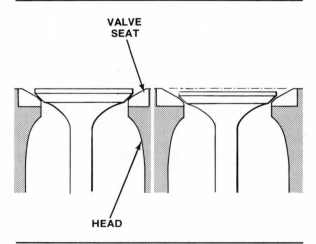

Figure 6-14. With the use of lead-free gasoline, valve seats began to wear so badly that they receded into the head.

head. The deposits can become so thick that they restrict the intake flow, figure 6-13.

Another area that can suffer from a lack of adequate lubrication is the valve face and seat. When the spring forces the valve face against the valve seat, there is a pounding action that tends to wear the seat. The valve does not wear as much because it is usually harder than the seat. This tendency of the seats to wear was unknown for years because gasoline contained tetraethyl lead. The lead acted as a lubricant on the valve and seat.

Figure 6-15. Valve with aluminized head.

When lead-free gasoline came out, valve seats started to wear so badly that the seat would actually recede into the head, figure 6-14. The receding of the seat reduces the valve spring pressure and eventually causes valve failure.

Valve Materials And Construction

As we learned in the section on valve cooling, the exhaust valve has to be able to withstand a lot more heat than the intake valve. For this reason, intake and exhaust valves are not always made of the same materials.

An exhaust valve is usually made of some type of heat resistant alloy steel. Many carmakers cover different parts of the valve with different materials. In the case of American Motors, the exhaust valve stem is a chrome alloy for wear, the head is chrome maganese nickel and the face is a combination of cobalt, chromium, and tungsten. This combination is commonly known by its most familiar brand name, **Stellite**.

Stellite is often used on valve faces because of its hardness. In fact, it has been used for so long that almost any hard-faced or premium valve is referred to as a "Stellite valve."

In most cases, intake valves are made of steel with aluminized heads, figure 6-15. Buick chrome flashes the valve stems for long wear.

The valve stem and head can be separate pieces welded together to form a valve.

Any small imperfection, or **stress riser**, in the valve could develop into a crack. Valves are ground to remove the stress risers, casting irregularities and machining marks that could cause disaster.

Polishing the valves is usually done only in high-performance shops. When a valve is polished it is less likely to accumulate carbon.

Valve Life

One of the main enemies of the valve is corrosion. Corrosion occurs when the metal in the valve combines with the oxygen or other gases or chemicals that pass by. Once the metal has

Valves, Springs, Guides and Seats

105

Figure 6-16. Corrosion occurs when the metal combines with oxygen. (Black and Decker)

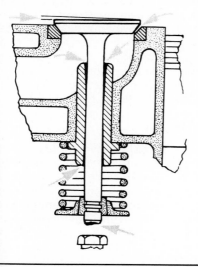

Figure 6-17. Wear on the valve stem or guide can cause the valve to seat in a cocked position. (Black and Decker)

corroded or rusted, it has very little strength and rapid wear occurs, figure 6-16.

Corrosion is difficult to fight. Metals that resist corrosion often do not resist wear.

One way to fight corrosion is to coat the head of the valve with aluminum oxide. Aluminum-oxide covered valves look as if the aluminum oxide has been sprayed on, but the process is done so that the aluminum actually fuses with the valve metal. Aluminized valves are used by many carmakers because of their longer life.

When we talked about valve materials and construction, you learned that the valve stems can also be coated with a special material to make them last longer.

If the valve stem or guide wears, the valve can come down on its seat in a cocked position, figure 6-17. The valve actually has to slide to reach its final position. This causes wear and leaking. For this reason, some manufacturers put a hard chromium plating on valve stems. The chromium finish has better wear characteristics.

■ Early Engine Valve Arrangement

Before the science of combustion chamber design was fully developed, more attention was paid to ease of valve actuation than the effect of location on the combustion chamber. The result was that the camshaft and valves were positioned alongside the engine cylinders. The camshaft lobes raised tappets which pushed on the valve stems and opened the valves. It was a simple arrangement with a minimum of parts, and it worked well on low compression engines.

But the flow patterns in an L-head combustion chamber bothered the engineers. It seemed that cylinder filling and scavenging would be better if the intake valve was opposite the exhaust valve. The T-head design, with intake and exhaust valves on opposite sides of the cylinder, and the F-head design, with one valve in the cylinder head, and the other in the block, were thought to be the answer to flow problems.

As compression ratios became higher and higher, with the engineers trying to extract increasing amounts of horsepower, it was obvious that leaving one valve in the block, or different arrangements of valves in the block, was not the way to go. Both valves were then placed in the cylinder head, resulting in the familiar overhead valve or I-head design universally used today.

Stellite: The brand name for a very hard alloy of mostly chromium and cobalt applied to valve faces for longer wear.

Stress Riser: A groove, scratch or other imperfection in the metal that could develop into a crack.

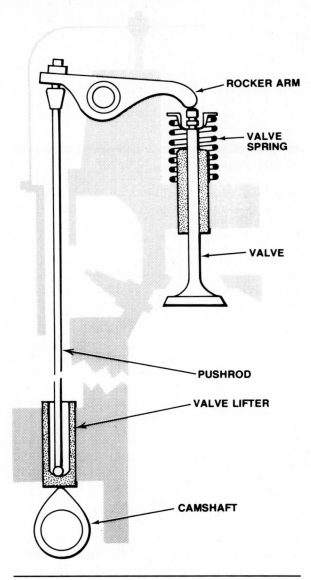

Figure 6-18. The job of the valve spring is twofold. It must hold the valve against the seat when the valve is closed and maintain tension in the valve train when the valve is open.

SPRINGS

The valve spring must hold the valve closed on its seat, figure 6-18. It must also maintain tension in the valve train when the valve is open.

As the camshaft lobe starts to open the valve, it sets in motion the components of the valve train. As the valve reaches the fully open position, the inertia of the valve train tends to continue to open the valve. If the inertia is great enough, it will continue opening the valve until the spring is completely compressed. At this point, the spring will attempt to close the valve

Figure 6-19. Dual springs are often used to reduce vibrations. (Audi)

with great force. Since the cam lobe has moved out from under the lifter, there is no resistance against the spring, and the parts of the valve train will come crashing together. This will quickly damage the engine. If hydraulic lifters are used, the sudden slack in the valve train will be taken up quickly by the lifter. This holds the valve open longer than necessary and the engine loses power. This phenomenon of valves opening through inertia is called **valve float**. When the valves float and the lifters take up the slack it is called **lifter pump-up**. These conditions occur when an engine is run much faster than the speed for which it was designed. A valve spring must exert enough tension to prevent valve float under normal operating conditions, but must not exert so much tension that the cam lobes and lifters begin to wear.

Spring Design

Springs appear to be very simple parts but in reality they must operate under very severe conditions. They are subject to vibrations or shock waves that travel through the spring like waves in the ocean. At certain engine speeds, these shock waves coincide with the opening push of the cam lobe and cause harmonic vibrations in the springs. On a test fixture, using a strobe light, these waves can actually be seen traveling through the spring coils.

When the vibrations become severe, the spring fails to exert a constant pressure on the valve train; this allows the valves to be open when they are supposed to be closed. It also creates play or excessive tension in the valve train.

Through experiments, engine designers have discovered that a simple valve spring does a marginal job on a modern engine.

Dual springs

Several modifications have been made to increase spring performance. One system frequently used is the dual spring, figure 6-19. It is a pair of springs, one inside the other. The two springs not only have different diameters, but slightly different lengths. Because of their differences, they vibrate at different speeds and their harmonic vibrations cancel each other.

Valves, Springs, Guides and Seats

Figure 6-20. A damper inside the spring is another way of reducing vibrations. (Buick)

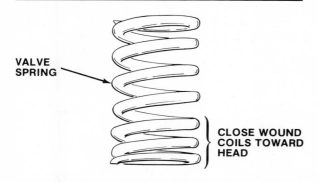

Figure 6-21. Some springs have the coils at one end closer together to reduce vibrations. (Buick)

Dampers

Another way to reduce harmonic vibrations is to use a damper inside the spring, figure 6-20. This is a flat wound coil that rubs against the inside of the spring. The rubbing tends to subdue the harmonics. Some springs are made with the coils closer together at one end, figure 6-21. When this type of spring is used to reduce harmonics, it must be installed according to the manufacturer's instructions, usually with the widely spaced coils toward the valve tip.

Variable rate spring

A spring with unequally spaced coils at one end is called a **variable rate spring**, figure 6-21. The more it is compressed, the harder it pushes for the distance it is compressed. Whereas an evenly spaced coil spring might increase its pressure 50 pounds for each tenth of an inch of compression, the variable rate spring will increase its pressure 50 pounds the first tenth of an inch; 60 pounds the second tenth of an inch; 70 pounds the third tenth of an inch and so forth. When the closely spaced coils touch each other, the rate of increase becomes less. This prevents excessive pressure on the tip of the cam lobe when the valve is fully open. Because the variable rate spring increases its pressure very quickly when it is opened, the pressure when the valve is closed can be reduced. This helps to prevent the valve from slamming onto its seat.

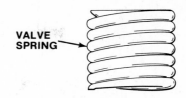

Figure 6-22. If the space between coils is not adequate, coil bind occurs.

When installing variable rate springs, always install them with the closely spaced coils against the cylinder head spring seat. This puts the lighter end of the spring next to the valve stem tip, and reduces some of the weight that must be removed in the valve train.

Coil bind

The amount of space between the spring coils determines how much the spring can compress. If the space is inadequate, the coils will jam against each other. This is called **coil bind**, figure 6-22.

With a properly designed spring, when the cam lobe lifts the valve fully open, some space will remain between the spring coils. However, if the engine is overspeeded and the valves float, the coils can bind. Coil bind puts a tremendous strain on the parts. Rocker arms and their supports can fracture and break.

Spring Material

Valve springs are made from wire. Metal is selected that will withstand heat without losing tension. The metal must be able to go through

---

Valve Float: The condition in which the valve continues to open or stays open after the cam lobe has moved from under the lifter. This happens when the inertia of the valve train at high speeds overcomes the valve spring tension.

Lifter Pump-up: The condition in which a hydraulic lifter adjusts or compensates for the additional play in the valve train created by valve float.

Variable Rate Spring: A spring that changes its rate of pressure increase as it is compressed. This is achieved by unequal spacing of the spring coils.

Coil Bind: The condition in which all the coils of a spring are touching each other so that no further compression is possible.

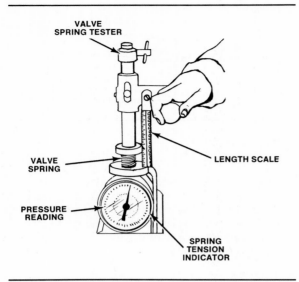

Figure 6-23. Be sure to check the valve spring length and strength with a tool like this one. (Silver Seal)

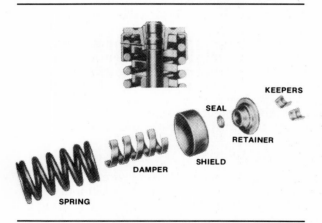

Figure 6-24. The spring retainer and split keepers hold the valve springs in place. (Buick)

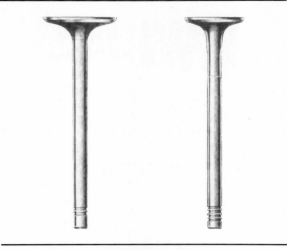

Figure 6-25. The keepers are designed to fit the number of grooves on the valve stem.

millions of compression and expansion cycles without losing tension. The better springs are made from special alloy steels. These include chromium and vanadium. Springs are usually shot-peened to blend in any stress risers and generally lessen tension in the spring, which lessens the possibility of cracking.

In spite of all precautions you may take, valve springs do break. The pieces of the spring may still exert some pressure on the valve so the cylinder will continue to operate, but at reduced power. It is possible for a broken valve spring to allow the valve to drop down and go through the piston.

Valve springs should be replaced not only when they are broken, but also when they are not up to specifications. When you are repairing or rebuilding an engine, check the valve spring length and strength with a valve spring tool, figure 6-23. Compare your measurements to those given by the car manufacturer.

Spring Installation

Valve springs are held in place by a spring seat, retainer, and two split keepers, figure 6-24. The spring seat can be a circular area cut out of the head to keep the spring from shifting sideways, or a washer with an inner ledge to locate the spring. The valve stem end of the spring is held by the retainer and split keepers. The retainer is put onto the valve spring and the valve spring is compressed; then the keepers are placed in the grooves near the valve stem tip. When the spring pressure is released, the keepers are wedged into the retainer. At the same time, the spring pressure pushes the valve against its seat.

Although all retainers and split keepers work on the same principle, they are not interchangeable. There are many different arrangements of grooves on the valve stem, figure 6-25. The keepers are designed to fit these grooves. Also, the retainers must be the correct diameter for the spring. The retainer is usually stepped to locate the end of the spring and keep the spring from sliding on the retainer. When dual springs are used, the retainer will have two ledges, one for each spring.

The distance between the retainer and the spring seat on the cylinder head determines the tension of the spring when it is installed. Refacing the valve or its seat will sink the valve farther into the seat. This moves the tip of the

Valves, Springs, Guides and Seats

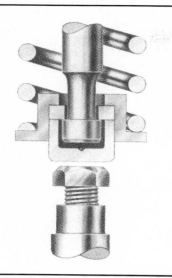

Figure 6-26. The release-type rotator allows the valve to move freely when the cam lobe starts its lift. (Ethyl)

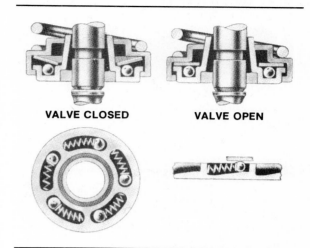

Figure 6-27. In a positive valve rotator, the rotator actually turns the valve. This type is called Rotocap. (Ethyl)

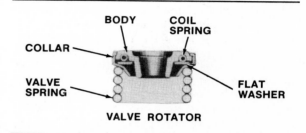

Figure 6-28. The Rotocoil valve rotator uses a coil spring to rotate the valve. (Buick)

valve and the retainer away from the spring seat, and reduces valve spring tension. The dimension from the spring seat to the bottom of the retainer, called **spring height**, must be measured whenever springs are installed. If the measurement is too great, spring spacer washers must be used between the spring and its seat to restore the tension.

Valve Rotators

The exhaust valve is especially prone to valve burning and wear because of the extremely high temperatures of the exhaust gases. If you could rotate the exhaust valve a little each time it opens, that little bit of friction caused by the rotation could help solve the problems of valve burning and sticking by scraping off the accumulation of carbon and emission deposits.

Many heavy-duty trucks, vehicles equipped for towing, and some cars are equipped with valve rotators. The rotators come in two different types: the release-type valve rotator and the positive valve rotator.

Release-type rotators allow the valve to move freely as the cam lobe starts its lift, figure 6-26. In this case, the valve turns slightly because of the vibration involved.

The positive valve rotator actually turns the valve several degrees every time the valve is opened. Small ramps inside the rotator cause the valve and spring to rotate. A design that uses small balls inside the rotator to turn the valve retainer, which turns the valve itself is called Rotocap, figure 6-27. Another design that uses a coil spring instead of balls is called Rotocoil, figure 6-28.

Carmakers such as Oldsmobile used valve rotators in their cars for a few years in the late 1950's, but phased them out because of the introduction of the hard alloy metals used to cover valve faces and seats. Another reason for the phasing out of automobile valve rotators was the advent of unleaded gasoline. Without the lead to protect the valve seat, the friction of the slight rotation increased seat wear and recession.

GUIDES

Valve stems are supported by valve guides, figure 6-29. There are two types of guides. A removable tube which is pressed or driven into

Spring Height: The dimension from the spring seat to the bottom of the retainer. It should be measured whenever springs are installed.

Figure 6-29. The valve stem is supported by the valve guide.

the head is called an **insert guide**, figure 6-29. The second type is called an **integral guide**. It is part of the engine casting and cannot be removed if worn, although it can be reconditioned. This process is covered in your Shop Manual.

Integral guides are repaired by knurling, by machining the guide to accept an insert type guide, or by tapping the guide bore and screwing in a special coil.

Insert guides are made of cast iron or silicon bronze. Although the silicon bronze is more expensive, it also has better wear characteristics. Insert type valve guides must be used in aluminum heads because the aluminum does not have enough durability.

Guide to valve stem clearances must be kept small to prevent oil from entering the combustion chamber or exhaust system through the space between the valve and its guide. Intake valve guide clearances are normally smaller than exhaust valve guide clearances, to prevent the intake port vacuum from pulling in oil. Intake valve stem to guide clearances normally range from .001" to .003" (.02 to .08 mm) while exhaust valves vary from .0015" to .0035" (.04 to .09 mm). These figures are representative of average ranges of tolerance; in all cases, the minimum possible clearance within the manufacturers' specifications is most desirable for maximum valve and guide life.

Guide Lubrication And Sealing

Valve stems and guides are lubricated by the oil that is pumped to the rocker arms. The oil runs onto the valve stem and guide.

Because of the large amount of oil that runs over the valve stem and its guide, some type of seal is necessary to prevent the oil from going through the guide and into the engine. Oil that runs down the intake valve stem is pulled into the combustion chamber by engine vacuum, where it is burned.

Oil that runs down an exhaust valve is pulled through the guide by a light vacuum in the exhaust. The vacuum is caused by the rush of exhaust gases past the end of the guide, figure

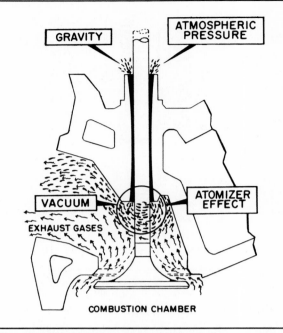

Figure 6-30. The oil that runs down the exhaust valve is pulled in by vacuum. (Perfect Circle)

6-30. This oil burns in the exhaust and goes out the tailpipe.

The first area that is sealed against oil is the spring retainer. Because the retainer is cup-shaped, it will collect oil and funnel it down the valve stem. To prevent this, a small square-section rubber ring is positioned in a groove on the valve stem, figure 6-31. If you examine a keeper and valve stem you will notice that there is one extra groove on those valves that use retainer seals. The upper groove or grooves are for the split keeper. The groove nearest the valve head is for the ring seal. When properly installed, the ring seal will prevent oil from funneling down through the retainer.

Car manufacturers use devices called deflectors or shields to keep the oil away from the upper end of the valve guide. On some engines, the shields are installed directly under the spring retainer, figure 6-33. The shield covers the top few coils of the spring. Oil that splashes on the shield runs down the outside of the spring and misses the end of the guide.

Some manufacturers use an umbrella-type shield that fits over the end of the valve guide, figure 6-32. Any oil that splashes on the valve stem above the guide will run off the shield instead of going between the valve stem and guide. This type of shield is sometimes responsible for too much oil going down the guides. If the deflector is a tight fit on the valve stem, it will go up and down with the valve. This is all

Valves, Springs, Guides and Seats

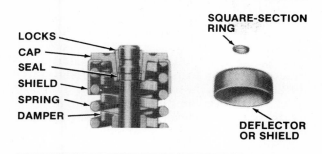

Figure 6-31. A small, square-section rubber ring is used on the valve stem to prevent oil seeping into the combustion chamber. (Buick)

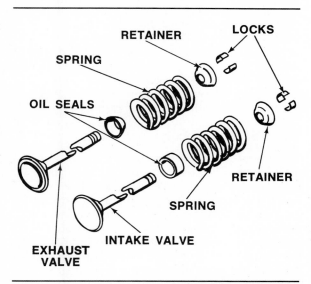

Figure 6-32. An umbrella-type shield is sometimes used on the end of the valve guide.

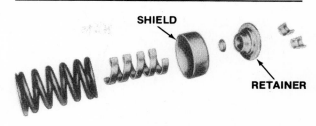

Figure 6-33. Sometimes the shields are installed directly under the spring retainer.

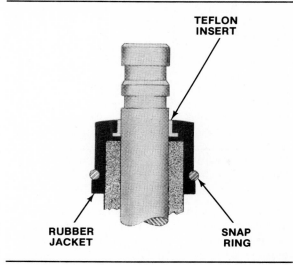

Figure 6-34. A Teflon insert is now used to prevent oil from getting into the guide. (Perfect Circle)

right, as long as the fit on the end of the valve guide is very loose. When the inside of the deflector rubs against the valve guide, it can create a pumping action that forces oil between the valve stem and guide. Ideally, the deflector should be a tight fit on the end of the valve guide and a sliding fit on the stem.

To prevent the pumping action of a deflector that is too tight, some manufacturers put lugs on the inside of the deflector. These lugs contact the valve guide, but the spaces between them eliminate the pumping action. On those deflectors without lugs, the fit of the stem and guide must be checked carefully.

In the last few years, car manufacturers have used oil shields on the exhaust valve and positive oil seals on the intake valves. These seals fit snugly against the top of the valve guide. A Teflon insert rubs on the valve stem and prevents any oil from getting into the guide, figure 6-34. Some of the guide seals are held in place with a light spring. Others go over the end of the valve guide and are held in place by a spring clip or by being pressed onto the guide. All of the positive guide seals require machining of the end of the valve guide. The valves must be removed from the engine to do this.

Car manufacturers put positive seals on intake valves for two reasons. The intake vacuum tends to draw oil down the intake valve, and the low viscosity of the new oils allows them to slip through more easily. Because of the large amount of oil necessary to lubricate overhead camshafts, manufacturers of these types of engines generally use positive seals on both intake and exhaust valves.

Insert Guide: A valve guide that is pressed, or driven into the cylinder head.

Integral Guide: A valve guide that is formed as part of the cylinder head.

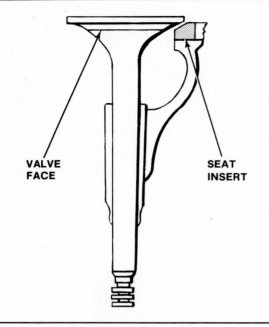

Figure 6-35. Hard seat inserts prolong seat and valve life.

High unexplained oil consumption can be caused by lack of oil seals on the exhaust valve guides. The rush of exhaust creates enough suction to draw oil down the guide. For maximum control, positive guide seals can be installed on both valves.

VALVE SEATS

When the valve is closed, the valve face rests against its seat, figure 6-35. If the seat and valve are ground to fit each other, the seal between the valve and seat will prevent any leakage. As we learned in the first part of this chapter, both face and seat are not ground at 45-degree angles. There is usually approximately one degree of difference in the angles. This difference is called the interference angle.

Valve seats are called integral seats when they are a part of the cylinder head casting.

Valve Seat Inserts

If a separate seat is inserted in a groove in the head, it is called a valve seat insert. The reason for using valve seat inserts is longer seat life. If a seat insert of alloy metal is used, it will withstand the high temperature with less wear. Valve seat inserts are harder than the head and are sometimes called hard seat inserts. Seat inserts must be used in aluminum heads because of the softness of the aluminum.

Hard seat inserts are the best, but also the most expensive way of prolonging seat life, figure 6-35. Installation of the seat inserts requires several machining operations, which raises production costs. In most cast iron production engines, the valve seats are the integral type. Seat inserts come in several grades of hardness, made of several different alloys. One of the hardest seats is made of Stellite or a combination of cobalt, chromium and tungsten.

Induction Hardening

Another method of combating seat wear is induction hardening. The seat is still an integral part of the head, but it is hardened by an electromagnet that heats the seat through induction. The induction method heats the valve seat to a temperature of approximately 1700° F (927° C). This gives the valve seat a more durable finish. Chrysler has adopted the use of induction hardening in its car engines.

Valve Seats And Ports

The intake valve seat is actually the end of the intake port. The same relationship holds true for the exhaust seat and port. The shape of the seat has a definite effect on the intake and exhaust mixture flow rate. The area of the seat that contacts the valve face must be located precisely to meet the valve. Above and below the actual circle of contact, a seat that protrudes into the airstream can slow the flow of gases and reduce the efficiency of the engine.

When you are rebuilding an engine, it is important to replace the valve seat insert in exactly the same location and at the same angle as the original equipment seat to ensure the same air-fuel mixture flow rate.

VALVE, SPRING, GUIDE, AND SEAT RELATIONSHIPS

The valve, spring, guide, and seat work as a team. If something is wrong with any one of them, it will affect the others. For example, it does little good to put new valves in worn guides. Clearance between the stem and guide may still be excessive, allowing too much oil consumption. The valve will wobble in its guide and not seat properly, figure 6-17. This will cause rapid wear and burning.

If a seat or valve is ground or cut to renew it, the valve stem will protrude through the head by the amount of metal that was removed. This will reduce the spring pressure on the valve when it is closed. With less spring pressure, the valve will burn easily. It will also change the angle between the tip of the valve stem and the

Valves, Springs, Guides and Seats

rocker arm. The original angle can be restored by grinding the end of the valve.

The height of the spring when the valve is installed should be within the manufacturer's specifications or the spring pressure will not be correct. If it is not, spacer washers can be installed under the spring to bring the spring to the correct, assembled height. The springs should also be checked for the correct tension on a spring testing machine, figure 6-23.

Review Questions

Choose the single most correct answer. Compare your answers to the correct answers on page 268.

1. Most engines now use:
   a. 30-degree valve faces
   b. 35-degree valve faces
   c. 40-degree valve faces
   d. 45-degree valve faces ✓

2. If the heat is not removed from the exhaust valves, they will:
   a. Leak
   b. Warp
   c. Burn
   d. All of the above ✓

3. In modern engines, valves are lubricated in their guides by oil:
   a. Vapor and splash
   b. Dripping from above ✓
   c. Injection
   d. Under pressure

4. The most efficient valve train is the:
   a. Overhead valve
   b. L-head
   c. Overhead cam valve ✓
   d. Hemi-head valve

5. In lubricating valves, which of the following wear points is also lubricated?
   a. Pushrod
   b. Rocker arm
   c. Guide
   d. All of the above ✓

6. The combination of chromium, cobalt, and tungsten used for valve faces is called:
   a. Chrome alloy
   b. Stellite ✓
   c. Coballite
   d. Chrome flashing

7. If the valve spring height is greater than specification, you should:
   a. Grind the valve stem
   b. Grind the valve seat
   c. Grind the valve face
   d. Install spacer washers on the spring seat ✓

8. Integral guides cannot be repaired by:
   a. Knurling
   b. Tapping the guide bore and screwing in a special insert
   c. Machining the guide to accept an insert-type guide
   d. Drilling out the guide and installing oversized valves ✓

9. Oil seals are used on:
   a. Only intake valves
   b. Only exhaust valves
   c. Both intake and exhaust valves ✓
   d. Neither intake nor exhaust valves

10. When rebuilding an engine it is important to replace the valve seat inserts in:
    a. A different location
    b. The same location but at a slightly different angle
    c. A different location at exactly the same angle
    d. Exactly the same location at the same angle ✓

11. 4-stroke engines use:
    a. Sodium valves
    b. Poppet valves ✓
    c. Rotary valves
    d. Reed valves

12. The greater mass of a large valve:
    a. Causes it to seal better when it hits the seat
    b. Reduces the possibility of valve float
    c. May cause it to bounce when it hits the seat ✓
    d. Reduces wear in the valve train

13. The length of time the valve stays open is determined by:
    a. The height of the cam lobe
    b. The speed of the camshaft
    c. The length of the cam lobe
    d. The angle from the beginning of one lobe ramp to the other ✓

14. The valves on an overhead cam engine are adjusted with:
    a. Shims
    b. Adjustable rocker arms
    c. Adjustable tappets
    d. All of the above ✓

15. Valve polishing is done to:
    a. Increase the flow of gases
    b. Eliminate stress risers
    c. Increase heat dissipation
    d. Reduce carbon accumulation ✓

16. To fight corrosion the head of the valve is often coated with:
    a. Chromium
    b. Molybdenum disulfide
    c. Aluminum oxide ✓
    d. Stellite

17. Lifter pump-up occurs:
    a. During valve float ✓
    b. If the oil pressure is too high
    c. When the lifter relief port is blocked
    d. During extreme engine load conditions

18. Seat wear can be reduced by using:
    a. Soft-faced valves
    b. Unleaded gasoline
    c. Aluminum valve seat inserts
    d. Induction hardening ✓

Chapter 7

Camshafts, Lifters or Followers, Pushrods, and Rocker Arms

CAMSHAFTS

The camshaft is a series of cams that are ground onto a long shaft, figure 7-1. There is one cam for each valve in the engine. As the camshaft rotates, the cams push on the valve lifters or cam followers. These, in turn, push on the valves, either directly or indirectly, through pushrods and rocker arms or simply through rocker arms.

The camshaft appears to be simple, but cam design determines the important factors of how much, how long, and when the valves open.

Cam design must take into account the fact that valves cannot open instantly. The components of the valve train, the lifter, pushrod, rocker arm, and valve all have mass, figure 7-2. If the cam lobe pushes the valve open too quickly, the force of lifting that mass at high rpm will be very great, and the cam lobe will wear rapidly. To prevent this, the valves must be opened gradually.

In addition, if the valve is lifted too quickly, the mass of the valve train will cause the valve to continue opening after the cam lobe has reached the full lift position. And, if the lobe does not allow the valve to close gradually, the lobe may drop out from under the lifter, leaving slack in the valve train. The valve spring will then slam the valve against its seat causing wear and possibly breaking the valve.

Each lobe on a camshaft is designed to gradually take up the clearance in the valve train, figure 7-3. The lobe lifts the valve rapidly. Near the fully open position, the lobe slows in its rate of lift. The idea is to slow down the rate of opening so that when the lobe reaches the fully open position the valve will be moving slowly.

The top of the lobe holds the valve in the fully open position. Then, as the top of the lobe passes out from under the lifter, the valve starts to close. Just before the valve reaches the fully closed position, the lobe reduces its rate of letdown so the valve closes very gently against the seat.

The cam lobes must open and close the valve at the proper time in relation to the position of the pistons. This is called valve timing. The camshaft must be correctly timed to the crankshaft or the engine may not run at all.

Camshaft Locations

Camshafts are located in the cylinder block or engine crankcase, or are mounted on the head, figure 7-4.

Camshafts, Lifters or Followers, Pushrods and Rocker Arms 115

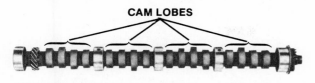

Figure 7-1. The camshaft combines a series of cam lobes on a long shaft. (TRW)

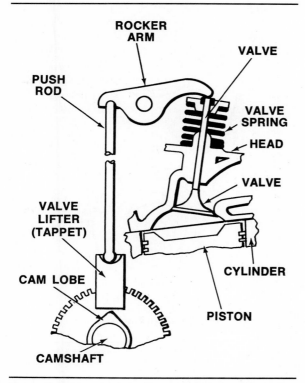

Figure 7-2. You see here the components of the valve train at work.

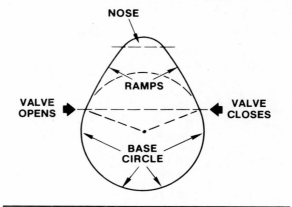

Figure 7-3. The cam lobe is designed so that the clearance in the valve train is taken up gradually.

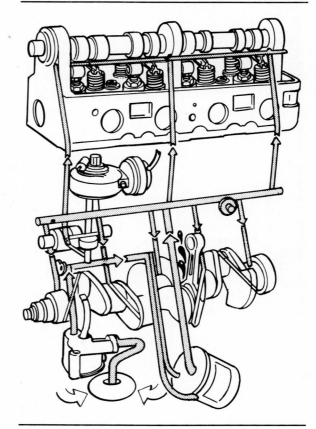

Figure 7-4. This Ford camshaft is mounted on the cylinder head. (Ford)

Camshaft Drive

The camshaft is driven from the crankshaft by gears, chains and sprockets, or cog wheels and **cog belts**. Whatever the method of drive, the camshaft must revolve at one half the speed of the crankshaft. That is, for every full revolution of the crankshaft, the camshaft must turn only one-half revolution. The reason for this is that each valve opens only once for every two crankshaft revolutions.

For example, the intake valve opens on the intake stroke. Then the piston must go through the compression, power, and exhaust strokes before getting to another intake stroke. These four strokes require two crankshaft revolutions, but if the camshaft is allowed to go through two revolutions in the same amount of time, the intake valve would open on the power stroke.

Cog Belt: A flat rubber belt with teeth that mesh with cog wheels.

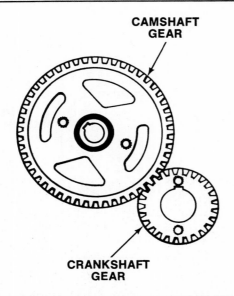

Figure 7-5. By making the camshaft gear twice as large as the crankshaft gear, the camshaft goes through one revolution for every two of the crankshaft.

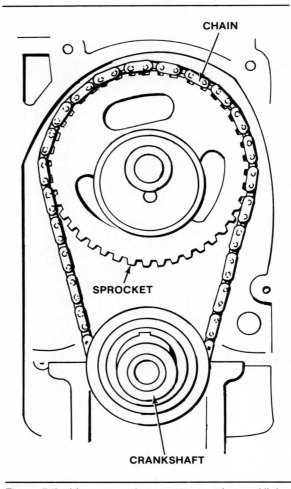

Figure 7-6. Many carmakers use a sprocket and link type chain drive. This is also known as a silent chain.

So the camshaft is designed to turn only one revolution while the crankshaft goes through two, figure 7-5.

Gears are the oldest type of camshaft drive, figure 7-5, but they have the disadvantage of being noisy. To eliminate noise, most carmakers use a sprocket and link-type chain drive, figure 7-6, also known as a **silent chain**. The word silent comes from a comparison with a roller chain, like those used on bicycles. The **roller chain** is stronger, but noisier.

The latest type of drive is the cog belt. It is quiet, and does not require any lubrication, figure 7-7. The cog belts are made from rubber and fabric, similar to a fan belt. Because of the cogs, or teeth, the cog belt canot slip. It keeps the camshaft in time with the crankshaft in this manner.

Tensioners

Both cog belts and silent chains often use **belt or chain tensioners** when used on overhead cam engines. On a silent chain, the tensioner is only to keep the slack side of the chain from flapping and making noise. On a cog belt, the tensioner actually determines the tension of the belt. The tensioner is spring-loaded. On engines with a belt tensioner, adjustment of the belt may not be needed. A new belt is put on, and the tensioner is allowed to press against the belt. It will maintain the proper tension as long as the belt lasts.

Engines without belt tensioners require tension adjustment. The Chevrolet Vega engine has a water pump with a sliding mounting, figure 7-7. The pump pulley is pushed against the belt with a torque wrench and then tightened in place to maintain the tension.

Accessory Drives

Camshafts do more than open valves. The camshaft also provides power to run other parts of the engine. Usually, one of the secondary jobs of the camshaft is to operate the ignition distributor. This is done by a gear that is mounted onto, or cast into, the shaft. **Helical gears** are often used, which allow the distributor shaft to operate at right angles to the camshaft, figure 7-9. Both the camshaft gear and the distributor gear have the same number and size of teeth so they turn at the same speed.

The camshaft may be used to drive the oil pump, figure 7-10. In most cases the oil pump is located in the engine directly below the distributor. Oil pumps are positively driven and are

Camshafts, Lifters or Followers, Pushrods and Rocker Arms

Figure 7-7. The cog belt is used on many domestic and import cars. It is made from fabric and rubber. This particular system is tensioned by adjusting the pump. (Chevrolet)

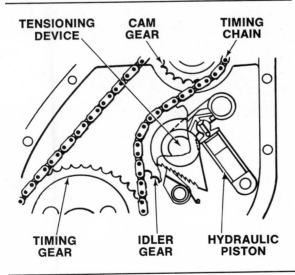

Figure 7-8. The tensioner on this engine presses against the chain. It is hydraulically activated and has a ratchet locking mechanism to keep the chain properly tensioned as it wears and stretches.

either the gear or rotor type, figure 7-11. The oil pump seldom suffers from a lack of lubrication because it handles the oil, but it can develop too much clearance and require replacement after a time.

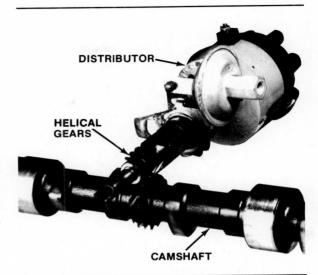

Figure 7-9. The helical gears allow the distributor shaft to operate at right angles to the camshaft.

The camshaft also may operate the fuel pump. This is done by an **eccentric** which is bolted to the front of the camshaft or cast into it. The eccentric changes the rotary camshaft motion to the reciprocating motion necessary to operate the fuel pump, figure 7-12.

On camshafts with many thousands of miles of wear, the eccentric usually is as good as new. Eccentrics bolted to the front of the camshaft can be made larger than the type cast into the middle of the camshaft, figure 7-13, because the cast-in type usually can't be bigger than the cam bearings or the camshaft can't be inserted

Silent Chain: A chain made of links that pivot around pins. Each link is a series of plates held together by the pins. The number of plates determines the width of the chain.

Roller Chain: A chain made of links, pins, and rollers. Each link is made of two plates, held together by pins, and separated by a roller around the pin. The width of the chain is determined by the length of the rollers.

Belt or Chain Tensioners: A wheel or pad, pressed against a belt or chain, to maintain its tension.

Helical Gears: Gears with teeth cut at an angle to the shaft instead of parallel to it.

Eccentric: A circle mounted off center on a shaft. It is used to convert rotary motion to reciprocating motion.

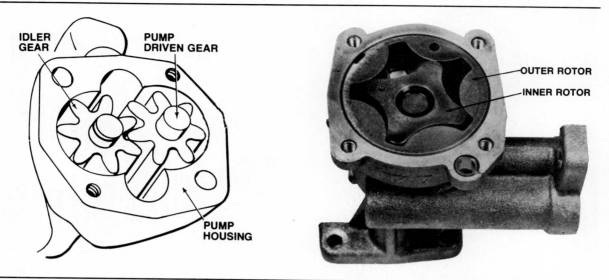

Figure 7-11. A gear type oil pump is shown on the left; a rotor type on the right.

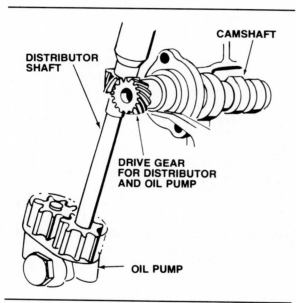

Figure 7-10. The oil pump may be driven by the camshaft.

Figure 7-12. An eccentric bolted to the front of the camshaft is used to operate the fuel pump.

into the block. The larger eccentric has a bigger bearing surface for the pump arm and is sometimes made with a slip ring to reduce friction between the arm and the eccentric.

Camshaft Materials

Camshafts usually are made from cast iron. They are brittle and must be handled with care. Cast steel, which is less brittle, is also used for camshafts.

Forged camshafts are made by some manufacturers and are extremely tough and long wearing. Alloy steels are used in forging which increases wear resistance. The most expensive camshafts are made from a steel billet. They have great resistance to wear and breakage and are mostly used in racing engines.

Cam Lobe Conditioning

Because the cam lobes receive such high loading from the valve train, several processes or treatments have been developed to reduce wear. Some camshafts have a hard-face overlay on the ramps and the tips of the lobes. Hard-face overlays are seldom found in production engines, but are used in racing engines.

Camshafts, Lifters or Followers, Pushrods and Rocker Arms

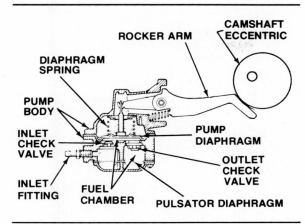

Figure 7-13. The fuel pump eccentric cast into the camshaft is often smaller than the one that is bolted onto the camshaft. (Pontiac)

Another method used to harden the lobes is called **carburizing**. It can only be done on steel with a specific carbon content. The metal is heated in an oven where the air is full of carbon. The surface of the lobe absorbs the carbon, making it slippery and long wearing. To prevent carburizing the entire camshaft and making it brittle, the entire shaft is copper-plated and the lobes and bearing surfaces are ground to remove the copper, allowing the carburizing process to work only on those areas.

Although long wear is a problem with camshafts, an even greater problem is scuffing of the lobes and lifters when a rebuilt engine is first started. Since the camshaft may be installed several hours or days before the engine is started, the lubricant can drain off the cam lobes in that time. In some cases, an engine may be put into storage for several months before being run. When it is started, a dry cam lobe running against a dry lifter will break through the hard outer surfaces and cause rapid wear.

Many camshaft manufacturers apply a manganese-phosphate coating on the cam. This slightly rough coating holds a film of oil and prevents scuffing during dry starts. Cams that are coated with this material have a dark gray appearance.

In spite of precautions by cam makers, there is still danger of wear from dry starts. For this reason, assembly lubricant is a must. Most camshafts are sold with an assembly lubricant, usually a heavy grease with additives, to work under extreme pressure.

Valve Timing

All valve timing events are measured in degrees of crankshaft rotation. At the end of each stroke, the piston is either at top dead center or bottom dead center. A stroke requires one half of a crankshaft rotation, or 180 degrees. Therefore, if a valve is described as opening at top dead center and closing at bottom dead center, the total time it is open will be 180 degrees of crankshaft rotation.

■ Donnell A. Sullivan

The career of Don Sullivan has been a long one marked with success after success in the automotive design field. Sullivan was born in 1904 in Port Huron, Michigan. At the age of 18, he began work on a steam turbocharger made out of parts from his mother's vacuum cleaner. He attached the turbo to a Model T roadster which made that vehicle the talk of Port Huron.

A few years later Sullivan went to work for the Ford Motor Company where his list of achievements is long and glowing.
- He was principal design engineer for Ford's 1932 V-8
- He developed the first over and under intake manifold for Ford's 1934 V-8
- Developed Ford's first 2-barrel intake manifold
- Holds the original patent for a 2-carburetor intake manifold for a V-8 engine
- Developed a new water pump system with the pump near the bottom of the cylinder block. Colder water was used in the system. This system was used on the 1937 and later model Ford V-8's
- Qualified a flat head Ford V-8-powered car for the 1940 Indianapolis 500
- Designed the Hutton-Sullivan high performance aluminum cylinder heads in 1946 for Ford flat head V-8's
- Designed the high performance camshaft and aluminum intake manifold for the 1957 Ford 312 cid engine

After his retirement from Ford in 1969, Sullivan went into business for himself as an independent design engineer and Ford consultant. Since that time he has designed the camshafts for A.J. Foyt's Indianapolis 500 winning race cars, the Oldsmobile diesel engine, Mercruiser marine engines, and the national champion motorcycle engines. In recent years, Sullivan has turned his talents to designing more fuel efficient camshafts.

Carburizing: A method of hardening camshaft lobes by heating the camshaft in a carbon atmosphere.

This open time, expressed in degrees of crankshaft rotation, is called the **duration** of the valve opening. Since valves usually open before the top dead center position and close after the bottom dead center position, the additional degrees must be added to 180 to get the total duration. For example, if an intake valve opens 20 degrees before top dead center, and closes 15 degrees after bottom dead center, the duration of the valve opening is the total 180, 20, and 15 degrees, or 215 degrees.

The same is true of an exhaust valve that opens at 25 degrees before bottom dead center, and closes at 18 degrees after top dead center. 180 plus 25 plus 18 equals 223 degrees of duration.

Another term used to describe valve timing is **overlap**. The intake and exhaust valves are usually designed to both be open for a short time when the piston is near top dead center on the exhaust stroke, which is also the beginning of the intake stroke. If an intake valve opens 12 degrees before top dead center, and an exhaust valve closes 12 degrees after top dead center, add the two figures to get the overlap, in this case 24 degrees.

Valve Lift

Another factor that affects engine operation is **valve lift**. The more the valve opens, the greater will be the flow of intake and exhaust gases and the higher the output will be. Valve lift is the actual distance that the valve lifts off its seat, measured in thousandths of an inch or hundredths of a millimeter.

The valve train is operated by the lobe of the cam. Each cam lobe has a specific amount of lift, called the **lobe lift**. This lobe lift can be measured by using a micrometer to measure the lobe, and subtracting the smallest measurement from the largest. It can also be measured with the engine assembled by rotating the cam and actually measuring how far the lifter body is raised by the lobe.

On an engine with the valves operated directly by the lifter or tappet, valve lift is the same as lobe lift, after subtracting any clearance present with mechanical tappets. On an engine with rocker arms, valve lift is usually greater than lobe lift because of the **rocker arm ratio**. One side of the rocker arm is longer than the other so that the valve lift is multiplied. This is explained in detail in the section on rocker arms.

There are many reasons for timing the valves to open at specific intervals of crankshaft rotation. The diagram, figure 7-14, shows how the intake valve opens when the piston is still mov-

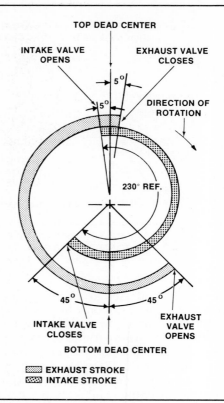

Figure 7-14. Typical valve timing diagram.

ing up on the exhaust stroke. However, this does not push exhaust into the intake manifold. The exhaust valve is still open when the intake valve opens, and the rush of exhaust gases out of the cylinder helps to pull in the intake mixture, even though the piston has not yet reached top dead center. Also notice that the exhaust valve is still open when the piston starts down on the intake stroke. This does not allow exhaust gases to be sucked back into the cylinder, because the outward rush of exhaust gases is still pulling the intake mixture into the cylinder.

The optimum opening and closing points of the intake and exhaust valves and the amount of valve overlap are found through experimentation on an engine dynomometer. A change of a few degrees one way or the other can have a big effect on performance, fuel mileage, and exhaust emissions.

The best time to open the exhaust valve is several degrees before the piston reaches bottom dead center on the power stroke. There is still some pressure in the cylinder from the expanding gases, and when the exhaust valve opens, the gases rush out into the exhaust system.

Similarly, the intake valve does not close precisely at bottom dead center on the intake stroke. Intake mixture is still flowing into the cylinder after the piston passes bottom dead center and starts to rise on the compression stroke. The cylinder will be filled with more air-fuel mixture if the intake valve closes several degrees after bottom dead center.

Engine speed has a direct bearing on cam design. If an engine is run slowly, the intake mixture and the exhaust gases move slowly. Keeping the valves open long or increasing the valve overlap in this case will not increase power. But if the engine is operated at high speed, the velocity of the gases is high. Keeping the valves open longer and increasing the valve overlap will more completely fill the cylinder and give more power.

Duration is also affected by the design speed of an engine. A low-speed engine may have a cam duration of 230 degrees, but the same engine designed for high speed may have a 300-degree duration. Intake duration and exhaust duration are usually very close. What works well to get the intake mixture in will also work well to get the exhaust gases out.

Don't accept a duration figure for a cam at face value. The duration figure given is meaningless unless the opening point is defined. If a cam has an opening specification for the intake valve of 15 degrees btdc, we must know if it just starts to open at that point, or if it is open a definite amount. It means nothing to say a cam opens at 20 degrees btdc if the opening rate is so slow that it takes ten degrees of rotation to lift the valve a few thousandths of an inch.

To avoid confusion on opening and closing points, many camshaft manufacturers have specified that valve duration time be computed only after the valve has opened 50 thousandths of an inch (1.27 mm). On the closing side, the valve is considered closed when it has only 50 thousandths of an inch (1.27 mm) left to close.

The duration of a cam lobe measured between the 50-thousandths (1.27 mm) lift points can be as much as 50 degrees less than that measured between the actual opening and closing points. The former is a more meaningful measurement of duration.

Cam Thrust

The longitudinal movement of the camshaft in its bearings is called **cam thrust**. The thrust can be forward or backward, depending on the loading of the gears that drive the cam or the distributor. On chain drive cams, the only thrust is from the distributor gears. The cam is pushed in the direction of the slant of the gears.

On most engines the cam thrust is controlled by a plate that bolts to the front of the block, figure 7-15. The front cover is close enough to the cam so that it can only move back and forth a few thousandths of an inch. The timing chain gives a certain amount of end thrust control. When it is new, it will not bend sideways easily. But a worn chain can allow quite a bit of cam movement which sometimes becomes so great that a cam lobe actually gets under the neighboring lifter.

Figure 7-15. Cam thrust is controlled by a thrust plate bolted to the front of the engine.

Camshaft Effect On Emissions, Performance, And Economy

Before the beginning of emission controls, camshafts in production engines were a compromise between performance, economy, and driveability. A camshaft with a lot of duration and lift is an excellent performer at high rpm. However, at idle and low rpm it might cause the engine to run rough and produce very little power.

On the other hand, a camshaft with less lift and duration will idle and run very smoothly at

---

Overlap: The period of crankshaft rotation in degrees during which both the intake and exhaust valves are open.

Duration: The number of crankshaft degrees that an intake or exhaust valve is open.

Lobe Lift: The amount of lift at the cam lobe.

Valve Lift: The amount of lift at the valve. It depends on lobe lift, rocker arm ratio, and the clearance in the valve train.

Rocker Arm Ratio: The ratio of the valve side of the rocker arm to the cam side, measured from the pivot.

Cam Thrust: The lengthwise movement of the camshaft to the front or rear of the engine.

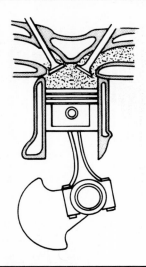

Figure 7-16. Valve overlap occurs when both the intake and exhaust valves are open and exhaust gases are pushed into the intake manifold.

low rpm, but at high rpm the engine would lack power.

For maximum efficiency, an engine should run at a constant speed so that the camshaft can be designed to give optimum breathing at that speed. The intake valve would close exactly when the full intake charge had entered the cylinder. The exhaust would open only after all the power had been applied to the piston.

When an engine is designed to run over a large range of speeds, the camshaft design must be a compromise. In modern engines, the compromise favors emissions, economy, and driveability at the expense of power.

Designing a camshaft to control emissions usually hurts both mileage and performance. Many camshafts today have more valve overlap than necessary for the most efficient operation of the engine, figure 7-16. The overlap is so great that some of the exhaust gases in the engine are actually pushed into the intake manifold through the open intake valve. Because most exhaust gases contain some unburned fuel, this gives the engine a second chance to burn the fuel. The result is less unburned fuel in the exhaust which reduces hydrocarbon emissions.

More valve overlap also reduces the production of oxides of nitrogen because the exhaust gases that stay in the cylinder to be burned a second time reduce the combustion temperature.

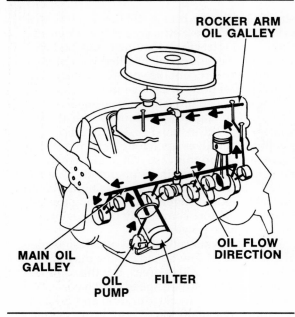

Figure 7-17. The camshaft receives its lubrication through the rocker arm galley as pictured here.

Camshaft Lubrication

Camshafts are lubricated by oil splash and full oil pressure, figure 7-17. The shaft bearings receive pressurized oil through passageways from the oil pump. Usually, these passages are fed from the same source as the crankshaft main bearings. Although worn camshaft bearings may not make any noticeable difference in the way an engine runs, they can make a big difference in oil pressure. Because camshaft bearings are usually fed from the same source as the crankshaft main bearings, a worn camshaft bearing can bleed off so much oil that the oil pressure at the main bearings drops dangerously low. This in turn, reduces the oil pressure at the connecting rod bearings and may eventually result in a rod knock. Therefore, camshaft bearings should be replaced along with all other bearings whenever an engine is rebuilt.

Overhead cam engines have oil fed under pressure directly to the cam lobes, internally or externally, figure 7-18. In the internal feed, oil is pumped into a passage inside the camshaft and allowed to come out at a hole in each cam lobe. In the external feed, oil squirts or drips on each lobe from a passageway or oil line near the camshaft.

In other engines, oil passes through the camshaft bearings and lifter bores, down the pushrods, off the crankshaft, and is whipped into a mist by the rotating parts. This oil mist thoroughly lubricates the cam lobes so that no direct oiling is needed.

Camshafts, Lifters or Followers, Pushrods and Rocker Arms

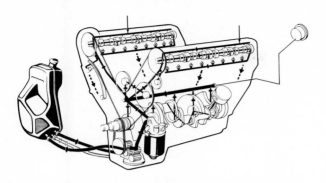

Figure 7-18. Oil can be fed to the lobes internally or externally as pictured here. (Mercedes-Benz)

VALVE LIFTERS, TAPPETS, AND CAM FOLLOWERS

Lifters, tappets, and cam followers all have the same purpose. They assist in transmitting the motion or the push of the cam lobe to the stem of the valve.

Valve stems expand considerably from heat. If the clearance between the cam lobe and the valve stem is not sufficient, the valve will be held open when the engine warms up and the valve stem expands.

Hydraulic Lifters

After many years of using adjustable tappets and adjustable rocker arms, the hydraulic lifter was invented, figure 7-19. It consists of an outer body, a plunger, and a check valve. A light spring inside the lifter keeps the plunger in constant contact with the end of the valve stem or pushrod. When the cam lobe starts to raise the lifter, the oil under the check valve is trapped. The plunger cannot move, and as the lifter is raised, the plunger also rises and opens the intake or exhaust valve. This is explained in more detail later.

Hydraulic lifters do not need adjustment because they automatically adjust for wear or play in the valve train. They always maintain zero clearance in the valve train without holding the valve open.

When hydraulic lifters are installed after engine work, the length of the valve train at each valve must be checked. If there is too much play, the lifter will not be compressed and will not operate. If there is too little play, the lifter may be completely collapsed, preventing the valve from closing. If an adjustable rocker arm is not provided, adjustable or different length pushrods must be used, or the valve tips

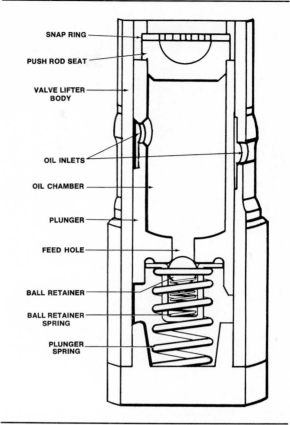

Figure 7-19. The hydraulic lifter consists of an outer body, a plunger and a check valve.

ground to adjust the lifter clearance. After the clearance is adjusted initially, it does not have to be readjusted until the next engine overhaul.

Tappets

Various methods are used to open valves, but they all use the cam follower, tappet, or hydraulic lifter. The **mushroom tappet** is one example, figure 7-20. It has a foot which is larger than the body of the tappet, and is used when the cam lobe is also larger than the body of the tappet. Because of the foot, the mushroom tappet can only be removed from the bottom of the lifter bore. Most tappets or lifters have a body of constant diameter over their entire length. These can be removed from either end of the lifter bore.

The **roller tappet** is a high-performance tappet. The roller reduces friction between the lobe and the lifter, figure 7-21. It also allows grinding of special cam lobe shapes, as it will follow virtually any shape lobe. Rollers require a special device to keep the roller square with the lobe. This is usually a link connecting two tappets together, or a guide bolted to the block.

Figure 7-20. In some engines the mushroom tappet is used to open the valves.

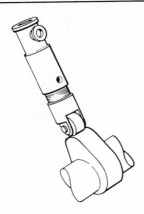

Figure 7-21. The roller tappet reduces friction between the lobe and the lifter.

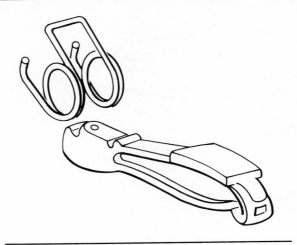

Figure 7-22. Some overhead camshaft engines use a cam follower.

Some overhead cam engines use cam followers, figure 7-22. One end rests on an adjustable pivot, while the other end rests on the valve stem. The clearance in the valve train is between the cam follower and the cam lobe. When hydraulic lifters are used in place of an adjustable pivot, they sit in a bore in the head and adjust the cam follower automatically. The follower has a leverage ratio similar to a rocker arm. The lift of the cam lobe is multiplied by this ratio to compute the valve lift.

Other overhead cam engines have tappets that operate directly off the cam lobe, figure 7-23. These are adjusted by turning a screw, or discs are placed either between the end of the tappet and the cam lobe or between the valve stem and the inside of the tappet's top surface. Adjustment is made by replacing the disc with one of a different thickness.

Tappets and lifters are made from cast iron. The lifter is hardened by heat treating so it will be long wearing. But lifters still wear out where the cam lobe rides. Whenever a camshaft is replaced with either a reground or new cam, the lifters should be replaced. Using worn lifters can damage a new cam in just a few thousand miles. The surface of a worn lifter may contact the cam lobe only at the high points between the wear grooves. This increases the pressure on the cam lobe face and quickly wears away the smooth lobe surface.

Hydraulic Lifters

Let's take a close look now at how a hydraulic lifter works, figure 7-24. The lifter rests in a bore. Oil passages supply oil under pressure from the engine oil pump. The body of the lifter is moved by the cam lobe, and slides up and down in the lifter bore.

Inside the lifter body is a plunger. Oil is forced through a hole in the side of the body, and through a second hole in the side of the plunger to the inside of the plunger. A check valve in the bottom of the plunger allows oil to pass out of the plunger into the lifter body, but not back into the plunger.

A light spring under the plunger pushes the lifter body against the cam and the plunger against the pushrod or cam follower. This removes all the play from the valve train, but does not lift the valve off the seat.

The space below the bottom of the plunger fills wih oil under pressure through the check valve. Because of the check valve, this oil can pass out of the lifter only through the very small clearance between the plunger and the lifter body.

When the cam lobe raises the lifter, the check valve closes and the plunger tries to compress the oil below it. But since oil cannot be compressed, as the lifter rises, the plunger rises with it, opening the valve.

Camshafts, Lifters or Followers, Pushrods and Rocker Arms

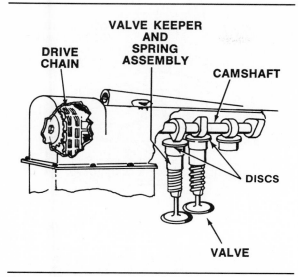

Figure 7-23. Some overhead cam engines have tappets that operate between the cam lobes and the valves. This particular engine has replaceable discs for adjustment.

Although some leakage occurs, because of the extremely small clearance between the body and the plunger, in the fraction of a second that a valve is opened and closed, the leakage is practically nothing. At an idle speed of 500 rpm, the valve is open less than nine hundreds of a second.

The pushrod seat is usually at the top of the plunger, figure 7-24. The seat has a hole in it in most designs. The hole allows oil to go up through the hollow pushrod to lubricate the valve stems and rockers. There may be a metering device under the pushrod seat, such as a small disc. This limits the oil pressure to the valves so that the stems are not flooded with oil.

Lifter operation

To prevent excessive wear in one spot, lifters are designed to rotate in their bores. One way of accomplishing this is by having the lifter bore slightly offset from the cam lobe. Instead of rubbing across the center of the lifter, the lobe rubs to one side. This makes the lifter rotate. When an engine is run with the rocker covers off, this rotation can be seen. The pushrods actually spin in their sockets from lifter rotation.

Another way to achieve lifter rotation is to grind the cam with a slight taper from one side of the lobe to the other. When the taper is combined with lifters which have convex bottoms, the pressure is applied offcenter and the lifter rotates.

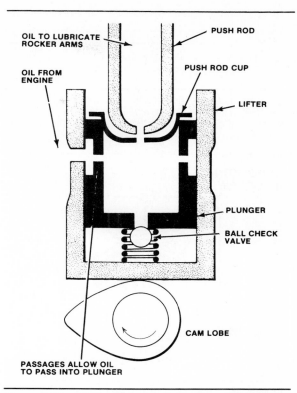

Figure 7-24. Oil from the engine flows under pressure into the oil passages of the hydraulic lifter. The oil lifts the body to remove play in the valve train.

PUSHRODS

A pushrod transmits the push from the lifter to the rocker arm, figure 7-30. Pushrods are usually hollow to reduce weight and to lubricate the rocker arms and the valve stems. Oil flows from the lifter through the hollow pushrod.

Pushrods are retained in the lifters and in rocker arms by sockets which are deep enough so that the pushrods cannot fall out.

Some engines do not have any provision for valve clearance adjustment. When the valves are refaced and the seats ground, the valves move into the head. This compresses the hydraulic lifters. If too many valve jobs have been done, the lifters will be compressed so much that they bottom out. The lifters will not adjust and the valves might be held open when the parts get hot and expand.

Mushroom Tappet: A tappet with a base larger in diameter than its body.

Roller Tappet: A tappet with a roller that rolls on the camshaft lobe.

Figure 7-25. Some older cars have adjustable pushrods. To make the necessary adjustment, just loosen the locknut, turn the threaded end, then tighten the locknut.

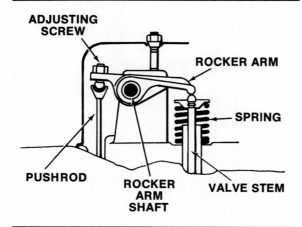

Figure 7-26. The rocker arm in an overhead valve engine.

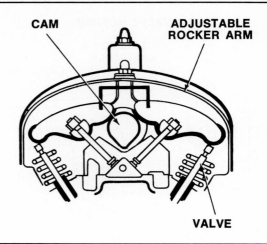

Figure 7-27. In an overhead cam engine, the rocker arm is operated by the camshaft with the pivot between the cam and the valve.

Pushrods are available in different lengths for engines which have no provision for valve adjustment. If shorter pushrods are not available, however, the tips of the valve stems must be ground. You may see a few older cars with adjustable pushrods. These have an adjustment screw at the top of the rod to change the length, figure 7-25.

When pushrods are used with rocker arms that pivot on a ball, it may be possible for the rocker arms to move from side to side. Many of these engines have a guide for the pushrod, often a plate fastened to the cylinder head casting.

Rocker arms that mount on a shaft do not need pushrod guides. The width of the shaft bushing keeps the rocker from twisting.

Pushrod ends require lubrication. The pushrod seats in the lifters and the rocker arms are usually lubricated by the oil that travels through the pushrods to the rockers. If the pushrod does not carry oil, then the lifter end gets oil from the lifter. The rocker arm end gets oil from the rocker.

ROCKER ARMS

Rocker arms are used in overhead valve engines, figure 7-26. A pushrod transfers the lift of the cam lobe to the top of the cylinder head when the camshaft is in the block. The rocker arm transfers this push down to the valve.

In an engine with an overhead camshaft, rocker arms are operated by the camshaft itself with a pivot between the cam and the valve stem, figure 7-27.

For many years all rocker arms were mounted on a heavy shaft which ran the full length of the cylinder head. It was supported by stands bolted to the head. The stands and rockers were arranged so that a rocker was against one side of a stand. A spring washer on the shaft put light pressure against the rocker arm and kept it against the stand. This kept the rocker arm aligned with the pushrod and the valve stem.

1955 was the beginning of the end for shaft-mounted rocker arms when Chevrolet brought out its famous small block V-8. Each rocker arm was mounted on a stud. A split ball acted as the pivot for the rocker. The split ball was held in place with a nut that could be screwed up or down the stud to adjust the valve clearance. There was no attempt to incorporate resistance to the tilting.

Some people doubted that the stud-mounted rockers would work, but they worked so well that this design and variations of this design have been used by many other engine makers, and are still being used today.

Some manufacturers use systems that may look different, but are basically the same as

Camshafts, Lifters or Followers, Pushrods and Rocker Arms

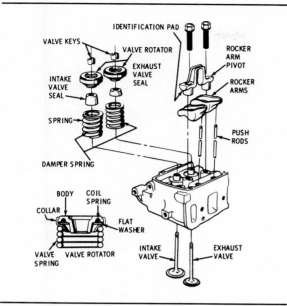

Figure 7-28. Oldsmobile rocker arms. (Oldsmobile)

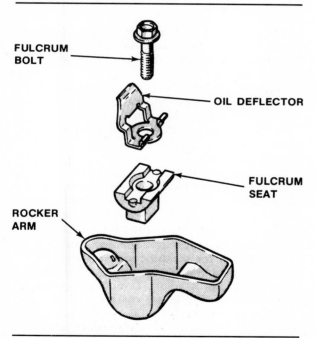

Figure 7-29. Ford rocker arms. (Ford)

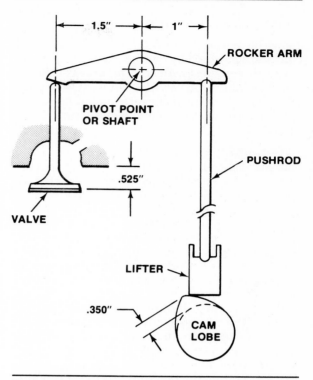

Figure 7-30. The distance on a rocker arm from the pivot to the pushrod seat is shorter than the distance from the pivot to the valve stem.

Rocker Arm Ratio

If you look closely at a rocker arm, you will see that the distance from the pivot point to the pushrod seat is shorter than the distance from the pivot point to the valve stem pad, figure 7-30. This lever action gives the rocker a mechanical advantage in the ratio of the long side to the short side. This advantage, or rocker arm ratio, can be anywhere from 1.1 to 1.6, but is usually around 1.5.

The rocker arm ratio directly affects valve lift. If the lift of the cam lobe is 0.35″ (8.9 mm) and the rocker ratio is 1.5, the actual lift at the valve will be 0.525″, or 13.35 mm, since .035 × 1.5 = .0525 (8.9 mm × 1.5 = 13.35 mm), figure 7-30.

The rocker arm ratio is designed to give the **desired valve lift while keeping cam lobe size to a minimum.** The positive rocker ratio does put more pressure on the cam lobe and makes it harder for the lobe to open the valve. But it is easy to deal with the increased pressure. Building all the lift into the cam lobe might require a larger lobe, larger bearings, and a redesign of the engine.

stud-mounted rocker arms, figures 7-28, 7-29. Instead of pivoting on a split ball, the rocker pivots on a split shaft. There is no stud. The split shaft fastens to the cylinder head with one or two capscrews. This is not quite as simple as the stud-mounted rocker, but it still eliminates the long, heavy rocker shaft formerly used.

Adjusting Rocker Arms

As we have learned, most modern cars and light trucks are powered by overhead cam or overhead valve engines with stud-mounted rocker arms. To adjust the valve lash or valve clearance you simply turn the rocker arm adjustment screw located at the end of the rocker arm, figure 7-26.

Rocker arm interference

When a valve job is done, the valves are refaced or replaced and the valve seats ground or cut. The valve stem will then stick out of the head more than it did before the valve job, which in turn will result in the valve stem end of the rocker arm being higher. On stud-mounted rockers, the rocker can be so high that the slot in the bottom hits the side of the mounting stud. If this happens, the valve will not close. It will leak and soon burn.

A rocker arm is designed to push on the center of the valve stem. If, through wear, the rocker pushes nearer to the edge of the valve stem, it may push the valve sideways and wear out the guide. In extreme cases, the rocker may move so far that the pushrod hits the side of the hole in the cylinder head.

Although these cases of rocker arm and pushrod interference are rare, they do happen. A simple visual check while cranking the engine is usually enough to see if anything is hitting that shouldn't be.

Overhead cam engines do not use pushrods, but do use rocker arms or cam followers. The rocker arm seat is in the top end of the lifter, which rests on the camshaft. Some engines with overhead cams use a cam follower on which the cam lobe pushes on the middle of the follower, with the valve on one end and the hydraulic lifter on the other. Oil pressure expands the hydraulic lifter until it pushes the rocker arm up against the cam lobe, removing the play in the valve train. Many overhead cam engines have cams which push directly on valve lifters to open the valves. This reduces the inertia of the valve train by completely eliminating the rockers and pushrods.

Materials and Lubrication

Rocker arms can be cast iron, pressed steel, or aluminum. The older, shaft-mounted rockers were made of cast iron which is relatively heavy. Now most rockers are made from pressed steel.

Rockers made from aluminum use hardened steel inserts at the valve and pushrod wear points.

Rocker arms get oil through hollow pushrods, through arm stands, or from an external line. When the oil supply is through the rocker stand, it may be through a drilled hole in the stand, or the oil may come up through the rocker stand mounting bolt. Since the bolt is hollow, it must be replaced in the stand that is positioned over the oil supply. In some cases the bolt shank is relieved to allow the oil space to travel up through the stand. The relieved bolt shank must also be replaced in the right stand.

ENGINE BALANCING SHAFTS

One feature of the heavy-duty diesel truck engine is the **balancing shaft**. It is a rotating shaft that serves no purpose other than to smooth out engine vibrations. Large, slow-speed engines sometimes develop unpleasant vibrations. These vibrations are called harmonics by engineers. The frequency of the harmonics can be measured by instruments. This is similar to the way your ear can differentiate between different sounds.

The balancing shaft sets up countervibrations that cancel out unwanted crankshaft vibrations. The result is a smoother engine, and the vehicle is more pleasant to drive.

The principle of balancing shafts has been known for many years. They were used on truck engines, but seldom on car engines. In 1976, the Dodge Colt came out with the 2000 cc Silent Shaft engine, figure 7-31, developed by Mitsubishi Motors Corporation of Japan.

An inline engine receives vibrations from three sources. The reciprocating piston and connecting rod give vertical vibrations, the piston pushes on the crankshaft and sets up a rolling force, and the burning in the combustion chamber causes vibrations and rolling forces.

A vibration can be cancelled by an equal vibration acting against it. To do this the Silent Shaft engines use two balance shafts, one on each side of the crankshaft, figure 7-31.

When comparing a V-8 and an inline 4-cylinder engine, there is a noticeable difference in vibration and noise. Yet there is almost as much difference between a regular 4-cylinder engine and a Silent Shaft 4-cylinder engine, the balancing shafts are that effective.

It is obvious by looking at the shafts that they are out of balance. But this is deliberate. They are out of balance so they can set up vibrations to cancel the three vibrations mentioned above. The shafts are driven by a chain and sprockets attached to the crankshaft, but separate from the camshaft drive.

Camshafts, Lifters or Followers, Pushrods and Rocker Arms

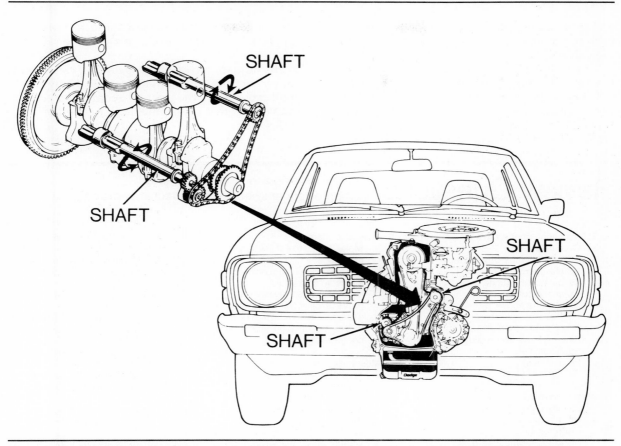

Figure 7-31. The 1976 Dodge Colt engine by Mitsubishi included a set of balancing shafts. (Mitsubishi)

To make the system work, one balance shaft has to turn in the opposite direction from the crankshaft. A clever method is used to accomplish this. The right shaft is driven through a set of gears to reverse its rotation. These gears are enclosed in a housing which also pumps oil to the engine. This eliminates the separate set of gears for the usual engine oil pump, and also makes the balance shaft reversing gears quieter because they are running in oil.

Because the balance shafts are individually out of balance, they must be timed when installed. Marks on the sprockets line up with a plated link on the drive chain. The two oil pump reversing gears also have marks that must be aligned.

Recently, a variation of the balance shaft principle has been applied to other Mitsubishi-Chrysler engines. A 1600 cc engine uses the same basic system, but with cog belts and cog wheels. The crankshaft has two cog wheels driving two separate cog belts. One short belt drives only the right balance shaft. Another longer belt drives the overhead camshaft, the left-hand balance shaft, and the oil pump. As in the chain drive system, the oil pump acts as the reversing mechanism for the balance shaft.

SUMMARY

The camshaft is a series of cams, or lobes, that rotate to open the engine valves. They may be located in the engine block or the cylinder head and are driven by gears, chain, or cog belt from the crankshaft.

The camshaft must be timed to the crankshaft rotation so that the valves open and close in time with piston movements. It is important to remember that when a camshaft is designed, lift, duration, overlap, and timing will all affect the engine's performance, emissions and fuel economy.

Balancing Shaft: A shaft that is used solely for the purpose of cancelling engine vibrations.

Tappets, cam followers, and rocker arms are used to transfer the motion of the cam lobes to the valves lifters.

When replacing any part of the valve train it is essential to take into consideration the original angles of each part in relation to the others. For instance, if you replace a pushrod with one that is slightly longer, you are changing the angle at which the rocker arm pushes against the valve stem. This, in turn, can affect the angle at which the valve face meets the valve seat.

By adjusting clearances, grinding valve stems or rocker arms, or changing pushrod length, you can achieve the original geometry of the valve train that is essential for it to operate efficiently.

Review Questions

Choose the single most correct answer. Compare your answers to the correct answers on page 268.

1. The camshaft opens valves:
   a. Gradually
   b. With an instantaneous thrust at exactly the right moment
   c. As fast as this can be done by a mechanical device
   d. None of the above

2. Most camshafts are made of:
   a. Aluminum-beryllium alloy
   b. Stainless steel
   c. Forged steel
   d. Cast iron or steel

3. Which of the following treatments reduces cam wear to a minimum?
   a. Hard-face overlays
   b. Carburizing
   c. Manganese-phosphate coating
   d. All of the above

4. If the intake valve opens 18 degrees before top dead center and closes 12 degrees after bottom dead center, the duration is:
   a. 186 degrees
   b. 210 degrees
   c. 222 degrees
   d. None of the above

5. Camshafts designed for high-speed operation:
   a. Usually provide comparatively long valve duration
   b. Make inlet valve duration much longer than exhaust valve duration
   c. Decrease valve overlap
   d. Have special gear ratios, to rotate faster than a regular camshaft

6. Hydraulic lifters:
   a. Make it easy to set the tappet clearance
   b. Are filled with hydraulic fluid similar to brake fluid
   c. Cannot rotate; they can only move up or down
   d. Transmit steady thrust to the valves, because oil cannot be compressed

7. Most pushrods:
   a. Are hollow and rotate when the engine is running
   b. Are adjustable
   c. Move freely in their guides, which are similar to valve guides
   d. Cannot bend

8. Balance shafts:
   a. Function by balancing engine output to minimize vibration in the transmission
   b. Cancel out harmonics
   c. Must rotate in the same direction as the crankshaft
   d. None of the above

9. Noise reduction is a major function of which of the following parts?
   a. Cog belt
   b. Helical gears
   c. Roller chain
   d. Valve guides

10. Rocker arms:
    a. Are usually mounted on a shaft
    b. Are unequal in length on each side of the pivot
    c. Pivot on a ball joint
    d. Provide a mechanical advantage anywhere from 2.1 to 2.6

11. Shorter pushrods are sometimes installed because:
    a. This increases engine power
    b. The original rods tend to spread out
    c. The original rods cannot be ground shorter
    d. Grinding valve seats changes the position of the valves

12. The camshaft operates the valves, and:
    a. Has no other function
    b. Usually drives the distributor shaft
    c. Always drives the oil pump
    d. Never drives the fuel pump

13. Camshafts are driven by:
    a. Sprockets and chains
    b. Cog wheels and cog belts
    c. Gears
    d. All of the above

14. On some engines, the camshaft operates the fuel pump:
    a. By a helical reduction gear
    b. Off one of the intake valve lobes
    c. By an eccentric
    d. By a concentric

15. Valve lift is equal to:
    a. Cam lift
    b. Cam lift plus rocker arm ratio
    c. Cam lift plus lifter clearance
    d. Cam lift times rocker arm ratio minus lifter clearance

Chapter 8

Crankshafts, Flywheels, Vibration Dampers, Pistons, and Rods

CRANKSHAFTS

The crankshaft converts reciprocating motion to rotary motion. The piston requires leverage to turn the crankshaft and the flywheel. The amount of leverage is equal to the distance from the centerline of the **crankpin** to the centerline of the crankshaft, figure 8-1. The illustrations show that this distance, called the throw, can be made longer or shorter to change the leverage. This distance determines the stroke of the piston.

The displacement of an engine in cubic inches, liters, or cubic centimeters is computed from its bore and stroke, figure 8-2. The crankshaft, with throws that measure one-half of the stroke, has a direct relationship to the size of the engine.

Burning the air-fuel mixture in the combustion chamber pushes the piston down on its stroke. The piston pushes on the piston pin, which is in the small end of the connecting rod. The connecting rod is the link between the piston and the crankshaft. Between the big end of the connecting rod and the crankpin of the crankshaft are insert-type bearings, figure 8-3. They are locked into the connecting rod in most designs. Their purpose is to allow the rod to turn easily on the crankpin without excessive wear.

Main Bearings

The crankshaft rotates in main bearings, figure 8-4. The main bearings are also the insert-type, similar to the connecting rod bearings. The simplest crankshaft with one throw requires two main bearings, one on each end of the shaft. If more cylinders are added and the crankshaft becomes longer, more main bearings may be necessary. Two main bearings will support the shaft, but may not keep it from bending.

Crankshaft Offset

The rotation of an automobile engine is almost always clockwise when looking at it from the front. The front of the engine is defined as the end opposite the flywheel. The design blueprints of most inline engines show that the centerline of the crankshaft main bearings does not line up with the centerline of the cylinders. When looking at the front of the engine, the crankshaft may be offset slightly to the left. Not

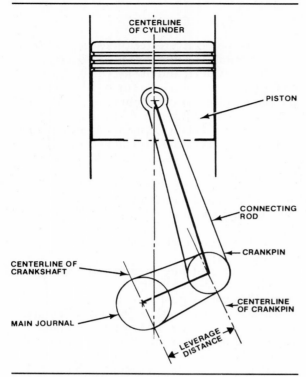

Figure 8-1. The distance from the crankpin centerline to the centerline of the crankshaft determines the leverage available to turn the crankshaft and flywheel.

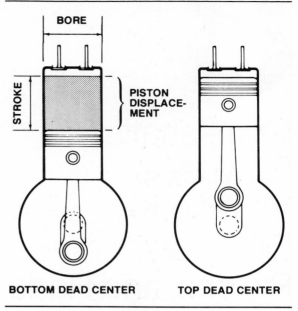

Figure 8-2. The displacement of an engine is directly related to its bore and stroke.

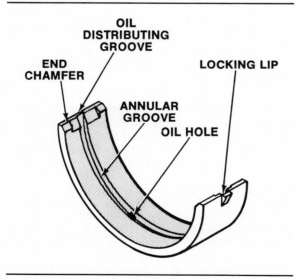

Figure 8-3. Insert bearings fit between the connecting rod and the crankpin.

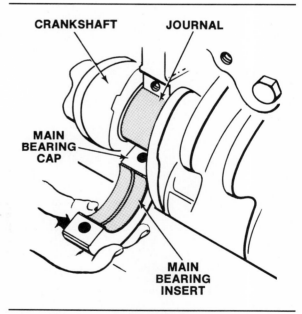

Figure 8-4. The crankshaft rotates in main bearings.

all engines use the offset, but its purpose is to reduce the angle of the connecting rod during the power stroke.

To understand this, look at the illustration of the crank throw that is halfway through the power stroke, figure 8-5. The piston is being pushed down by the burning mixture, but the rod is at such an extreme angle that the piston is being forced into the side of the cylinder. If the crankshaft is offset, the angle of the rod will be reduced. This will let the piston push without as much side thrust, and there will be less pressure on the connecting rod bearing.

Crankshaft offset is sometimes called cylinder offset, since the cylinders are offset from the

Crankshafts, Flywheels, Vibration Dampers, Pistons, and Rods

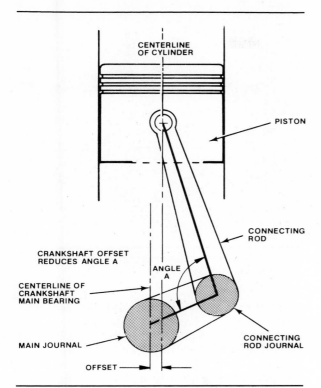

Figure 8-5. This crank throw is halfway through the power stroke. You can see that the piston is being pushed down by the burning mixture, but because of the angle of the rod, the piston is forced against the cylinder.

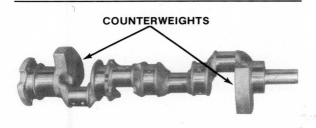

Figure 8-6. Engineers add counterweights to the crankshaft to balance it. (Pontiac)

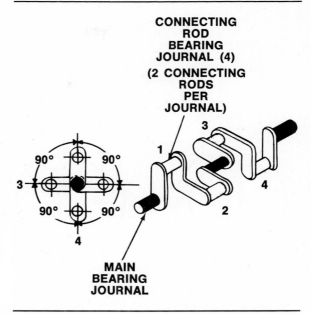

Figure 8-7. The arrangement of throws around the crankshaft makes it possible for the firing impulses to occur at equal intervals.

centerline of the crankshaft. When offsets were first used, they were as great as three-quarters of an inch (17 mm), but modern engines use an offset of only about one sixteenth of an inch (1.5 mm), or none at all.

Counterweights

Most modern crankshafts are counterweighted, figure 8-6. The **counterweights** are cast in one piece with the shaft. When looking at a crankshaft, it appears that the counterweights are heavier than they need to be. Each pair of counterweights is opposite a rod journal, but the counterweights are much heavier than the journal.

This is because the weight of the bearings, connecting rods, and pistons must also be balanced. When these components are assembled, the engine will be almost perfectly balanced.

Crank Throws And Firing Impulses

The arrangement of rod journals or throws around a crankshaft follows a definite pattern. The throws are arranged so the firing impulses are at equal intervals, figure 8-7, which makes the engine run smoother and vibrate less. We will begin our discussion with a single cylinder engine, and add cylinders to see how they are spaced.

Crankshaft Offset: The distance that the centerline of the crankshaft is offset from the centerline of the cylinders.

Counterweights: The weights opposite the rod journals on a crankshaft that balance the weight of the journal, bearing, connecting rod, and piston assembly.

Crankpin: Connecting rod crankshaft journal.

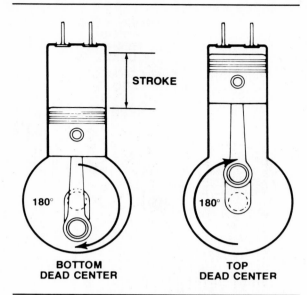

Figure 8-8. Every 360 degrees, the one throw on a single cylinder engine returns to top dead center.

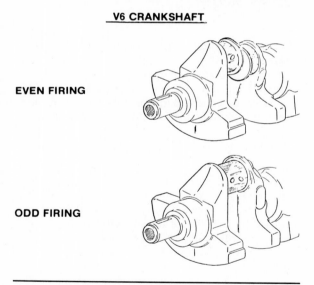

Figure 8-10. On the even firing Buick V-6, there is no common crankpin. The pins are staggered. (Buick)

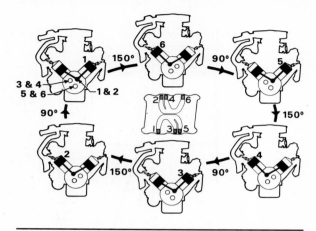

Figure 8-9. The firing impulses on this Buick V-6 were unequally spaced. (Buick)

The crankshaft for a single cylinder engine has only one throw. Every 360 degrees of rotation, it returns to top dead center, figure 8-8. Every 720 degrees of rotation there is a firing impulse in a 4-stroke engine. Now add three cylinders to create a 4-cylinder engine with four crank throws. We then divide 720 by four to determine the crankshaft throw spacing for this engine. This gives us a firing impulse every 180 degrees.

A V-8 engine also has four crank throws, figure 8-7, but has two connecting rods per crank throw. Each throw has a single crankpin that carries two connecting rods, side by side.

A V-8 must be designed so the angle between the cylinder banks is the same as the angle between the crankshaft throws, or a multiple thereof. If it is not, the firing impulses will not be equally spaced.

Actually, this rule applies to any arrangement of V-type, radial, or opposed cylinder engines.

The original Buick V-6 engine has unequally spaced firing impulses, figure 8-9. It used a 90-degree block, with a 120-degree crankshaft, and three crankpins. The two cylinders that were attached to the same crankpin fired in succession. After firing one cylinder, the crankshaft had to rotate only 90 degrees to reach top dead center on the opposite cylinder because 90 degrees was the angle between the cylinder banks. After the second cylinder fired, the engine had to rotate 150 degrees to get to the next cylinder. The engine fired from every other crankshaft throw. With 120 degrees between throws, going to the second throw took 240 degrees of rotation. The engine fired this way continuously, alternating between 90-degree and 150-degree firing intervals. It was known as the uneven firing V-6.

Buick eventually decided to make the V-6 even firing so it would run more smoothly. They did this by splitting the crankpins. Instead of a common crankpin for each pair of cylinders, the pins are staggered on the even firing V-6, figure 8-10. The separation between adjacent crankpins on the same throw is 30 degrees.

Another exception to the evenly spaced firing impulse engines is the V-6 introduced by Chevrolet in 1978. This engine uses a 90-degree V-8 block with two cylinders cut off, and has

Crankshafts, Flywheels, Vibration Dampers, Pistons, and Rods

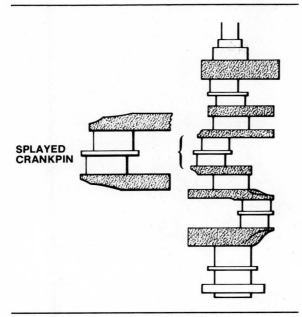

Figure 8-11. The Chevrolet V-6 has the pins separated.

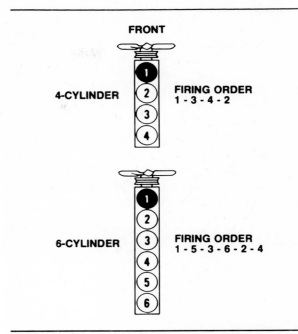

Figure 8-13. Cylinder numbering of an inline engine.

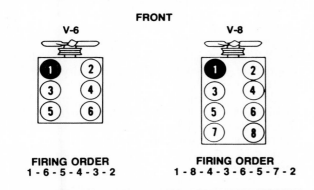

Figure 8-12. American Motors, Chrysler, and most General Motors V-type engines are numbered this way.

alternating 108-degree and 132-degree firing intervals. Instead of having split crankpins, as on the Buick even-firing V-6, the Chevrolet engine has the pins separated, so that each connecting rod has its own throw, figure 8-11.

Firing order

As described in Chapter One, each cylinder in an engine is numbered. The numbers are usually cast into the intake manifold near the cylinder intake port. In some cases, the cylinder numbers may appear only in the service literature.

Cylinder numbering sequence varies from manufacturer to manufacturer. In recent years, General Motors numbered their cylinders with the odd numbers on the left and the even numbers on the right, figure 8-12. But earlier GM engines were different. Ford usually has numbers 1 through 4 on the right and 5 through 8 on the left. Chrysler numbers their engines the same as GM, with the odd numbers on the left and the even on the right. Virtually all inline engines are numbered in sequence from front to rear (flywheel end), figure 8-13.

When cylinder numbering is combined with the arrangement of the crankshaft throws, a firing order results. The firing order is part of the design of the engine and never changes, figures 8-12 and 8-13.

Firing orders are always listed starting with cylinder number one. A typical firing order for a 4-cylinder engine is 1-3-4-2. Inline 6-cylinder engines have the firing order 1-5-3-6-2-4. V-8 engines have several different firing orders, mainly because the cylinder numbering sequence varies.

Materials And Construction

Crankshafts are made of steel or iron, and different types are available depending on the strength required.

A **billet crankshaft** is cut from a solid piece of steel. It is used in racing engines when a special

---

Billet Crankshaft: A crankshaft cut on a lathe from a solid piece of steel.

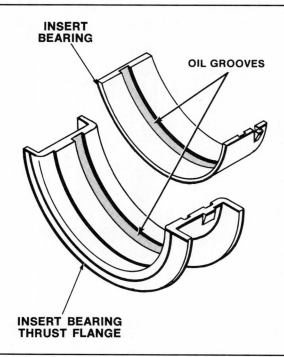

Figure 8-14. Crankshaft main bearings get their oil through grooves in the bearing.

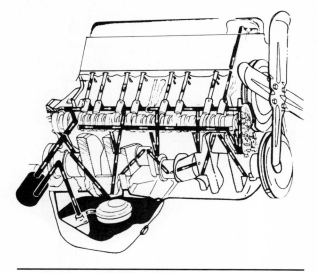

Figure 8-15. Oil flows to the connecting rod bearings through passageways in the crankshaft. (Chevrolet)

stroke is required and a forged crank is not available. The forged crankshaft is the strongest and is used by carmakers who feel the extra strength justifies the added cost.

A nodular cast iron crankshaft is a compromise between a forged and a cast crankshaft. This type of crankshaft is cast so that the graphite separates into little spheres or nodules. Another term for nodular iron is spherical graphite iron. The structure of the little spheres dispersed through the iron in great quantity gives the nodular iron its strength. Normal cast iron has graphite in little flakes.

Crankshafts used in racing engines or heavy-duty truck engines are sometimes treated to harden them for longer wear. The hardness of the shaft and the bearings must be closely matched. A heavy-duty bearing is harder than a normal bearing. Putting a hard bearing in an engine with a normal shaft will cause excessive wear on the shaft.

There are two basic methods of hardening crankshafts. A strong magnetic field can be used to heat the shaft only in the bearing areas. This is called **induction hardening**. The other way is to submerge the shaft in an acid bath. This is called **nitriding**. Induction hardening goes deeper than nitriding. An induction-hardened shaft can be reground without breaking through the hard surface into the softer material below. A nitrided shaft should be nitrided again after regrinding, because the grinding may remove the hard surface.

Racing crankshafts are sometimes hard chrome plated. This puts an extremely hard and long wearing surface on the crankshaft journals. A relatively soft bearing is used and changed frequently. This combination gives very long life to the crankshaft. A disadvantage is that when the shaft finally does wear, it cannot be rechromed unless it is ground undersize. There is no way of removing just the chrome or adding chrome to restore the original surface.

Lubrication

The crankshaft is lubricated by oil under pressure. The oil is pumped through passageways in the block to a hole in the crankshaft main bearings. The oil then passes into a groove in the bearing insert, figure 8-14. This groove is aligned with a hole in the crankshaft journal. Oil flows into the hole and through passageways in the crankshaft to the connecting rod bearings, figure 8-15. At the same time, the oil is forced outward over the main bearing surface. This provides a thin film of oil to protect the bearing and the crankshaft from wear.

The oil film on the bearings is strong enough to prevent the bearings and crank journal from touching each other. However, under heavy loads or during starting, there is a possibility that the oil film will break down and allow the parts to touch. Modern engines have long crankshaft and bearing life. Many engines, disassembled after high mileage, have no perceptible wear on the crankshaft.

Crankshafts, Flywheels, Vibration Dampers, Pistons, and Rods

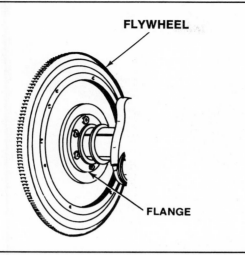

Figure 8-16. A flywheel bolts onto a flange at the end of the crankshaft.

FLYWHEELS AND FLEXPLATES

A flywheel must be heavy so its momentum will keep the crankshaft rotating until the engine is in position for the next firing impulse. It also adds mass to the engine so that it will run and accelerate smoothly. An engine without a flywheel could accelerate so fast that the inertia of the valve train components would bend pushrods and break rocker arms.

A flywheel is made of cast iron or steel and bolts to a flange on the end of the crankshaft, figure 8-16. The rear surface of the flywheel is machined smooth to accept the clutch friction disc. A clutch pressure plate is bolted over the clutch disc onto the flywheel. The pressure plate squeezes the disc against the flywheel, engaging the engine with the transmission. When the clutch pedal is depressed, the pressure plate is forced back from the disc, allowing the disc to turn freely and disengaging the engine from the transmission.

A ring gear is mounted on the circumference of the flywheel, to provide for starter engagement.

On an engine equipped with an automatic transmission, the heavy torque converter acts as a rotating mass for the engine, so a flywheel is not needed. Instead, a flexplate is used in place of the flywheel, figure 8-17. It is very light, and its only function is to provide a mount for the ring gear and take up any thrust movement from the crankshaft. The flexplate is named for its ability to absorb this thrust. Instead of the torque converter being thrust in and out of the transmission, the flexplate cushions this movement by flexing.

VIBRATION DAMPERS

A vibration damper is not used to make the engine run more smoothly. Nor is it used to eliminate vibrations that the driver or the passengers might feel. The purpose of the vibration damper or **harmonic balancer** is to prevent the crankshaft from breaking.

When the crankshaft throw receives the force of the piston on the power stroke, the crankshaft twists slightly. This twisting may be so minute that it is not measureable. It may only strain the crankshaft, causing stress in the metal without exterior movement. But the heavy crankshaft resists the strain, and after the initial push from the piston and connecting rod, pushes back. In effect, the piston tries to bend the crankshaft, and the crankshaft tries to straighten itself.

When the crankshaft pushes back, it overreacts and then tries to straighten itself in the other direction. These oscillations take place for several cycles and finally die out, similar to a tuning fork.

Every time there is a firing impulse in the engine, the crankshaft gets a push and starts vibrating again. At certain speeds, the pushes received by the crankshaft occur simultaneously with the efforts of the crankshaft to straighten itself. This increases the magnitude of the oscillations. At certain speeds the vibration can become so great that the shaft will break.

This vibration is called **torsional vibration**. It can also cause heavy wear on timing chains. The sprocket at the front of the crankshaft constantly oscillates against the chain. Torsional vibration rarely causes any problems at the flywheel end of the engine because the heavy flywheel dampens the oscillations.

To lessen the torsional vibration, a vibration damper or harmonic balancer is used, figure

Induction Hardening: The use of a strong electromagnet to heat small areas on a large piece of metal. The area is then cooled rapidly to harden it. The process is used on crankshaft journals and valve seats.

Nitriding: Hardening of metal by dipping it into an acid bath.

Torsional Vibration: The vibration set up by the piston thrust against the crankshaft causing it to twist slightly, and the crankshaft's efforts to straighten itself.

Harmonic Balancer: A vibration damper.

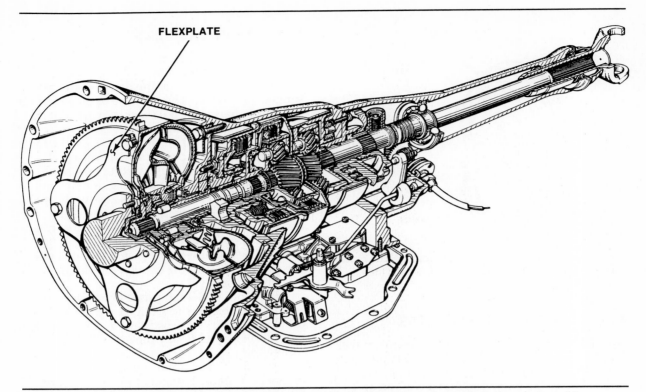

Figure 8-17. The flexplate is sometimes used instead of a flywheel on automatic transmission engines. It is attached to the torque converter here. (Chevrolet)

8-18. It is usually the front pulley on the crankshaft or has the front pulley bolted to it. The damper is made of an outer heavy ring, a rubber ring, and an inner hub, figure 8-19. When the crankshaft oscillates, the balancer oscillates with it. When it gets to the end of the oscillation, the outer ring keeps going and twists the rubber slightly. This happens at the end or beginning of each oscillation. The outer weight ring stretches the rubber and dampens the torsional vibration in the crankshaft.

Vibration dampers must be tuned to the crankshaft.

A 4-cylinder engine usually does not require a damper. The crankshaft is short, and the number of impulses is less than an engine with more cylinders. An inline 6-cylinder engine always requires some kind of damper. A V-8 engine with a timing chain ordinarily uses a damper. When a V-8 engine uses timing gears instead of a sprocket and chain, the gears act as a damper.

Dampers can wear out. The outer ring constantly stretches the rubber ring, and will sometimes lose its bond to the rubber. It may rotate freely or even come off. When this happens, the damper must be replaced.

Vibration dampers are also made of a combination of springs and friction material between the outer and inner members. This type can also wear out. If the springs are loose and rattle, the damper has lost its effectiveness and should be replaced.

CONNECTING RODS

The connecting rod is the link between the piston and the crankshaft, figure 8-20. It has a small end and a large end. The large end has a hole that fits the crankshaft rod journal.

Connecting rods have evolved considerably since the beginning of the automobile engine. The first connecting rods were very crude. The small end of the rod was split and clamped into the piston pin with a bolt. The large end was split with a removeable cap held on by two bolts.

To reduce wear on the crankshaft, babbit was poured into the large end of the rod, and was scraped by hand until the end of the rod fit the crankpin. To adjust the fit of the bearing, shims were used between the cap and the rod. Fitting bearings was an art requiring skill and experience. If the bearings were too tight, they overheated; if they were too loose, they pounded out.

As manufacturing techniques improved, connecting rods were made with removable bearing

Crankshafts, Flywheels, Vibration Dampers, Pistons, and Rods

Figure 8-18. A vibration damper or harmonic balancer reduces vibrations.

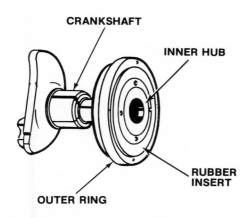

Figure 8-19. The damper is made of rubber.

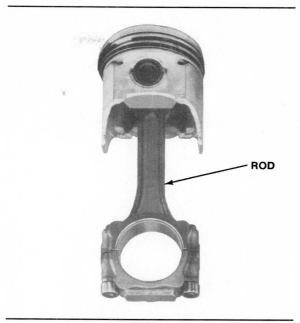

Figure 8-20. The connecting rod links the piston to the crankshaft. (Oldsmobile)

inserts. The big end of the rod was a smooth steel surface with a small cutout at one side. The bearings were thin shells with a tang that fitted into the cutout. Removable bearing inserts did away with babbit and shims. The crank journal was ground to a precise size and the bearings were made to fit it. No scraping or fitting was required. When the shaft wore, it was ground undersize so that undersize bearings would fit.

Some engines used a full-floating rod bearing. There were no tangs on the bearing insert, and no notches in the rod. As the crankshaft rotated, the bearing insert was free to rotate with the shaft or remain stationary with the large end of the rod. On Ford V-8 engines from 1936 until 1948, the floating design used a single bearing insert for each pair of connecting rods. Wear rates were difficult to predict because the rotation of the insert was variable. The insert might rotate with one rod, with the other rod, or with the shaft. In modern designs, the bear-

■ The Non-Rotating Crankshaft

Crankshafts always rotate, right? Wrong! It isn't any more necessary for a crankshaft to rotate than it is for a carburetor to be on top of an engine. In an air-cooled engine there is no heavy cylinder block or liquid cooling system. The structure to which the cylinders are attached could be allowed to rotate, while the crankshaft remained stationary. It may sound crazy, but it has actually been done. During the first World War, the rotary aircraft engine was built that way. It was manufactured by several companies, and powered planes from France, England, and Germany. Because the crankshaft was the only stationary part of the engine, fuel had to be pumped through the crankshaft to get to the cylinders. As odd as it was, the rotary was a great success. It was the first successful air-cooled engine, which isn't surprising, considering the breeze that those whirling cylinders must have stirred up. The rotary proved that air cooling would work, and paved the way for the modern air-cooled radial engine.

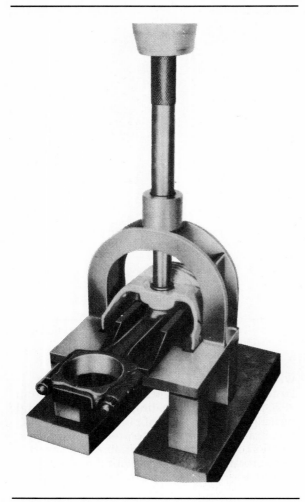

Figure 8-21. The pin is forced or press fit into the piston. (Chevrolet)

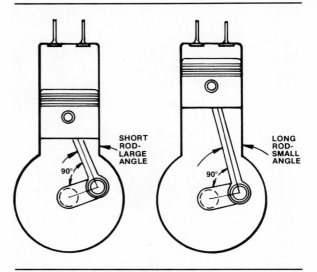

Figure 8-22. The longer connecting rod allows the piston to push more on the rod instead of against the cylinder wall.

ing inserts are locked in place and are firmly supported by the connecting rod.

At the same time, the small end of the rod was also changing. Some rods were made with a bushing in the small end. The bushing allowed the piston pin to turn in the rod. Because the pin also turned in the piston, these were known as **full-floating pins**.

Rod bearings were a weak point in many engines, but today, rod bearings seldom fail. When they do, it is usually because of dirt, improper assembly, or lack of oil.

Now most connecting rods have pins that are a press fit. The small end of the rod is slightly smaller than the pin. The piston, pin, and the rod are placed in a press and the pin forced into the rod until it is centered, figure 8-21. The press fit prevents the pin from touching the cylinder wall, but does not affect the movement of the piston on the pin.

The length of the connecting rod is determined by the crankshaft stroke, the length of the cylinder, the piston pin location, and the desired compression ratio.

The Effects Of Rod Length

The length of the rod affects the way the piston moves. When the piston is at top dead center, there are several degrees of movement of the crankshaft throw during which the piston does not move. A long rod increases the angle that the crankshaft can move without moving the piston, and decelerates the piston more slowly as it approaches top dead center. When the piston starts to leave top dead center, it also accelerates more slowly with a long rod.

With the piston at top dead center for a longer period of time, the low pressure in the exhaust manifold will pull in more intake mixture. Also, the longer rod does not tend to push the mixture back into the intake manifold because it is moving more slowly.

When a longer rod is at a 90-degree angle with the crankshaft throw, the angle between the side of the piston and the rod is less. This allows the piston to push more on the rod instead of against the cylinder wall, figure 8-22.

A shorter rod accelerates and decelerates the piston faster near top dead center. This is an advantage in getting the piston out of the way of the valves, but because the short rod accelerates the piston faster, it puts more stress on the piston and rod. A rod that is too short may crack and break.

Crankshafts, Flywheels, Vibration Dampers, Pistons, and Rods

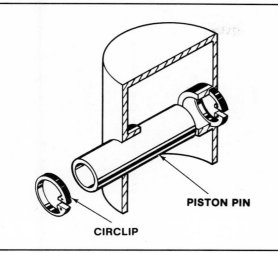

Figure 8-23. The pin is held in place with circlips.

Materials and Construction

Connecting rods are usually made in an "H" section. This shape has good resistance to bending and is easy to manufacture. Rods are usually made of steel. The better rods contain nickel, chrome, or other elements to make them tougher. Rods are generally forged because casting does not have the required strength.

Rods used in racing engines are sometimes made of aluminum. Aluminum is much lighter than steel and allows higher speeds and greater acceleration. However, aluminum fatigues easily, so the rods must be changed frequently to prevent breakage.

Titanium connecting rods have also been used in some engines. They are stronger, although heavier, than aluminum, but a little lighter than steel. They also have a problem with fatigue.

Sometimes rods are **shot-peened** to make them stronger and more resistant to cracking. The shot-peening smoothes out flaws in the material which might develop into cracks.

When a full-floating pin is used in the small end of a rod, a bushing is pressed into it. The pin is retained with **circlips** or locks at each end of the pin hole in the piston, figure 8-23.

When press-fit pins are used, the pin is pressed into the rod metal without any bushing. Some other methods of retaining piston pins are illustrated in figure 8-24.

In the large end of the rod, the replaceable inserts are located by tangs that fit into notches, figure 8-25. To prevent movement of the bearing in the rod and to make sure it is tight against the rod, the bearings are slightly larger than the rod. This is called bearing crush, figure 8-26. Bearings are covered completely in Chapter Nine.

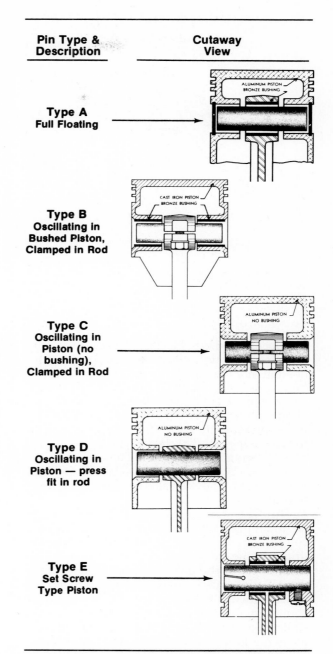

Figure 8-24. Other methods of retaining piston pins. (Sunnen)

Full-Floating Pin: A piston pin that is free to move in both the piston and the rod.

Shot-Peening: The process of shooting small steel balls against metal to compress and strengthen it.

Circlips: A round spring-steel device that fits into a groove at the end of the piston pin hole to lock in the pin.

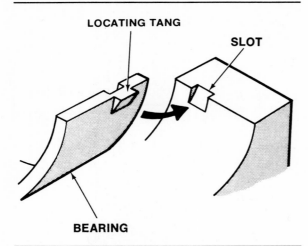

Figure 8-25. Replaceable inserts are located by a tang in a notch.

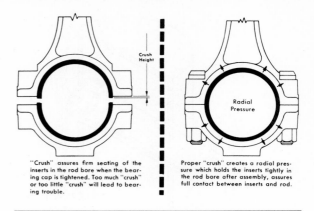

Figure 8-26. Bearing crush. (Sunnen)

Lubrication

Connecting rods are lubricated by oil splash or oil under pressure. When the piston pin is full floating, the pin end of the rod requires lubrication. Holes in the top of the rod allow oil to go between the bushing and the pin. There can also be a hole drilled the length of the rod. Oil under pressure then runs from the large end of the rod to the pin. This latter type of rod is used only in racing engines today. They are called drilled rods because the hole is drilled after the rod is forged. They can also be called pressure fit rods.

The big end of the rod is always lubricated by oil under pressure in modern engines. The oil comes from the engine oil pump through holes drilled in the crankshaft.

Connecting rod oil is also used to lubricate the cylinder walls in many engines. Either a **squirt hole** or a notch in the cap allows the oil to squirt out, figure 8-27. Inline engines have a squirt hole above the rod bolt so it will squirt on its own cylinder. In V-type engines, the hole is a notch between the cap and the rod. It squirts on the opposite cylinder in the other bank. The rod must be installed in the correct position or the squirt hole cannot do its job. Refer to the shop manuals to determine the correct position for the rod and squirt hole.

PISTONS

Pistons receive the full force of the burning mixture. They transmit this force to the connecting rod, which pushes on the crankshaft to turn the engine.

All the parts of the piston have specific names. The crown is the top of the piston that forms part of the combustion chamber. The crown can be domed, flat, or dished. This is discussed in more detail on the following pages.

Below the crown are the ring grooves in which the rings rest. The part of the piston between the grooves is called a land. Lands are always smaller in diameter than the piston skirt to keep them from touching the cylinder wall.

The areas where the piston pin fits into the piston are called **pin bosses**. Sometimes strengthening webs are cast into the walls of the piston to connect them with the pin bosses.

Rings on most modern pistons are above the pin. Usually there are three rings, two compression rings and one oil control ring. The two compression rings on top prevent gases from getting past the piston, figure 8-28. The oil control ring on the bottom scrapes off oil splashed on the cylinder walls, so the oil will not get into the combustion chamber.

Older engines were often made with four piston rings. The fourth ring was below the piston pin. It was strictly for oil control. Some modern industrial engines that run at slow speeds still use four piston rings. Because of their high compression ratios, most diesel engines have four or more rings to give added oil control and control of blowby.

Construction And Material

Pistons in early cars were iron or steel. They were heavy, but it did not matter because the engine did not turn very fast. As engines began turning faster and producing more horsepower, the inertia of the heavy pistons put added strain on other parts, which began to break.

Since the piston must stop and reverse direction at the end of each stroke, a heavy piston puts a greater load on the bearings. As engine speeds increased, the need for a lighter piston became more evident.

Crankshafts, Flywheels, Vibration Dampers, Pistons, and Rods

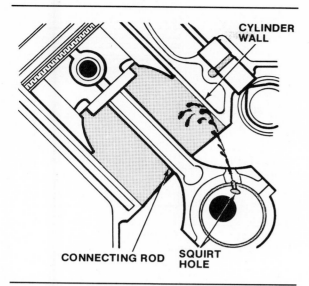

Figure 8-27. The squirt hole allows the oil to spurt out and lubricate the cylinder walls.

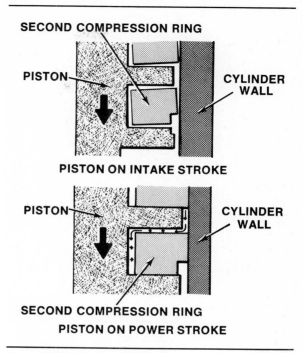

Figure 8-28. The two rings on top prevent gases from escaping past the piston.

Iron pistons were made from thinner metal, but they were not durable and frequently broke.

Modern pistons are made of aluminum alloy. Aluminum is about one-third the weight of iron. It conducts heat more than four times faster than iron. This means that the heat of combustion will be conducted into the cylinder walls and into the oil much faster.

Aluminum alloy pistons are approximately 90 percent aluminum with the other ten percent being traces of metals such as copper, zinc and chromium.

Many pistons are cast instead of being forged. Casting is less expensive, but does not produce as strong a piston.

Piston finish

The finish on pistons varies with the manufacturer. Most pistons have an oil-retaining finish to resist **scuffing**. Scuffing is mainly a problem when breaking in an engine. Until the high spots on the cylinder wall and on the piston are worn smooth, the piston can rub hard enough to transfer metal to the cylinder wall. Once this happens, metal continues to come off the piston until it is ruined, figure 8-29.

Scuffing can be caused by distorted or overheated cylinders. The overheating results from a poorly maintained cooling system. The radiator coolant does not boil, but hot spots occur from rust and the clogging of the passage around the cylinder.

To ensure that the piston skirt will hold enough oil to prevent scuffing, some pistons come with a fairly rough finish on the skirt. The finish may be obtained by blasting with small

■ Flywheel Engines

The inertia properties of the flywheel have always fascinated inventors. Many experimental vehicles have been built with no motive power other than a big flywheel. A large heavy flywheel could be brought up to speed at a charging station. Then the vehicle would use the power of the rotating flywheel to propel itself wherever the driver wanted to go. As the flywheel slowed, a warning light would come on. The driver would then try to get to a charging station before the flywheel stopped. Because of the limited period of power available from the flywheel, they were usually used on buses. A bus normally travels a fixed route, so it was easy to predict how far the bus could go before it "ran out of flywheel."

The main advantage of the vehicle was no pollution and very little noise.

Squirt Hole: A hole in the side of a connecting rod that causes oil to squirt onto the cylinder wall.

Pin Bosses: The areas surrounding the holes in the piston for the piston pin.

Scuffing: The transfer of metal between two rubbing parts; caused by lack of lubrication.

Figure 8-29. The piston can sometimes rub hard enough to transfer metal to the surface of the cylinder wall. (Sunnen)

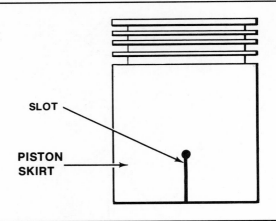

Figure 8-30. The split skirt is one way to control piston expansion.

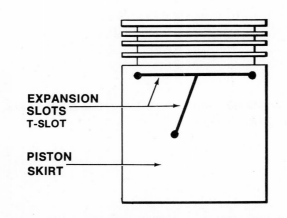

Figure 8-31. T-slot pistons also control expansion.

beads or by turning the piston in a lathe. The turning cuts a very fine thread on the skirt that collects oil and lubricates the skirt.

Other pistons are ground smooth, and then tin plated or anodized. The plating gives a surface that can rub off to prevent scuffing.

Piston Expansion

With the invention of aluminum pistons, the weight of parts was reduced by half. Unfortunately aluminum expands at twice the rate of cast iron, so if a piston is fitted to a cast iron cylinder with a slip fit at room temperature, when the engine is hot it will be so tight that it cannot move.

The solution to this problem is to make the piston smaller to give it a loose fit at room temperature. Then, when it warms up, it will fit perfectly. But a loose piston rattles when it is cold, and it takes the piston several minutes to warm up and expand.

Several methods of controlling piston expansion have been tried. One of the first experiments was the split skirt, figure 8-30. By cutting a slot in the skirt, the piston could expand and fill the slot. This increased the pressure on the cylinder wall to some extent, but not enough to stop the engine.

Unfortunately, split skirts weaken the pistons, and sometimes cause them to collapse. A collapsed piston is one which has a shrunken skirt. The skirt shrinks enough to cause **piston slap**, and the piston has to be replaced.

T-slot pistons, figure 8-31, are another method of controlling expansion. The top of the "T" prevents heat transfer from the dome to the skirt. The vertical line of the "T" acted just like a split skirt.

Further design and experimentation showed that piston expansion could be controlled without splitting the skirt. First of all, the skirt is narrowed so it only touches the cylinder wall at right angles to the pin. Next, a slot is cut between the head of the piston and the skirt, figure 8-32. The slot prevents heat transfer from the very hot head. If the skirt does not get as hot, it does not expand as much.

In addition, pistons were cam ground. Instead of being ground in a perfectly round shape, they were ground in an ellipse, figure 8-33. The skirt is about 0.010" (0.25 mm) less in diameter at a 45-degree angle to the pin than it was at a 90-degree angle. Also, the skirt is tapered. The top of the skirt is slightly smaller than the bottom because the top is closer to the flame. Because of its proximity to the flame, the top of the piston skirt will get hotter and expand more.

Crankshafts, Flywheels, Vibration Dampers, Pistons, and Rods

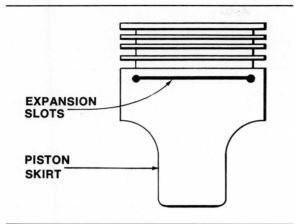

Figure 8-32. By narrowing the skirt and cutting a slot between the piston head and the skirt, expansion can be controlled.

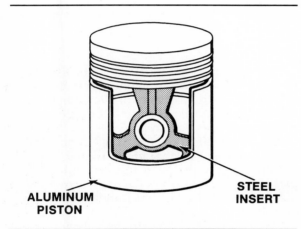

Figure 8-34. Steel piston insert.

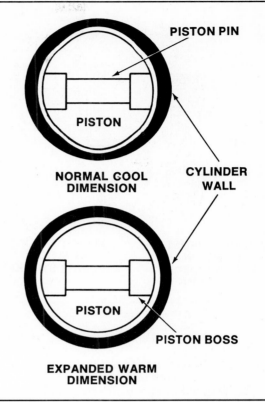

Figure 8-33. Cam ground pistons have an elliptical shape.

Some pistons are cast with struts or belts of steel inside the aluminum to control expansion, figure 8-34. A strut surrounds the pin hole in each side of the piston and prevents it from expanding at the skirt faces. Instead, expansion takes place along the pin.

Piston Speed

During engine operation, **piston speed** is a prime factor in the life of the piston.

Piston speed is the average velocity of the piston at a given rpm. It is usually expressed in feet per minute (meters per minute). For example, an engine with a stroke of three inches (76 mm) will move one of its pistons three inches (76 mm) down and three inches (76 mm) up on every revolution. Therefore piston movement on that engine would be six inches (152 mm) during each engine revolution. During ten revolutions, the total movement would be 60 inches (1524 mm) or five feet (1.5 m). If it took one minute for the engine to turn the ten revolutions, then the piston speed would be five feet (1.5 m) per minute.

At 3,000 rpm, that same engine would move one piston 18,000 inches (457,200 mm), or 1500 feet (457 m) each minute.

As the stroke gets longer, the piston travels farther, and piston speed increases. Piston speeds of 3000 feet (914 m) per minute and higher are common in racing engines, but the piston speed of a passenger car stays around 1500 feet (457 m) per minute.

When the first automobiles were made with their heavy pistons, the bearings could not withstand much pressure and lubrication was marginal. It was considered quite an accomplishment to design an engine that would run for a

---

Piston Slap: A noise caused by the piston skirt hitting against the cylinder wall.

Piston Speed: The average speed of the piston, in feet or meters per minute, at a specified engine rpm.

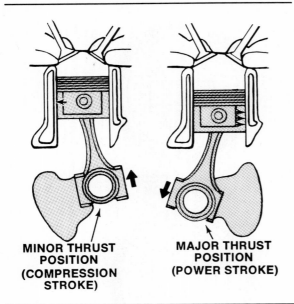

Figure 8-35. The major thrust face receives the thrust of the power stroke.

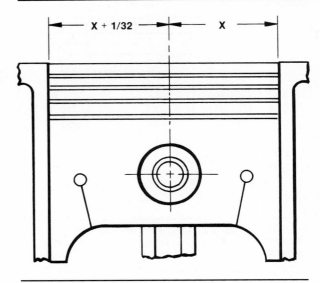

Figure 8-36. The piston pin is offset to prevent piston slap.

reasonable length of time at a piston speed of 1500 feet (457 m) per minute.

Piston speed can be misleading because it is only the average piston velocity; it does not indicate maximum piston speed which can be much greater than the average speed. There is also the acceleration of the piston to consider. Every time the piston gets to top or bottom dead center, it stops. Immediately afterwards, it accelerates to full speed. If the engine is going fast enough, this sudden acceleration will pull the piston pin right out of the piston. This is one reason why overrevving an engine can be so damaging.

Piston Offset

Thrust forces on a piston are caused by the angle of the connecting rod. When the piston is pushed down on the power stroke, the angle of the rod forces the piston against one side of the cylinder. The power stroke generates much more force on the piston than the compression stroke, so the side of the piston that receives the thrust on the power stroke is called the **major thrust face**, figure 8-35.

At top dead center, between the compression and power strokes, the thrust on the piston changes from the minor to the major thrust face or side. If there is any clearance between the piston and the cylinder, the piston will slap against the wall. This makes noise and can damage the piston. To prevent slap, the piston **pin may be offset up to 1/16" (1.6 mm) toward the major thrust side, figure 8-36**.

Offsetting the pin causes the piston skirt to tilt as the thrust shifts from one side to the other. With a slight amount of tilt, the bottom of the skirt touches first against the cylinder wall, which eliminates the slapping.

Pins can be offset to either the left or right, as viewed from the front of the engine, but are usually offset to the right. Pistons with offset pins are marked with a notch or an arrow to indicate the position in which they must be installed in the engine.

When the offset is toward the right side of the piston, it increases the angularity of the rod. This puts more thrust pressure on the piston and reduces engine torque output. Even so, designers favor offsetting to the right. It results in the piston tipping so the bottom of the skirt presses against the cylinder wall first. If the pin was offset to the left, the top of the skirt would hit first and the piston might slap anyway. The top of the skirt is in line with the pin and receives the full force of the thrust. When the bottom of the skirt touches first, it flexes slightly and allows the top to come in slowly without slapping.

Piston Crown Shape

Pistons frequently have crowns that protrude into the combustion chamber, figure 8-37. These protrusions promote swirling of the air-fuel mixture and increase burning efficiency. A recent example of this is the 1981 Ford Escort piston, figure 8-38.

Pistons sometimes have valve clearance pockets cut into the crown. If only the edge of the valve comes close to the piston, then the piston

Crankshafts, Flywheels, Vibration Dampers, Pistons, and Rods

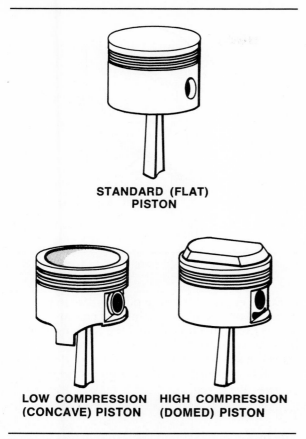

Figure 8-37. Some pistons have crowns that protrude into the combustion chamber.

Figure 8-38. The Ford Escort piston. (Ford)

Figure 8-39. In a dished piston, most of the combustion chamber is often in the piston crown. (Pontiac)

clearance or **eyebrow** will be just a small, scooped out area.

Sometimes the crown of the piston is shaped like a bowl or dished. Most of the combustion chamber is actually in the bowl, figure 8-39. Dishing pistons is one way of lowering compression to control oxides of nitrogen emissions.

Piston Clearance

Pistons are fitted to cylinders with clearances from 0.001″ (0.025 mm) to as much as 0.010″ (0.25 mm). The measurement for the diameter of the piston is taken at a specific place on the skirt. The exact place is recommended by the piston manufacturer or in the factory shop manual, figure 8-40.

When rebuilding an engine, the clearance is established by boring or honing the cylinder to fit the piston being used. When engines are assembled at the factory, several different sizes of pistons are available. The pistons are selected to fit each individual bore, which eliminates having to change the size of the cylinder bore.

PISTON PINS

Piston pins are sometimes called **wristpins** because they have somewhat the same action as the human wrist. Piston pins are usually hollow to reduce their weight, figure 8-41. They con-

Major Thrust Face: The side of the piston that pushes against the cylinder wall during the power stroke.

Piston Eyebrow: Small pockets cut into the top of pistons to provide clearance for the valve.

Wristpin: The hollow metal rod that holds the piston to the connecting rod; also called a piston pin.

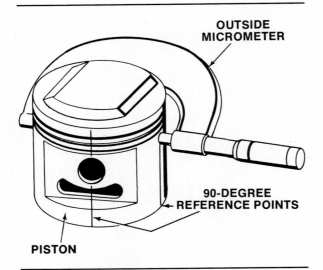

Figure 8-40. The measurement for the diameter of the piston skirt is taken at the largest diameter or in a direction perpendicular to the pin bosses.

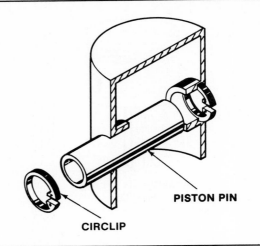

Figure 8-41. Hollow piston pin.

nect the piston to the connecting rod but allow the rod to rotate as the crankshaft turns.

Pins are made from steel. They must be strong to withstand the full force of the piston pushing on the rod. The surface of the pin has a high polish. It runs freely on the aluminum pin boss without scoring or gouging.

Through the years, piston pins have been secured in many different ways. The modern way, used in most engines today, is to press fit the pin into the rod. The pin is free to turn in the piston, but fits so tightly in the rod that it cannot shift and hit the cylinder wall.

Scoring of the cylinder wall from loose pins was a problem before the adoption of press-fit pins. When small circlips are used at the pin ends, they sometimes come loose. There have been pistons with screws that held the pin rigid in the piston, but the pin was free to turn in the end of the rod, figure 8-24, type E.

The steel pin fits directly into the aluminum pin boss. When the pin is made to turn in the rod, a bronze bushing must be used in the rod. Pins that can move in both the rod and the piston are called full-floating pins.

Lubrication

Pistons and piston pins usually get their lubrication from oil splash. Oil is splashed on the cylinder wall by the crankshaft. This lubricates the pistons and rings. The oil ring scrapes this oil from the cylinder wall. From there it drains through holes drilled in the top of the pin boss to lubricate the pin. The end of the pin next to the cylinder wall also gets oil because it is below the oil ring.

Some engines have full oil pressure to the pins. Oil from the connecting rod bearings comes through a hole drilled the full length of the connecting rod. This full pressure oil lubricates the pin where it turns in the rod; it does not lubricate the pin in the piston except by splash.

PISTON RINGS

The piston in the cylinder is sandwiched between combustion and compression above and splashing oil below. Piston rings are used to keep combustion gases up and oil down. If combustion gases are allowed to leak into the crankcase, the engine loses power. The force of the expanding, burning air-fuel mixture is wasted if it escapes alongside the piston instead of pushing on the piston. If oil is allowed to get into the combustion chamber it will oxidize and turn to carbon from the heat. The carbon can obstruct the intake and exhaust gas flow and raise compression, which causes detonation. Glowing particles of carbon can cause pre-ignition of the air-fuel mixture.

As described earlier in this chapter, there are two types of piston rings: compression rings are designed to keep gas pressure above the piston from escaping and oil control rings are designed to scrape oil from the cylinder wall and return it to the crankcase, figure 8-42.

Pistons on modern engines normally have three rings, figue 8-43, all above the pin. It was common in older engines to have an oil ring below the pin or at the bottom of the piston skirt. In modern ring design, it is not necessary to have more than three rings in passenger car engines. The additional rings do not seal significantly better and they increase friction, which consumes horsepower.

Crankshafts, Flywheels, Vibration Dampers, Pistons, and Rods

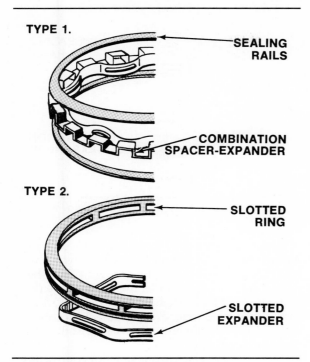

Figure 8-42. Two types of oil control rings.

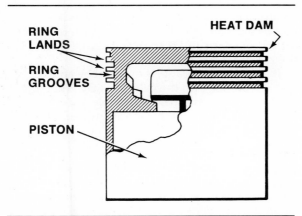

Figure 8-44. The ring lands support the bottom and top of the ring.

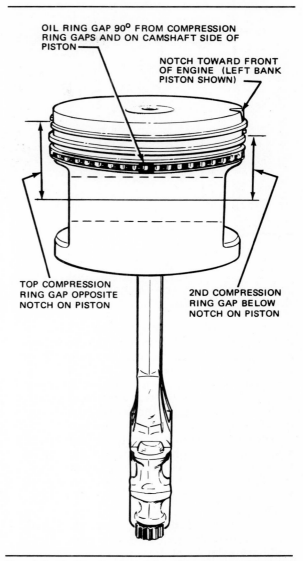

Figure 8-43. The top two rings of the three used on the piston are the compression rings.

Friction is actually the major problem with piston rings. It is possible to make a ring that seals so completely that it will not allow any gases or oil to pass. However, such a ring would fit so tightly against the cylinder bore that the piston could not be moved. Modern rings are a compromise. They leak small amounts of combustion gas into the crankcase, and they allow some oil to get into the combustion chamber. But as long as the leakage is below the acceptable limit, the ring is considered good.

How Rings Operate

Rings are mounted in grooves on the piston. Between the grooves are the lands, which support the bottom and top of the ring, figure 8-44. A ring is made so that when it is inserted in the piston, and the piston is inserted into the cylinder, tension forces the ring against the wall of the cylinder. On a compression ring, the tension does not do all the sealing. When the tremendous pressure of the burning mixture passes alongside the top of the piston and hits the ring, it increases the ring pressure by getting into the space behind the ring and pushing it against the cylinder wall, figure 8-45. With the extra pressure created by combustion, the ring makes a very good seal. At other times during

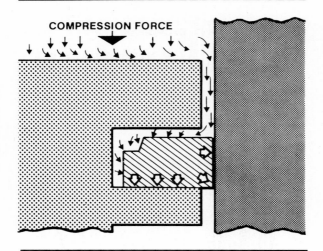

Figure 8-45. Combustion chamber pressure forces the ring against the cylinder wall and bottom side of the groove. (Perfect Circle, Dana Corp.)

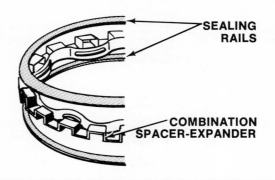

Figure 8-46. The three parts of the oil control ring.

the piston stroke, the ring does not seal as tightly, which reduces friction. But slight leakage on other than the power stroke does very little harm.

The second compression ring on the piston acts as a backup for the top compression ring. It seals most of the gases that manage to escape past the top ring.

The compression ring ends are separated by a small gap. Although the gap does allow some leakage, it is necessary to allow for expansion of the ring. Without the gap, the ring ends would butt together and buckle the ring when the engine got hot.

The third ring is the oil control ring, figure 8-46. It is usually made in three pieces. The top and bottom layers of the ring are steel rails and between them is an expander. The expander keeps the rails separated and pushes them out against the cylinder wall. The expander is perforated and slotted. Any oil that is scraped off the wall by the top rail, goes between the rails, through the expander, and back into the crankcase. Most of the oil is scraped off by the bottom rail. The top rail only removes the oil that the bottom rail cannot handle.

The rails both have square cut ends with a gap between them; this is to allow for expansion as the engine heats up. The expander between the rails does not touch the cylinder wall. In most cases it does not touch the back of the piston groove either. In fact, none of the rings are supposed to touch the back of the piston grooves. This allows them to ride over any imperfections in the cylinder wall without affecting the piston. In this way, the rings can follow the cylinder wall for the best seal.

Rings operate under extremely difficult conditions. When the piston is moving down, the ring is against the top of the groove. When the piston reverses direction at the bottom, the ring is slammed against the bottom of the groove. This occurs twice every revolution. This constant hammering eventually wears out the ring grooves in the piston.

The rings are also responsible for conducting some of the heat from the piston, figure 8-47. The source of heat is on top of the piston. As the heat moves down, it contacts the rings and moves into the cylinder wall. From there it goes into the coolant and is transferred to the air at the radiator.

If a ring is too heavy, it may not be able to stay in contact with the cylinder wall when the piston reverses at top of bottom dead center. The ring actually vibrates in the groove. This is called **ring flutter**.

Different types of rings

Rings are made in dozens of different shapes. All of them have been designed to do a better job than previous designs for a particular application. Some rings have a tapered face with the sharp point at the bottom. The contact area with the cylinder wall is very small, and the ring seats quickly. The pointed edge also scrapes off any oil on the cylinder wall that leaks past the oil ring.

The seating of rings means that they have worn so they fit snugly. If the wall is smooth and the ring fits closely, very little material has to be worn away.

Some rings are designed to twist when they touch the cylinder wall. The face is square, but the twisting rocks the edge of the face into contact with the wall. Twist rings have a bevel on the inside upper corner, figure 8-49. The bevel is sometimes in the form of a groove. Putting the bevel or groove on the outside of the ring reduces the area of the ring face. This

Crankshafts, Flywheels, Vibration Dampers, Pistons, and Rods

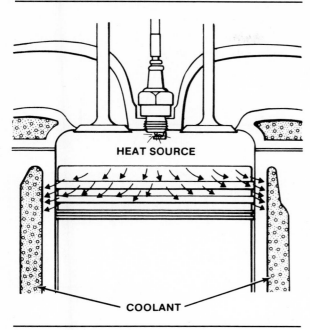

Figure 8-47. The rings conduct heat from the piston.

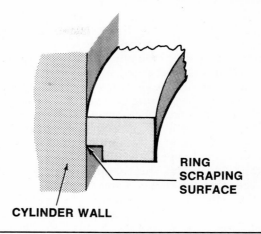

Figure 8-48. The sharp edge of the ring scrapes oil off the cylinder wall.

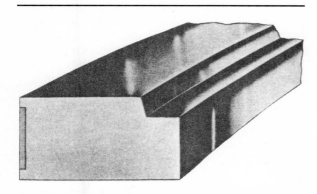

Figure 8-49. Twist rings have a bevel on the upper inside corner. (Perfect Circle, Dana Corp.)

increases the pressure against the cylinder wall. The ring also twists to make the lower edge dig into the wall, figure 8-48.

Twist and taper face rings can be combined. The sharp edge at the bottom of the ring is forced into the wall by the twisting action.

Materials and Construction

Rings are made from special cast iron called ductile-cast iron or nodular iron because of the way it is heated and cooled. Cast iron is not normally flexible, but when made with a small cross section such as in a piston ring, it flexes easily. However, it will break easily if not handled with care. Chrome-faced, molybdenum oxide-faced, and stainless steel rings are also available.

Cast iron rings are not used as original equipment on cars because they have a relatively short service life compared to chrome-faced rings. They frequently are used during engine overhaul because of their ability to seat quickly and easily. A ring is seated when the surface of the ring wears until it conforms to the shape of the cylinder bore to form a good seal. If the cylinders are not bored oversize during the overhaul, it is possible to deglaze the cylinders and install cast iron rings.

Chrome rings are used because of their wearing ability and resistance to abrasion from dirt. However, the cylinder bore must be honed to the proper fit and finish or the rings will not seat properly. A new process has recently been developed to help chrome rings seat more easily. This consists of reversing the electro-plating polarity during the last few moments of the plating process. This causes the outer layer of the chrome to become porous which allows the ring to retain oil as well as seat more readily during the break in process.

Molybdenum oxide or "moly" rings, as they are often called, are extremely resistant to scuffing. They work very well under peak loads when there are hot spots on the cylinder walls or when there is minimum lubrication. The disadvantage of moly rings is their short life compared to chrome.

Ring Flutter: Rapid up and down movement of a piston ring that breaks the seal between the ring and the cylinder wall.

Stainless steel rings are not used in production engines because of their expense. They are designed to work under extreme temperature and load conditions usually found only in supercharged and turbocharged engines.

The different types of rings described are used only in the top ring groove. The second ring is always made of cast iron because it does not have to function in as severe an environment as the top ring. Most piston rings have a sprayed-on phosphate coating on the top and bottom of the ring. This coating absorbs oil and ensures that the ring will wear into its groove without sticking. When a ring sticks in the groove, it cannot expand against the cylinder wall. This increases blowby or oil consumption. Blowby refers to the combustion gases that blow past the rings. Although every engine has some blowby, too much will increase crankcase pressure and force oil into the air cleaner or out the crankcase breather.

Lubrication

Rings are lubricated by splashed oil from the crankshaft. If the rings seal perfectly, the middle ring does not get any oil. But all the rings leak enough so that they are well lubricated.

However, dry starts with cold oil do present a lubrication problem. The engine must run a few minutes before the oil is warm enough to be thrown onto the cylinder wall in any quantity. To avoid scuffing pistons and rings, an engine should be allowed to warm up at a speed fast enough to allow oil to be thrown from the crankshaft.

CRANKSHAFT, ROD, PISTON, AND PIN RELATIONSHIP

The piston is halfway through its stroke before the crankshaft gets to the 90-degree position. At 90 degrees, the piston has completed 55 to 57 percent of its stroke. If you analyze the drawings, figure 8-50, you will see that the connecting rod is at a different angle when leaving top dead center than when leaving bottom dead center. It is this angle that causes the piston to travel farther in the first half of the crankshaft stroke than in the second.

Because the piston travels farther in the top half of the stroke, it must move faster. It is moving fastest when the crankshaft throw is at an angle of 90 degrees to the rod. This position occurs when the crankshaft is about 80 degrees from top dead center. From that point on, the piston slows down until it reaches the end of its stroke.

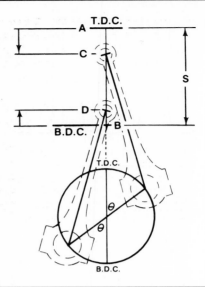

Figure 8-50. The connecting rod is at a different angle when leaving top dead center than when leaving bottom dead center.

Although the piston speed is comparatively low, the maximum velocity of the piston is very high. In an engine with a 4-inch (101.5 mm) stroke and an 8-inch (203 mm) rod, running at 4000 rpm, the piston moves eight inches (203 mm) with each revolution. At 4000 rpm, this amounts to a piston speed of 2666 feet (810 m) per minute. But at the point of maximum velocity, the piston is actually moving at a rate of 4200 feet (1280 m) per minute.

The acceleration must also be considered. The piston does not gradually increase to that velocity. It reaches it with only 80 degrees of crankshaft movement. At 4000 rpm the crankshaft takes less than 0.004 of a second to move that far. The inertia force on the piston during acceleration is over 1000 times its weight, which can cause bearings to break up and pistons to crack.

The acceleration of the piston is directly affected by the length of the connecting rod. A short rod increases acceleration. A short rod also increases the angle of the rod when the crankshaft throw is halfway through its stroke. This reduces the torque output of the engine. The ideal relationship seems to be about two to one, with the rod twice as long as the stroke.

Another factor in the geometry of rod angles is the offset of the piston pin. They are usually offset to the major thrust side, which is the right side of the engine on engines that rotate clockwise. But the rod angle will be decreased and more torque will be produced if the offset is to the left.

Crankshafts, Flywheels, Vibration Dampers, Pistons, and Rods

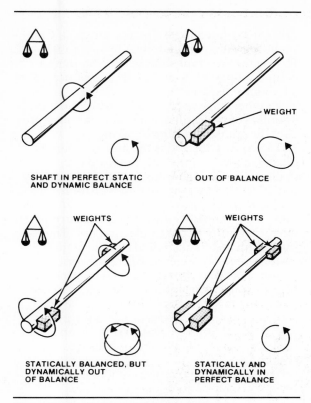

Figure 8-51. Static and dynamic balance.

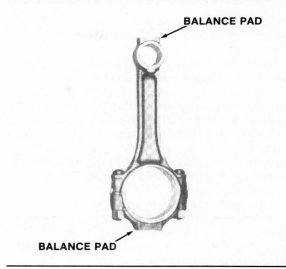

Figure 8-52. There are balance pads at each end of the connecting rod. (Sunnen)

The offset of the crankshaft is also a consideration. To reduce the rod angle, the crankshaft must be offset to the right side. In an inline engine this is a straightforward measurement, but in a V-8 engine, moving the crankshaft to one side makes the pistons on that side rise a little higher, while those on the other side do not rise quite as high. This is the reason that most V-8 engines today do not have crankshaft offset.

ENGINE BALANCE

Dynamic unbalance exists in a rotating part when the weight is not equally distributed around the pivot. An automotive part such as a crankshaft can also be in static unbalance.

To run smoothly, a part must be in dynamic and static balance. Any long part that rotates, such as a crankshaft or a driveshaft, can be in perfect static balance, but not in dynamic balance.

To illustrate, let us imagine a long shaft that is in perfect balance. Add a small weight to one side of the shaft at one end. Add another weight to the other side of the other end, exactly opposite the first weight. If both weights are equal, the shaft will still be in static balance. But dynamically it will be badly out of balance, because the two weights are not directly opposite each other, figure 8-51. Each weight tries to swing out as it rotates with the shaft. This makes the shaft badly out of balance, and causes vibration that can be felt, or seen.

A crankshaft is balanced by counterweights. Extremely heavy weights are cast onto the shaft to give it both static and dynamic balance. The lower part of the connecting rod and the bearings rotate with the crankshaft throw. To balance these added weights, the crankshaft counterweights must be heavier than if all they had to balance was the weight of the throw itself.

The reciprocating parts of the engine must also be balanced. Crankshafts are usually made so that when one piston is at top dead center, another piston is at bottom dead center. The two pistons then balance each other. But if one piston is heavier than the other, it will cause an imbalance. To avoid this, pistons are made so that they weigh within a few grams of each other.

New engines are balanced at the factory closely enough to prevent vibration, but that is all. Custom engine builders provide engine balancing service that is much closer to perfect balance. Each piston is weighed. If any are heavier than others, they are trimmed to make them equal within a fraction of a gram. Weight is usually removed from areas under the piston crown or near the pin bosses from which metal can be removed without weakening the piston.

Connecting rods are made with a balance pad at the pin end and another pad on the cap, figure 8-52. Ordinary balancing is done by making all the rods the same weight by grinding the pads. A better job is done with a double scale, which measures the weight of each end of the rod. The pads are then ground so that each rod weighs the same and each end weighs the same.

The weight of the rings and bearings is also taken into account when everything is balanced. Bob weights, equal to one-half of the reciprocating weight, are attached to the crankshaft for balancing. Then the shaft is spun in a balancer that detects unbalance and its location. If any counterweight is heavy, it is drilled to remove metal. If light, a hole is drilled and some heavy metal is inserted in the hole. The hole is then welded shut.

Flywheels and clutches are also balanced this way. When finished, a fully balanced engine runs smoothly. However, such precision balancing is usually done only on racing engines which operate at high rpm.

SUMMARY

The job of the crankshaft is to convert reciprocating motion to rotary motion. The crank is supported by the main bearings and is counterweighted to balance the weight of the rods and pistons. The spacing of throws on the crankshaft is determined by the number of cylinders, their arrangement, and whether the firing intervals are designed to be equal or unequal. Crankshafts are either forged steel or cast iron. To resist wear, crankshafts can be hardened, treated, or plated. Crankshafts are lubricated by oil under pressure.

Flywheels are used to smooth out the firing impulses of the engine. Vibration dampers prevent the crankshaft from breaking.

The connecting rod is the link between the piston and the crankshaft. Rods are usually made of steel and the length of the connecting rod affects the movement of the piston.

The modern piston is made from aluminum. The piston transmits the force of the burning air-fuel mixture to the connecting rod. The rod pushes on the crankshaft to run the engine. Piston speed is a function of engine rpm. The piston speed has a great deal to do with how long the piston lasts. Most passenger cars use pistons with three rings. The top two are designed to keep gas pressure above the piston. The third ring, the oil control ring, scrapes oil from the cylinder walls and returns it to the crankcase.

To run smoothly, any automotive part must be in both static and dynamic balance.

Engine Bearings

Figure 9-7. The ability to change shape slightly with the shape of the crankshaft is called conformability. (Federal-Mogul)

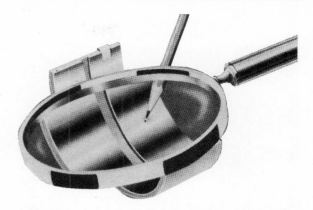

Figure 9-8. Embedability allows a particle to sink into the bearing material. (Federal-Mogul)

Figure 9-9. Many modern bearings are made in layers. (Federal-Mogul)

backing shell so it can be used as a replaceable insert.

Bearing Materials Development

Babbitt, a mixture of tin and lead, was the original bearing material. It was invented in 1839 by Isaac Babbitt. It was called tin base babbitt because 89 percent of the material was tin. As we have already learned there is also a lead-based babbitt.

When babbitt was invented there were no automobiles. The babbitt was used in stationary power plants, steam engines, and many other places where shafts turned in bearings.

The lead-based babbitt was developed next and used by railroads for wheel bearings. From this use, lead-based babbitt was developed for automotive engines. It was used for many years in connecting rods and main bearings, either poured in place or on a steel-backed insert.

The latest bearings are made of copper alloys, aluminum alloys or multiple layers of alloys and metals.

Most modern bearings consist of three, four, or even five layers of materials, figure 9-9. The back may be steel with a copper, tin, or lead alloy lining. There may be a barrier plate between the lining and the overplate to prevent the overplate from reacting chemically with the lining material. To prevent corrosion on the shelf or in the engine, the entire bearing is plated with a thin coating of tin, figure 9-9.

Each bearing material has its own uses and advantages.

Tin base babbitt

This material is regarded as a good, all-around bearing material. It operates well under such handicaps as poor lubrication and misalignment.

Lead base babbitt

Good conformability, load-carrying ability and embedability are the strong points of this bearing material.

Copper alloys

Bearings with copper alloy linings came into use to accomodate the increased speed and load of the automobile. This lining has good fatigue resistance and fair embedability and conformability.

Multi-layer bearings

In a multi-layer bearing, the overlay is usually put on top of a strong lining to resist high

---

Compatibility: The ability of a bearing lining to allow friction without excessive wear.

Conformability: The ability of a bearing lining to conform to irregularities in a bearing journal.

Embedability: The ability to absorb particles into the lining of the bearing.

Figure 9-10. The web or saddle supports the main bearings.

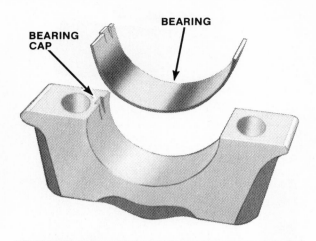

Figure 9-12. The bearing cap holds the bearing in place. (Federal-Mogul)

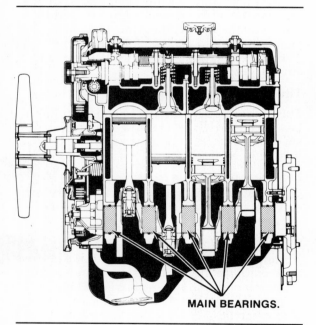

Figure 9-11. Modern 4-cylinder engines usually have five main bearings. (Chrysler)

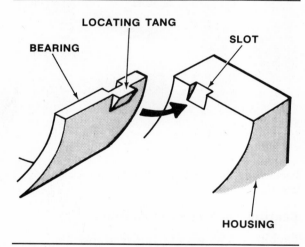

Figure 9-13. The tang and slot prevents the bearing from spinning in its bore.

loading. These are often used for heavy-duty service. Some of the combinations are:
  Steel back with bronze inner layer, and lead or tin base babbitt.
  Steel back with copper nickel inner layer and lead alloy overlay.
  Steel back with a copper alloy inner layer, a barrier plate, a lead alloy overplate and flash tin over the entire bearing.
  Steel back with an aluminum alloy lining, a lead alloy overplate, and a flash tin plate over all.
  Steel back with an aluminum alloy lining and a flash tin plate over the entire bearing.

Aluminum alloy bearing

The aluminum alloy bearing is a combination of tin, copper, nickel and aluminum and is cast as one piece. The finished bearing is covered with a thin plate of tin to help during the break-in period. This bearing is made with a heavy wall because of the tendency of aluminum to expand at high temperatures. You cannot use this bearing as a replacement for a thin wall bearing. Aluminum bearings are sometimes used with hardened crankshafts for extended life.

Engine Bearings

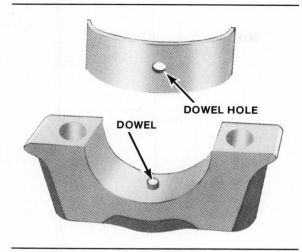

Figure 9-14. In some cases a dowel in the bore holds the bearing in place. (Federal-Mogul)

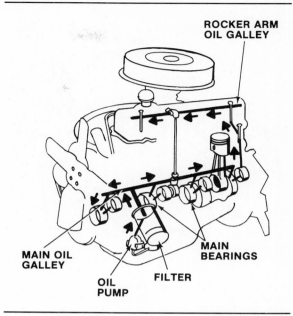

Figure 9-15. The main bearings receive oil through this lubrication system.

BEARING CONSTRUCTION AND INSTALLATION

There are many ways to construct crankshaft main and connecting rod bearings. The main bearings must have support. This support is supplied by **webs** or **bearing saddles** cast into the block or crankcase, figure 9-10. Ideally these webs should be alongside each cylinder or pair of cylinders. Opposite V-8 cylinders are so close together that their connecting rods are on the same crankpin. There is no room to put a main bearing on both sides of each cylinder. So the main bearings are located alongside each pair of the four pairs of cylinders in a V-8. This gives a total of five main bearings.

On 6-cylinder inline engines, the ideal number of main bearings is seven. This puts a main bearing alongside each connecting rod bearing. Some older 6-cylinder inline engines have only four main bearings.

In the inline 4-cylinder engines of today, you will generally find five main bearings, figure 9-11.

Crankshaft and connecting rod bearings are held in place by the bearing cap, figure 9-12. To install the bearings, half of the bearing is inserted in the web and the other half in the cap. The bearing is then lubricated and the crankshaft laid in the web. Finally the main bearing caps are placed over the shaft journals and the bolts inserted and tightened. Connecting rod bearings are put into the rod and cap. The rod is then placed on the crank journal, and the cap slipped onto the rod and tightened.

When the nuts or bolts are tightened, the bearing is firmly against the supporting metal. But the bearing must be prevented from spinning in its **bore** or shifting sideways. On most bearings this is done by a tang and slot, figure 9-13. Each bearing half has a small tang at one end which fits into a slot in the bore. Because the tangs are offset from each other, the edge of the tang bears against the lip of the other half of the bore. This prevents the bearing from rotating in its bore. Lateral motion of the bearing is also prevented by the tang.

Some engines have a locating dowel in the bearing bore, figure 9-14. The dowel fits into a hole in the bearing shell and prevents movement in any direction.

Lubrication

Every bearing gets its lubrication through holes in the crankshaft or crankcase. The main bearings receive oil directly from the oil pump, figure 9-15. The oil enters the bearing through a hole, usually in the center of the upper half of

---

Bearing Saddle or Web: The machined area which supports the connecting rod or main bearing insert.

Bearing Bore: The full circle machined surface that supports the back of the bearing. It may consist of two saddles bolted together, as in rod or main bearings, or a bored hole, as in most camshafts.

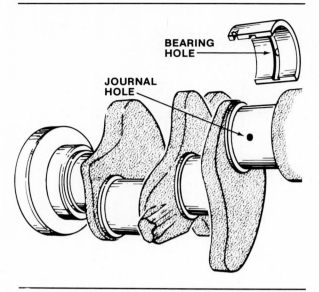

Figure 9-16. When the hole in the main bearing aligns with the hole in the journal, oil passes through. (Federal-Mogul)

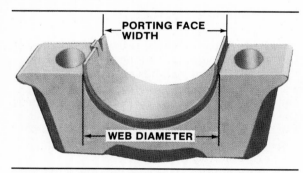

Figure 9-18. Bearing spread allows the bearing to snap into place. (Federal-Mogul)

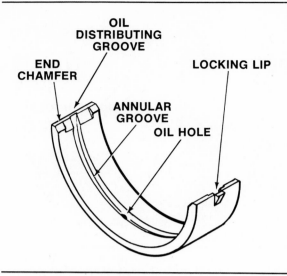

Figure 9-17. In many bearings there is a groove down the middle for oil to flow around the main journal.

the bearing. A matching hole in the center of the main journal allows oil to go through a drilled passage inside the crankshaft to the connecting rod journals. If there is one hole in the main bearing and one in the journal, the holes will align only once in a revolution, figure 9-16. Some journals are cross drilled, so the holes align twice every revolution. But this still shuts off the oil supply when the holes do not align. For this reason, many bearing inserts have a groove down the center of the bearing that allows the oil to flow around the main journal and maintains pressure on the rod journal hole at all times, figure 9-17. In some designs the groove is only in the top half of the insert. With a cross-drilled journal, a half circle groove will still apply pressure at all times. Any groove, however, still takes away from the bearing area.

Connecting rod bearings are usually not as wide as main bearings. They do not need grooves to distribute oil. A hole in the journal is enough to keep the bearings supplied with oil. Some crankshafts have cross drilled rod journals so the rod bearing receives oil at two points.

Bearing Measurements

The distance that the parting face of a bearing is greater than the diameter of its web is called the **bearing spread**, figure 9-18. Making the bearing spread dimension from 0.005" to 0.030" (0.125 to 0.750 mm) ensures that it will be tight against its bore. The bearing snaps into place when inserted and will not fall out during handling.

The distance that the bearing is higher than the web is called **bearing crush**, figure 9-19. When the rod or main bearing cap is tightened, the bearing is actually "crushed" into the bore slightly. The crush makes sure the bearing is tight in its housing. Bearings without enough crush show signs of movement in the web when the engine is torn down. Movement between the bearing and its bore will cause wear and looseness.

OIL CLEARANCE

The gap or space between a bearing and its shaft or journal is called the **oil clearance**, figure 9-20. A certain amount of space between the bearing and the shaft must be allowed for oil. A bearing too close to the shaft won't have sufficient oil clearance and will not last as long as one that has the correct amount of oil clearance.

Oil clearance depends on the type of bearing material and the shaft size. For example, a two-

Engine Bearings

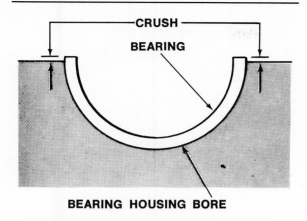

Figure 9-19. Bearing crush is the distance that the bearing is higher than the web. (Federal-Mogul)

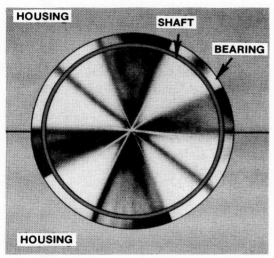

Figure 9-20. Oil clearance is the gap or space between the bearing and the shaft. (Federal-Mogul)

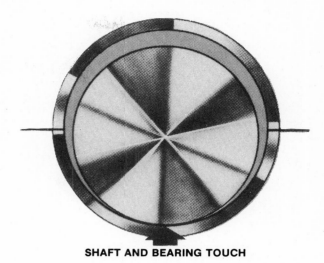

Figure 9-21. When the engine is stopped for awhile, the shaft and bearings touch. (Federal-Mogul)

inch shaft using babbitt bearings may require 0.001″ (0.025 mm) of oil clearance, while the same shaft using copper alloy bearings may require 0.002″ (0.050 mm) of oil clearance. A multi-layered copper alloy bearing may require only 0.001″ (0.025 mm) of oil clearance.

Bearing clearance is the same as oil clearance. It is usually specified as a range. For example, rod bearing clearance may be specified as .0015″ to .003″ (.035 mm to .075 mm), with a maximum clearance in service of .0035″ (.90 mm).

Bearing clearance is determined by engine designers according to the diameter of the shaft and the bearing material. The greater the diameter of the shaft, the more bearing clearance it requires. Specifications usually call for more main bearing clearance than rod bearing clearance. The reason is that main bearings are larger than rod bearings, and therefore need additional clearance.

When an engine is running, the oil film keeps the shaft centered in the bearing except during heavy load conditions. When the piston exerts maximum force on the rod, the rod bearing and shaft may touch momentarily. The bearings and shaft also touch when the engine is not running, figure 9-21.

When the engine is not under heavy load or is not at rest, there is a continuous oil film surrounding the shaft. This oil film is not formed by the pressure of the oil. The turning shaft actually rolls up on a wedge of oil, figure 9-22. It does this almost instantly whenever the engine is started. However, cold starting will cause some wear because the bearing and shaft are touching at the first instant of movement, figure 9-21.

Bearing Spread: The distance that the parting face of a bearing is wider than the diameter of its saddle or web.

Bearing Crush: The distance that a bearing is higher than its saddle or web.

Oil Clearance or Bearing Clearance: The gap or space between a bearing and its shaft that fills with oil when the engine is running.

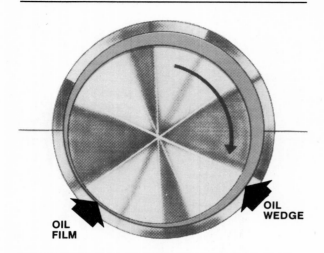

Figure 9-22. The shaft rolls up on a wedge of oil. (Federal-Mogul)

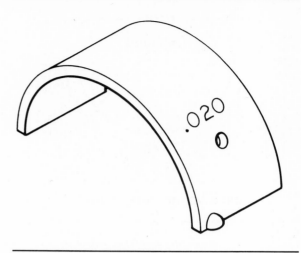

Figure 9-23. The bearing may be marked for oversize or undersize.

The amount of oil pressure, the bearing clearance, and bearing life are related. If the bearing clearance is excessive, oil will pass through the bearing more quickly, and oil pressure will be reduced. Oil throwoff from the spinning crankshaft is also increased when the clearance is excessive. This results in rings overloaded with oil and high oil consumption. Extremely high oil pressure will abrade the bearing, and wash the bearing material away.

UNDERSIZE AND OVERSIZE BEARINGS

On a modern engine, when the bearings and shaft wear, the bearings are thrown away and replaced with new ones. Bearings are manufactured .010″ (.25 mm), .020″ (.50 mm), and .030″ (.75 mm) under the standard size. An **undersize bearing** is really thicker than a standard bearing. The name undersize is not a description of the bearing, but of the shaft it is made to fit.

When a crankshaft is worn, it is removed from the engine and ground to a standard undersize on a crank grinding machine. If the surface of the journal is still rough, out-of-round or tapered, it is ground to the next available undersize. After the crank is reground, it is marked for the size bearing it requires. If the correct undersize bearings are used, the crank will fit and operate efficiently.

Sometimes a problem occurs with the main bearing bore when it is machined at the factory. To correct this, the factory may bore the web one bearing size oversize. This allows them to use a block which would otherwise have to be scrapped.

The **oversize bearing** will be the standard size on the inside, but the outside will be larger. This makes no difference to engine operation. But when the bearings are changed, the engine rebuilder must be careful to spot the oversize or undersize bearing. In most cases, the bearing is marked for undersize or oversize, figure 9-23. But do not take the markings at face value; measure for yourself to determine the bearing size.

Oversize bearings are not normally made for connecting rods. Rod caps can be ground and the rod remachined to accept a standard size bearing.

Camshaft Bearings

The camshaft is supported by several bearings. The load on these bearings comes from the pressure generated by the valve springs every time a valve opens and closes. This pressure is fairly constant because of the consistent motion of the valves.

Camshaft bearings normally last at least as long as the crankshaft bearings. In some cases, a set of cam bearings will outlast two sets of main bearings. Because camshaft bearings seldom cause trouble, they are sometimes overlooked when an engine is torn down for repairs.

Remember, cam bearings can cause trouble if they are worn enough to allow too much oil flow. This is because the same oil passageway feeds both the cam bearings and the crankshaft main bearings. If the cam bearings bleed off excessive quantities of oil, the crankshaft may be starved for oil because of a drop in oil pressure. Cam bearings should always be replaced when rebuilding an engine.

Engine Bearings

Because the load on cam bearings is not as severe as the load on crankshaft bearings, the cam bearing material does not take as much load. Many cam bearings are made with a steel back and babbitt lining. In some engines, the camshaft runs directly on the metal of the block or head without any insert bearings.

SUMMARY

Bearings are used to reduce wear between rubbing or sliding parts. In most cases bearings need the assistance of a lubricant to keep wear to a minimum.

Anti-friction bearings use balls or rollers to act as a buffer between parts.

Friction bearings are named because friction is a result of the part sliding against the bearing. It is necessary for a friction bearing to have the following characteristics: Compatibility, conformability, embedability, ability to withstand loads, a resistance to acids formed in the engine, and heat conductivity.

Bearings can be made of various materials. One of the most popular bearing linings is babbitt. Babbitt is a mixture of soft metals. Babbitt can be either tin or lead based. Many modern engine bearings are constructed in layers to provide as many qualities as possible.

The bearing receives its lubrication through holes in the crankshaft. The space between the bearing and its shaft which fills with oil is called the oil clearance. Mechanics also refer to this area as the bearing clearance.

The bearing can be retained in its seat by tangs, dowels, or allowed to float.

In cases where the crankshaft is ground undersize or the web ground oversize, you can replace the bearings with undersize or oversize bearings. In both cases, the name undersize or oversize does not describe the bearing, but the crank journal or web.

■ What Is The Metric System?

The entire metric system is based on the meter. Originally, French scientists divided the distance between the North Pole and the equator into ten million equal sections. Each section was called a meter. Now, however, scientists have given the meter a new, more complex (but more accurate) definition: One meter equals the length of 1,650,763.73 wave lengths of the red-orange radiation line produced by an atom called Krypton 86!

The base unit can be divided or multiplied by 10, 100, or 1,000. The prefixes on the names of the units tell which factor (number) to use when multiplying or dividing. These are the most common prefixes, although there are others:

micro = 1/1,000,000	deka = 10
milli = 1/1,000	hecto = 100
centi = 1/100	kilo = 1,000
deci = 1/10	mega = 1,000,000

For example, 1 kilometer equals 1,000 meters, and 1 centimeter equals 1/100 (one one-hundredth) of a meter. Another example: 1,000 millimeters equals 100 centimeters equals 10 decimeters equals 1 meter. These prefixes apply to the other units as well, such as the gram (kilogram, centigram, milligram, etc.).

Camshaft Bearing: The bearings that support the camshaft in the engine block or cylinder head.

Undersize Bearing: A bearing made thicker with a smaller inside diameter to fit an undersize crankshaft.

Oversize Bearing: A bearing made thicker with a larger outside diameter to fit an oversize web.

Review Questions

Choose the single most correct answer.
Compare your answers to the correct answers on page 268.

1. There are two types of bearings, friction bearings and:
   a. Ball bearings
   b. Roller bearings
   c. Anti-friction bearings
   d. None of the above

2. A ball bearing is a/an _____ bearing.
   a. Anti-friction
   b. Friction
   c. Frictionless
   d. None of the above

3. Which of the terms below does NOT apply to engine bearings:
   a. Embedability
   b. Roundability
   c. Conformability
   d. Compatibility

4. The bearings that support the crankshaft in the block are called:
   a. Main bearings
   b. Web bearings
   c. Support bearings
   d. Connector bearings

5. On a six-cylinder engine the ideal number of main bearings is:
   a. 12
   b. 6
   c. 4
   d. None of the above

6. The gap between the bearing and the shaft or journal is called the:
   a. Bearing clearance
   b. Oil clearance
   c. Air gap
   d. Both A and B

7. Extremely high oil pressure can:
   a. Wash away the bearing material
   b. Cause the bearing to shift
   c. Raise the engine temperature
   d. None of the above

8. The distance that the bearing is higher than the web is called:
   a. The float
   b. Bearing height
   c. Bearing crush
   d. Bearing level

9. The material used in bearings is called:
   a. Alloy
   b. Babbitt
   c. Layers
   d. None of the above

10. Most bearing wear occurs when the engine is:
    a. Started when cold
    b. Started when hot
    c. Shut off
    d. Idling

11. A worn cam bearing can cause:
    a. Cam wobble
    b. Low oil pressure
    c. High oil pressure
    d. Camshaft knock

12. The greater the diameter of a shaft:
    a. The greater the oil pressure
    b. The lower the oil pressure
    c. The more bearing clearance required
    d. The thicker the bearing required

13. Main bearings are usually:
    a. The same size as the rod bearings
    b. Smaller than the rod bearings
    c. Harder than the rod bearings
    d. Larger than the rod bearings

14. Friction bearings are always made of a _____ material than the shaft on which they ride.
    a. Lighter
    b. Stronger
    c. Harder
    d. Different

15. Bearings are made in three different oversizes:
    a. .010, .020, .030
    b. .015, .025, .035
    c. .001, .002, .003
    d. None of the above

Chapter 10

Engine Lubrication and Ventilation

The lubrication system circulates motor oil throughout the engine to do a number of jobs. This chapter will tell you:
- The purpose of motor oil
- How it is rated
- How additives help the oil do specific jobs
- How the lubrication system works
- The relationship between lubrication and performance, economy, and emission control.

PURPOSES OF MOTOR OIL

Motor oil in a car engine does five major jobs:
1. It reduces friction between moving parts, which lessens both wear and heat.
2. It acts as a coolant, removing heat from the metal of the engine.
3. It carries dirt particles away from moving surfaces, cleaning the engine.
4. It helps seal the combustion chamber by forming a film around the valve guides and between the piston rings and the cylinder wall.
5. It acts as a shock absorber, cushioning engine parts to protect them from the force of combustion.

All of these jobs help keep the engine running smoothly and efficiently. If the motor oil fails to do any one of these, engine performance may be reduced and the engine might be damaged.

MOTOR OIL COMPOSITION AND ADDITIVES

Motor oil is a byproduct of petroleum, as are gasoline, kerosene, and other types of lubricants. To change or improve the performance of a motor oil, manufacturers blend in chemical additives.

Motor Oil Composition

Petroleum-based motor oils contain mostly hydrogen and carbon. They are hydrocarbon compounds, as is gasoline.

Motor Oil Additives

The purpose of a motor oil additive can be to:
1. Replace a property of the oil that was lost during refinement.
2. Strengthen a natural quality already in the oil.
3. Add a property that the oil did not naturally have.

A few of the common motor oil additives and their jobs are described in the following paragraphs.

Oxygen tends to combine chemically with

Figure 10-1. Sludge deposits in an engine.

New System		Old System	New System		Old System
SA	=	ML	SF	=	none
SB	=	MM	CA	=	DG
SC	=	MS (1964)	CB	=	DM
SD	=	MS (1968)	CC	=	DM
SE	=	none	CD	=	DS

Figure 10-2. An older rating method had fewer categories and used different letters. It is no longer used. This is how that system corresponds to the "S" and "C" system now used.

hot motor oil. This is called oxidation, and it can leave hard carbon and **varnish** deposits in the engine. **Antioxidants** reduce this high-temperature problem.

In any engine, some of the combustion chamber gases get past the piston rings and enter the crankcase. This is called **blowby**. Blowby gases contain water vapor and acids that will rust or corrode engine parts. Rust and corrosion preventives are added to motor oil to neutralize acids and reduce the bad effects of blowby gases.

The oil in an engine is constantly being churned by moving parts. This can mix air with the oil and cause it to foam. Oil foam does not protect moving parts as well as liquid oil. Foam inhibitors in a motor oil will reduce foaming. Both the water vapor and the fuel in blowby gases tend to mix with cold oil and cause **sludge**, figure 10-1. This thick black deposit clogs oil passages and increases engine friction. **Detergent-dispersants** reduce sludge formation and keep sludge particles suspended in the oil, to be removed when the oil and filter are changed.

Viscosity is the tendency of a liquid to keep from flowing. Some additives help an oil to flow under wide temperature ranges. These are called "viscosity index improvers."

MOTOR OIL DESIGNATIONS

Because engines and operating conditions vary greatly, oil refiners blend and sell different types of motor oil. For example, the oil used in a diesel truck is different than that used in a high-performance car engine. Both engines need lubrication, but most oils cannot meet all the requirements of both engines.

The three designations applied to motor oil are:
• Service classification
• Viscosity grades
• Energy conserving

API Service Classification

The **API Service Classification** is defined by the American Petroleum Institute (API). An oil is used in standardized laboratory engines and the oil's performance is monitored during specific operating conditions. The areas monitored include resistance to oxidation, corrosion, wear, and carbon formation. On the basis of the oil's performance, a service classification is assigned.

Oils for use in gasoline engines have an "S" followed by a letter from "A" through "F" indicating the service level:

• SA has no performance requirements. It is for use under very mild conditions, and might contain only one or two mild additives.

• SB provides antioxidant and antiscuff protection and resistance to corrosion, and is also for use under mild conditions.

• SC is suitable for use in car engines, and some truck engines, manufactured from 1964 through 1967. It contains additives to protect against high- and low-temperature deposits, wear, rust, and corrosion.

• SD is suitable for use in engines manufactured from 1968 to 1970, and some 1971 engines. It provides more protection than SC, and can be substituted for an SC grade oil.

• SE should be used in some 1971 engines, all 1972 through 1979 engines, and some 1980 engines. It can be substituted for SC and SD grades. It will lubricate and protect under severe conditions.

• SF is the latest service classification, for use in some 1980 and all later gasoline engine cars. It can be substituted for SC, SD, or SE grades, and provides better engine protection than those grades.

Oils for use in diesel engines have a "C" followed by a letter from "A" through "D". The CC and CD grades are used in diesel car engines.

Figure 10-2 shows an older service classification no longer in use.

An oil may carry more than one API service

Engine Lubrication and Ventilation

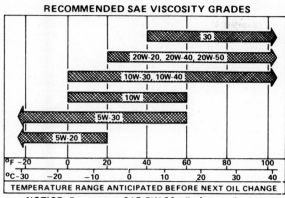

Figure 10-3. Viscosity chart for 1981 Chevrolet car engines. (Chevrolet)

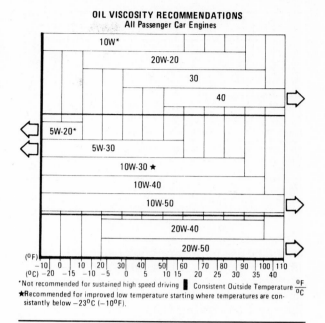

Figure 10-4. Viscosity chart for 1981 Ford car engines. (Ford)

classification. This means that the oil has passed the tests for two or more categories, such as an SE-CC oil.

SAE Viscosity Grades

All motor oils also receive viscosity grades from the Society of Automotive Engineers (SAE). A thick oil that does not flow easily has a high viscosity. A thin oil that does flow easily has a low viscosity. The **SAE viscosity grade** indicates the oil's viscosity at a specific temperature. Motor oil viscosity numbers are 5, 10, 15, 20, 30, 40, and 50. The low numbers represent low-viscosity (thin) oils. The high numbers represent high-viscosity (thick) oils.

Oils are tested at 0°F (−18°C) to determine their cold viscosity. The grades given during the cold test are SAE 5W, 10W, 15W, and 20W. The "W" means that oil viscosity was determined at 0°F (−18°C). Oils are tested at 212°F (100°C) to determine their hot viscosity. The grades given during the hot test are SAE 20, 30, 40, and 50.

When an oil has only one SAE viscosity number it is called a **single-grade** or **straight-grade** oil. These oils can be used safely when the engine operating temperature does not vary greatly. However, if the engine operates under widely varying temperatures, it may benefit from a **multigrade** or **multiviscosity** motor oil.

These oils are tested at both 0° and 212°F (−18° and 100°C), and have two or more viscosity numbers. For example, an SAE 20W-40 oil has a viscosity of an SAE 20W when tested cold and a viscosity of an SAE 40 when tested hot. It will protect an engine at either temperature extreme.

Varnish: An undesirable deposit, usually on the engine pistons, formed by oxidation of fuel and of motor oil.

Antioxidants: Chemicals or compounds added to motor oil to reduce oil oxidation, which leaves carbon and varnish in the engine.

Blowby: Combustion gases that get past the piston rings into the crankcase; this includes water vapor, acids, and unburned fuel.

Sludge: A thick, black deposit caused by the mixing of blowby gases and oil.

Detergent-Dispersant: A chemical added to motor oil to break down and disperse sludge and other undesirable particles picked up by the oil.

Viscosity: The tendency of a liquid such as oil to resist flowing.

API Service Classification: A system of letters signifying an oil's performance; assigned by the American Petroleum Institute.

SAE Viscosity Grade: A system of numbers signifying an oil's viscosity at a specific temperature; assigned by the Society of Automotive Engineers.

Single Grade (Straight Grade): An oil that has been tested at only one temperature, and so has only one SAE viscosity number.

Multigrade (Multiviscosity): An oil that has been tested at more than one temperature, and so has more than one viscosity number.

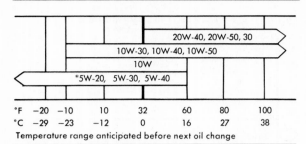

Figure 10-5. Viscosity chart for 1981 Chrysler car engines. (Chrysler)

Lowest Temperature Anticipated	Recommended Single Viscosity	Recommended Multi-Viscosity
Above + 40° F + 5° C	SAE 30 or SAE 40	SAE 10W-30, 20W-40, or 10W-40
Above 0° F −18° C	SAE 20W-20	SAE 10W-30 or 10W-40
Below 0° F −18° C	SAE 10W*	SAE 5W-20 or 5W-30

*Sustained high speeds (above 55 mph, 88 km/h) should be avoided when using SAE 10W engine oil since oil consumption may be greater under this condition.

Figure 10-6. Viscosity chart for 1981 AMC car engines. (AMC)

Carmakers specify the exact kind of oil that should be used under certain operating conditions. Figures 10-3, 10-4, 10-5, and 10-6 show oil charts prepared by different manufacturers.

Energy Conserving Oils

As part of the effort to conserve energy, the oil industry is developing new oils that reduce the friction between lubricated surfaces in an engine, thus providing more miles driven per gallon of fuel used.

This can be accomplished by using a lower viscosity base oil and by the use of friction reducing additives such as graphite, molybdenum disulfide (moly), or other suspended materials, or by the use of oil-soluble organic compounds.

These oils are identified by printing on the container such phrases as "fuel efficient", "energy saving", and "energy conserving".

Synthetic Motor Oils

Synthetic-base lubricants have been used in aviation and special-purpose engines for decades. Since the late 1970's, **synthetic motor oils** have attracted a lot of attention and created controversy.

Before we look at the controversy, we must define "synthetic base."Most so-called synthetic oils are at least partly derived from petroleum. Of the four major classes of synthetic-base stocks, two are 100-percent petroleum based; one is 60- to 80-percent derived from a petroleum base; and the fourth is 10- to 30-percent derived from a petroleum base. The differences between most conventional and synthetic motor oils are in the refining or preparation process and the additives used.

During the refining of conventional oils, no attempt is made to change the molecular structure of the crude oil. To make a synthetic oil, chemists carefully combine certain smaller molecules that come from the crude oil and from other sources. The smaller molecules are joined to form the larger molecules needed for motor oil. Chemists can change the synthetic stock by controlling the molecular combinations. After preparation, additives are blended with the synthetic stock just as with conventional stock, although synthetic oils tend to have more additives than do conventional oils.

The SAE and API have not yet set any standards for synthetic oils. However, oil manufacturers have made some performance claims, including statements that synthetics can:

- Last up to 50,000 miles or more (compared to the 6,000- to 12,000-mile replacement intervals for conventional oils)
- Reduce wear on moving parts
- Increase fuel mileage by decreasing engine friction
- Protect an engine throughout a greater-than-normal temperature range.

Much research has been done to prove or disprove these statements. Synthetic oils do tend to have less viscosity change with temperature change. Engine wear tests have yielded conflicting results, with some engine parts having increased wear with synthetic oil use. The extended drain interval has not been claimed for all synthetic oils.

The energy used to prepare synthetic oils makes them two to six times more expensive than conventional oils. If an extended drain interval is followed, the higher original price is partially offset. However, the carmakers have not accepted the performance claims of synthetics, and continue to require the same service intervals for synthetics as for conventional oils to maintain new-car warranty.

Nevertheless, carmakers are lengthening the oil change interval for conventional motor oils.

Engine Lubrication and Ventilation

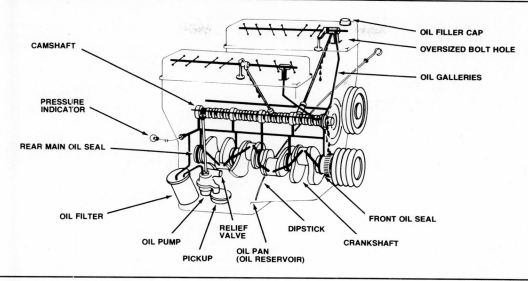

Figure 10-7. The engine oiling system, showing the oil galleries and the oil lines.

The 1971 Dodge, for example, required 12 oil changes in 50,000 miles of operation. By contrast, the 1981 Dodge required only six oil changes in the same number of miles.

ENGINE OILING SYSTEM AND PRESSURE REQUIREMENTS

All modern car engines have a pressurized lubrication system. The parts of this system, figure 10-7, are the:
- Oil reservoir and its ventilation
- Oil pump and pickup
- Pressure relief valve
- Filter
- Galleries and lines
- Indicators.

These parts are described in the following paragraphs.

Reservoir And Ventilation

There must be enough oil in the engine to circulate throughout the system, plus some reserve so that the oil can cool before being recirculated. All of this oil is kept in the engine oil pan, or sump. Because the oil pan is at the bottom of the engine, the oil drains into the pan after passing through the engine. The flow of air past the pan when the car is moving helps cool the oil.

Engine oil pan capacities vary greatly. 4-cylinder engines can hold 3 to 5 quarts of oil. Most V-type engines hold 4 to 5 quarts. Some high-performance and diesel V-type engines require 6 or 7 quarts. Oil is put in the crankcase through a capped oil filler hole at the top of the engine.

We have seen that blowby gases enter the crankcase from the combustion chambers. The gases can mix with motor oil and cause engine

■ Multiple API Classifications

It is possible for an oil manufacturer to blend an oil that can pass both the SF and the CC or CD performance tests. The result would be rated as an SF/CC or SF/CD oil able to protect a gasoline engine as well as a diesel engine.

General Motors, however, cautions the public that for its engines the above information is true only since the development of the SF Service Classification. A look at the accompanying chart will show recommendations for their gasoline and diesel engines before and after the improvement in wear proection from the SE to the SF Service Classification.

MODEL YEARS 1978 TO 1981 ENGINE OIL RECOMMENDATIONS

	ENGINE	
PREFERRED CATEGORIES	GASOLINE	DIESEL
SF	YES	NO
SF/CC	YES	YES
SF/CD	YES	YES
OTHER CATEGORIES		
SE	YES(1981's NO)	NO
SE/CC	YES(1981's NO)	YES(1978's NO)
SE/CD	NO	NO(1978's YES)
SA, SB, SC, SD	NO	NO
CA, CB, CC, CD	NO	NO

Synthetic Motor Oils: Lubricants formed by artificially combining molecules of petroleum and other materials.

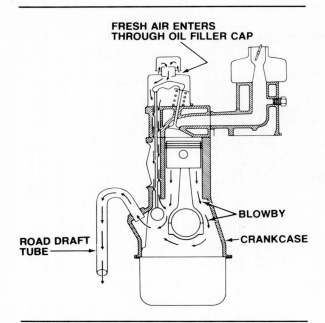

Figure 10-8. An open crankcase ventilation system using a road draft tube.

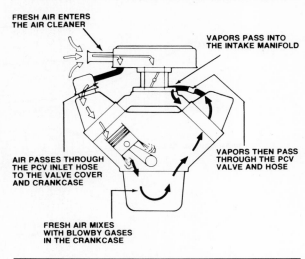

Figure 10-9. The latest PCV systems are totally closed.

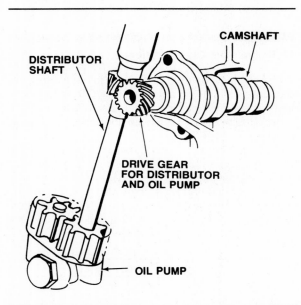

Figure 10-10. The oil pump is driven by the camshaft.

damage. Because the combustion chamber gases are under very high pressure, they increase the pressure within the crankcase. If the crankcase is not vented, the pressure will force oil out of the engine at loosely sealed points such as the oil filler cap and the junction of the oilpan and block, among other points. This is messy, wasteful, and can be dangerous as a fire hazard or if the oil gets on the brake or clutch materials. To prevent this, the crankcase is ventilated.

On older cars, the oil pan was vented directly to the atmosphere through a **road draft tube**, figure 10-8. Until the 1960's, this was the most common type of crankcase ventilation.

All late-model cars have **positive crankcase ventilation** (PCV) systems, figure 10-9. In a PCV system, clean, filtered air is drawn into the crankcase, and crankcase vapors are recycled to the intake manifold. This keeps blowby from polluting the air and provides better crankcase ventilation than the road draft tube. Crankcase ventilation and PCV systems are explained in detail in the Engine Performance Diagnosis and Tune-Up text.

Oil Pump And Pickup

The oil pump is a mechanical device that forces motor oil to circulate through the engine. On most engines, the pump is driven by the camshaft through an extension of the distributor shaft, figure 10-10. Some engines may have a crankshaft-driven oil pump, and most overhead-cam engines have a separate engine shaft to drive the oil pump and the distributor.

Two types of oil pumps are in use today:
- The gear type
- The rotor type.

The most common pump is the gear type, figure 10-11. One gear is driven by gears and a shaft from the camshaft and is called the drive gear. When the drive gear turns it forces the second gear, called the idler gear, to turn. As the two gears turn, the oil between the gear teeth is carried along. At the point where the

Engine Lubrication and Ventilation

173

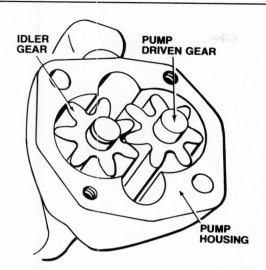

Figure 10-11. A gear type oil pump.

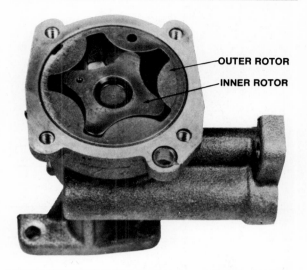

Figure 10-12. A rotor type oil pump.

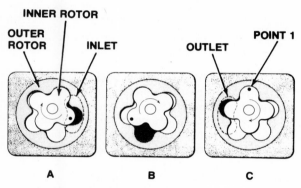

A. Oil is picked up in lobe of outer rotor.
B. Oil is moved in lobe of outer rotor to outlet.
C. Oil is forced out of outlet because the inner and outer rotors mesh too tightly at point 1 and the oil cannot pass through.

Figure 10-13. The principle of operation of the rotor type oil pump.

two gears mesh, there is very little room for oil, so the oil is forced out of this area under pressure.

The rotor type pump works on the same principle of carrying oil from a large area into a smaller area to create pressure. An inner rotor is mounted offcenter within an outer rotor, figure 10-12. The inner rotor is driven through gears by the camshaft. The outer rotor is driven by the inner rotor. As the rotors turn, they carry oil from the areas of large clearance to the areas of small clearance, figure 10-13, forcing the oil to flow from the pump under pressure.

The oil entering the pump comes from the oil pan. An oil pickup tube extends from the pump to the bottom of the pan. A screen at the bottom end of the pickup tube keeps sludge and large particles from entering the oil pump and damaging it, or clogging the oil lines and galleries.

Pressure Relief Valve

The oil leaving the pump passes through a pressure relief valve, or pressure regulating valve. The valve limits the maximum engine oil pressure. It consists of a spring-loaded ball or piston set into an opening in the valve body, figure 10-14. Oil pressure forces the ball or piston to move against spring tension and open the hole in the valve body. Some oil can escape

■ Splash Oiling

Some older car engines and small gasoline engines do not have an oil pump or a pressurized oiling system. Instead, they depend on the motion of engine parts to splash oil from the oil pan into the engine.

Special scoops on the connecting rod big ends scoop up the oil and direct it toward critical spots. Some engines may rely on splash oiling for the lower half of the engine and use an oil pump to supply oil to the upper half of the engine.

Road Draft Tube: The earliest type of crankcase ventilation; it vented blowby gases to the atmosphere.

Positive Crankcase Ventilation (PCV): Late-model crankcase ventilation systems that return blowby gases to the combustion chambers.

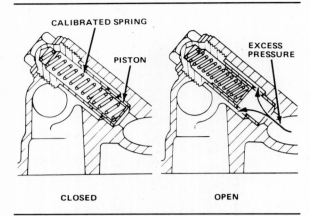

Figue 10-14. The oil pressure relief valve.

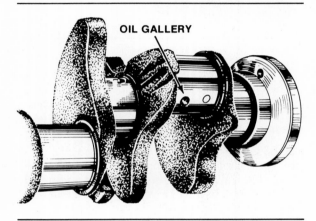

Figure 10-16. The crankshaft has oil galleries drilled in it.

through this hole to decrease the overall system pressure.

The pressure relief valve is usually built into the oil pump housing. The oil that escapes through the relief hole is sent back to the inlet side of the pump. This arrangement of oil flow from the relief valve reduces oil foaming and agitation so that the pump puts out a steady stream of oil.

Filter

Oil leaving the pressure relief valve flows through the oil filter before reaching the rest of the system. The filter is made of paper or cloth fibers that will pass liquid oil but trap dirt.

If the filter becomes plugged with dirt, oil flow is restricted.

To prevent a clogged filter from completely stopping oil flow and damaging the engine, a bypass valve is built into the filter or the filter housing. It allows oil to flow around the filter instead of through it when the oil is too thick to go through the filter because of its low tempera-

Figure 10-15. A spin-on, completely disposable oil filter.

ture, or when the filter outlet pressure drops. Oil filters must be replaced at specific intervals to prevent clogging. Older cars have replaceable oil filter elements inside permanent housings. Most modern engines use spin-on filters that are completely disposable, figure 10-15.

Galleries And Lines

A modern engine has many areas that must have a constant supply of oil under pressure. To ensure that they do, the lubrication system includes a network of passages that direct oil to these parts of the engine, figure 10-7. These passages, called oil galleries, can be drilled through the block, the head, and the crankshaft. They can also be separate tubes, called oil lines, connected to the engine. Not all engines have separate oil lines, but all do have oil galleries.

From the filter, oil flows to a large main gallery. Inline engines and some V-type engines have one main gallery. Other V-type engines have two main galleries, one for each bank of cylinders. From the main gallery, smaller passages direct oil to the camshaft bearings and to the crankshaft main bearings.

Oil must also reach the crankshaft connecting rod bearings. This is done by drilling holes through the crankshaft, figure 10-16, so that oil at the main bearings can also flow to the rod bearings. The caps that are bolted around the rod bearings have small holes that can line up

Engine Lubrication and Ventilation

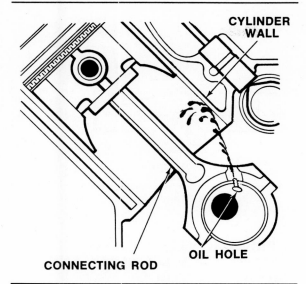

Figure 10-17. Some oil spurts through the connecting rod cap and onto the cylinder walls.

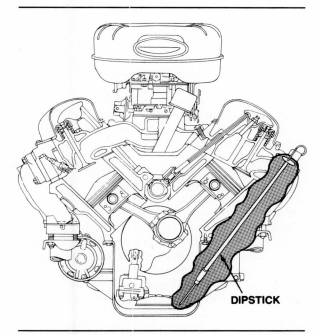

Figure 10-18. The engine dipstick.

with similar holes in the crankshaft and the rod bearings. When the holes in the cap and bearing align with the hole in the crankshaft, figure 10-17, a small stream of oil squirts through and hits the lower cylinder wall. This helps lubricate the cylinder wall and piston.

Different engine designs have different ways of getting oil from the main gallery to the head and valve assemblies. There can be galleries drilled through the block and head, or the oil can travel through hollow valve pushrods.

Also, an enlarged head-bolt hole can allow oil flow. From the valve assembly, the oil drains down through the engine into the oil pan. The oil drain holes can be placed so that the dripping oil helps lubricate the camshaft.

Indicators

To check the oil level when the engine is at rest, a dipstick is installed in the oil pan. To keep the driver informed of engine oil pressure while the engine is running, carmakers equip cars with a low oil pressure warning lamp or a gauge that indicates the pressure at all times.

■ Bypass Oil Filters

Before the full-flow oil filtering system was invented, engines that used oil filters had what is called a bypass system. The filter was mounted on a bracket attached to the engine or anywhere in the engine compartment. Oil lines were connected to a tapped hole in the side of the engine block, and to a drain on the pan or block. The oil was fed under pressure to the filter, and allowed to drain back into the oil pan after filtration.

Since the oil did not have to go through the engine filter before getting to the bearings, a piece of dirt could, theoretically, circulate through the oil system indefinitely until it happened to get into the filter and be trapped. This catch-as-catch-can system was discontinued in favor of the full-flow system.

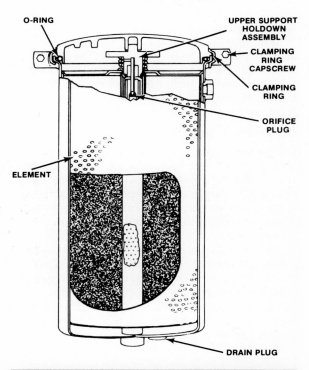

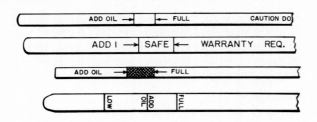

Figure 10-19. Typical dipstick markings.

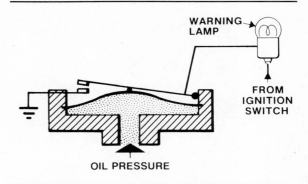

Figure 10-20. A diagram of a low oil pressure warning lamp.

Dipstick
When the engine is at rest, almost all of the oil drains into the oil pan. All engines have a measuring rod, called a dipstick, that extends from the outside of the engine into the pan, figure 10-18. The dipstick has markings on it that indicate the maximum and minimum oil levels for that engine. When you pull the dipstick out of the pan, a film of oil can be seen on the stick. The level of the film relative to the markings on the stick indicates how much oil is in the pan, figure 10-19.

Oil pressure warning lamp
The oil pressure warning lamp lights when the oil pressure is less than a set amount. This happens during cranking and when there is a problem in the lubricating system.

Warning lamps light when a set of electrical contacts close, figure 10-20. The contacts are controlled by a movable diaphragm that is exposed to engine oil pressure. The contacts close when oil pressure is low, and the lamp lights. When pressure increases, the diaphragm moves and the contacts open, turning off the lamp.

The contacts and diaphragm are contained within a sending unit. The sending unit is threaded into the engine block or into the oil filter housing.

Oil pressure gauge
Oil pressure gauges are operated electrically or mechanically. An electric gauge has a sending unit similar to the warning lamp sending unit. A movable diaphragm varies the current flow through the gauge in proportion to oil pressure. The current flow determines the position of the gauge needle.

Mechanical gauges, figure 10-21, have a tube running from the engine to the gauge. When there is engine oil pressure, oil is forced into the tube and to the gauge. The pressure at the gauge end of the tube is the same as the engine oil pressure, and determines the position of the gauge needle.

LUBRICATION EFFECTS ON PERFORMANCE, ECONOMY, AND EMISSION CONTROL
The lubrication system can have direct, measurable effects on performance, economy, and exhaust emissions.

Performance
An engine's performance is related to its mechanical condition. When an engine is new, the moving parts fit together very closely. The distance between two moving parts is called clearance. As the engine wears, the clearance increases. When an engine wears, or gets "loose," it does not operate as efficiently as it once did, and performance suffers.

The lubrication system helps keep engine wear to a minimum. Using the proper API rated oil will reduce engine wear because the oil is formulated to match the engine's requirements. Changing the oil and filter at or before recommended intervals will minimize engine wear by replacing used additives and by keeping harmful particles out of circulation.

The SAE oil viscosity rating will affect engine wear and performance as well. If oil with too low a viscosity is used during high-speed or high-temperature operation, the oil film between moving parts may become so thin that it breaks down, allowing metal-to-metal contact and rapid wear. If oil with too high a viscosity is used during low-temperature operation, power will be wasted in forcing engine parts to overcome the resistance of the thick oil. In some cases, a combination of cold temperature and high-viscosity oil will prevent the starting system from cranking the engine at all.

Engine Lubrication and Ventilation

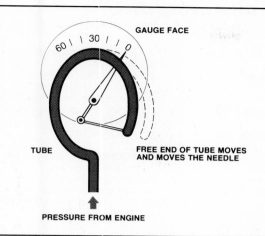

Figure 10-21. A typical mechanical oil pressure gauge.

Economy

Economy is also affected by engine mechanical condition. Camshaft and valve assembly wear, for instance, will change valve timing. Other factors such as the amount of power wasted as blowby will also affect economy.

Some manufacturers of synthetic oils claim that the increased slipperiness of their oil will decrease engine friction and increase fuel economy. Although these claims have been substantiated in most cases, the savings in fuel economy must be weighed against the higher price of the oil.

Emission Control

A small amount of oil is normally present in the combustion chamber when the air-fuel mixture is ignited. The oil must be there to lubricate the cylinder walls and to help seal the chamber. Because oil does not burn as easily as gasoline, some of the oil leaves with the exhaust gases as unburned hydrocarbons (HC), one of the gasoline engine's main pollutants.

When the engine oil is old and diluted by blowby gases, or when the engine is worn and clearances have increased, more oil than normal will enter the combustion chamber. The oil can come down through the valve assembly or get past the piston rings. Once the oil is in the chamber it is a source of HC emissions.

SUMMARY

Motor oil has five major jobs in an engine: reducing friction, cooling, cleaning, sealing, and absorbing shock. Additives mixed with the oil help it to do these jobs. Two oil rating systems are generally used: the API service classification and the SAE viscosity rating. The API number rates an oil's performance in a laboratory engine, and the SAE viscosity number rates the oil's thickness. An oil may have one or more API ratings. An oil's service classification and viscosity rating must be matched to an engine's requirements for best performance, economy, and emission control.

Synthetic motor oils are usually made from a petroleum base, but have chemicals added for greater control over the motor oil's qualities and performance.

An engine lubrication system includes the pan, filter, pump, oil galleries or oil lines (or both), dipstick, and pressure warning lamp or pressure gauge. The oil is stored in the oil pan. The pump takes it from the pan, pressurizes it, and sends it through the oil galleries or oil lines to lubricate the moving parts of the engine. The filter traps dirt and other particles that the oil holds in suspension, and so the oil and filter must be changed periodically. A dipstick is used to measure the oil level in the pan when the engine is not running, and a pressure warning lamp (for low oil pressure) or a pressure gauge (for continuous pressure reading) is used when the engine is running.

The main purpose of engine lubrication is control of mechanical wear. Excessive wear will hurt performance, economy, and emission control. The use of the proper oil, and regular oil and filter changes, will keep engine wear to a minimum.

Review Questions

Choose the single most correct answer.
Compare your answers to the correct answers on page 268.

1. Which of the following is *not* a primary job done by motor oil:
   a. Cooling the engine
   b. Reducing friction
   c. Reducing exhaust emissions
   d. Cleaning the engine

2. Additives in motor oil can:
   a. Replace qualities lost during refinement
   b. Strengthen natural qualities
   c. Add qualities not naturally present
   d. All of the above.

3. Blowby gases come from the:
   a. Crankcase
   b. Combustion chamber
   c. Oil filter
   d. Carburetor

4. The oil used in a 1972-79 passenger-car gasoline engine should have an API service classification of:
   a. SD
   b. CD
   c. CC
   d. SE

5. The "W" in an SAE viscosity grade indicates that the oil:
   a. Has been tested at 0° F
   b. Is for winter use
   c. Has no additives
   d. None of the above

6. The highest viscosity number given to motor oils is:
   a. SAE 40
   b. SAE 50
   c. SAE 100
   d. SAE 70

7. The term "multigrade" means that an oil:
   a. Has been given two API service classifications
   b. Has many additives
   c. Has been tested for viscosity when hot and cold
   d. Can be used in gasoline or diesel engines

8. Most synthetic motor oils:
   a. Are partially petroleum based
   b. Cost less than conventional oils
   c. Reduce engine operating temperatures
   d. Can be safely mixed with conventional oils

9. If the crankcase is not ventilated:
   a. Exhaust emissions will increase
   b. The oil will foam and not protect the engine
   c. Too much oil will enter the combustion chambers
   d. Blowby gases will pressurize it and force oil out

10. The oil pump is usually driven by the:
    a. Crankshaft
    b. Camshaft
    c. Fuel Pump
    d. Timing gears

11. The oil pressure relief valve:
    a. Is usually mounted in the oil pump housing
    b. Limits the maximum pressure in the oiling system
    c. Contains a spring-loaded ball or piston
    d. All of the above

12. When an oil filter becomes clogged:
    a. A bypass valve allows dirty oil to lubricate the engine
    b. The oil pump must slow down
    c. Engine operating temperature increases
    d. All of the above

13. To direct oil to critical points, the oiling system includes:
    a. Internal galleries
    b. Connecting rod squirt holes
    c. Holes drilled in the crankshaft
    d. All of the above

14. Most dipsticks will show you:
    a. What grade of oil to use
    b. If the oil is oxidized
    c. The maximum and minimum oil levels
    d. None of the above

15. Oil pressure warning lamps have a _____ installed in the engine:
    a. Sending unit
    b. Lamp bulb
    c. Oil tube
    d. Pressure gauge

16. Lubrication's greatest effect on performance, economy, and emission control is:
    a. Making the engine run cooler
    b. Keeping the compression ratio steady
    c. Controlling engine mechanical wear
    d. Increasing engine horsepower

Chapter 11

Gaskets, Fasteners, Seals, and Sealants

GASKETS

A gasket is used between two pieces in a stationary joint to prevent leaks, figure 11-1. Such joints are found between the cylinder head and the block, between the oil pan and the block, and between the intake manifold and the heads. Smaller gaskets are used at the front cover, the fuel pump, the water pump, the thermostat elbow, and many other places.

If it were possible to fit two pieces together perfectly so that air or liquids could not pass between them, a gasket would not be necessary. But a gasketless joint is difficult to make and might develop a leak in service. Even so, there are a few gasketless joints in an automobile engine.

The purpose of a gasket, then, is to fill in the spaces between two parts. A gasket must be soft enough to fill the minute grooves and holes when the joint is bolted together, but strong enough so that it will not tear or break. If the clamping force is sufficient, the joint will not leak.

Gasket Types

Gaskets are divided into two major types, hard and soft. Hard gaskets are made of metal or a combination of metal and a softer material. Soft gaskets are made of soft, compressible substances such as cork or rubber.

Hard gaskets

These gaskets are used between the cylinder head and the block. The edge of the gasket exposed to combustion is coated with metal to prevent burnout. Metal is also used around the coolant holes in the head gasket, figure 11-2.

Head gaskets are sometimes a **sandwich type** of construction with steel or copper touching the head and block, and asbestos in between, figure 11-3. Some early gaskets used steel only around the holes and combustion chamber; between the holes, the asbestos was exposed. Today some head gaskets are made with a soft material touching the head and block, and a steel core. The steel core has curled projections that help support the soft facing. Some steel core gaskets have an additional steel facing on one side to act as a heat shield.

An **embossed steel gasket**, sometimes called a shim steel gasket, is frequently put on a new engine. This is a sheet of steel with a raised pattern embossed onto it, figure 11-4. The pattern is in the form of circles or shapes that match the area of the combustion chamber and the coolant and bolt holes. An embossed steel gasket is fine on a new engine put together under carefully controlled conditions. If the engine is

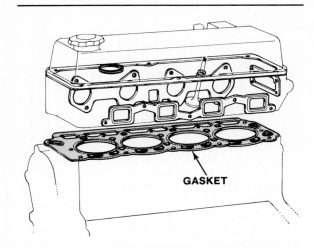

Figure 11-1. The gasket is used between two parts.

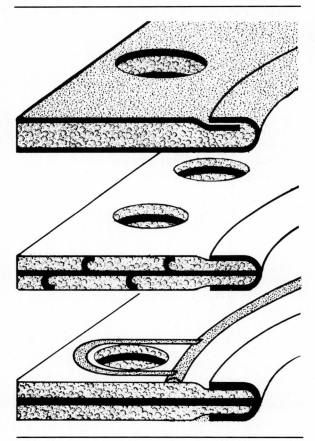

Figure 11-3. A sandwich-type gasket is made in layers. (Fel-Pro)

torn down in the field, it can be repaired using a new embossed steel gasket, but if the head or block is slightly warped, it may leak. Dirty or rusty bolt threads may reduce the clamping force or make it unequal, which could cause a

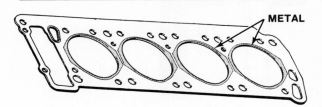

Figure 11-2. A head gasket has metal around the coolant holes and where the gasket may be exposed to combustion.

Figure 11-4. The embossed steel gasket is sometimes caled the shim steel gasket. (Fel-Pro)

leak. Therefore, the use of a sandwich type gasket is often recommended to avoid leaks.

Soft gaskets
Soft gaskets are used on fuel pumps, front covers, oil pans, rocker covers, and almost any place two parts form a joint.

One of the oldest gasket materials still in use today is cork, figure 11-5. Cork gaskets perform well, but they shrink in storage and are fragile.

Paper and composition gaskets, made from a combination of materials, are also used. They are more flexible than cork, and not as easily broken.

Rubber gaskets are probably the best, although the most expensive, kind of gasket available. These gaskets are made of a synthetic rubber and are so tough that they are not affected by oil, grease, or water. Gaskets are also made of ground up pieces of cork and rubber. These have an advantage over the pure rubber gasket because they do not slide out of place easily. Rubber gaskets must be cemented in place or they will squeeze out when the bolts are tightened, figure 11-6.

Gaskets for fuel pump flanges, front covers, and water pumps are sometimes made from paper or cardboard. These work fine, especially if they are coated with a sealing compound before installation to prevent seepage. If the flanges are clean, the sealer may not be re-

Gaskets, Fasteners, Seals, and Sealants

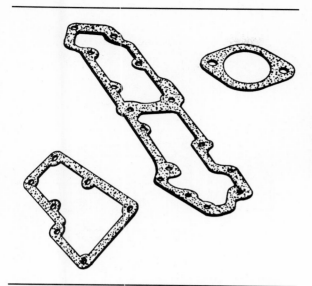

Figure 11-5. Cork gaskets are one of the oldest type of gaskets still in use.

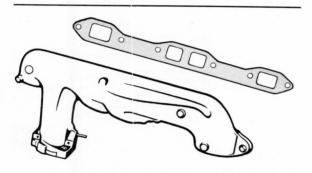

Figure 11-7. Because the exhaust manifold gasket is exposed to extreme heat, it is made of perforated steel and asbestos.

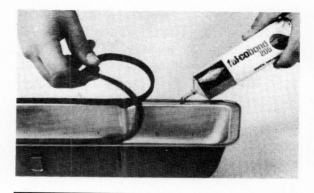

Figure 11-6. You must use cement with rubber gaskets.

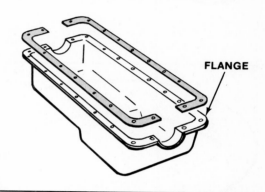

Figure 11-8. Some parts have a lip or flange to help position the gasket.

Gasket Retention

Gaskets are usually held in place by bolts or studs that go through the mating surface or flange of one part, through the gasket, and into the mating surface of the other part. A lip or protrusion around the flange or mating surface may exist to help position the gasket, figure 11-8. A flange made of sheet metal, such as a rocker cover or oil pan, is easily distorted by overtightening. If the metal around the bolt holes is crushed out, when the cover is tightened the gasket will be very tight around the bolts, but very loose everywhere else. Covers bent like this should be straightened before installation.

Sandwich type gasket: A gasket with layers of different materials, such as asbestos and steel.

Embossed steel gasket: A gasket made from thin sheet steel covered with a raised design.

quired. Seals and sealants will be discussed later in this chapter.

A gasket for an exhaust manifold must be different from other gaskets because of the high temperatures to which it is exposed. Some exhaust gaskets are made from embossed steel; others have a steel perforated sheet on one side and exposed asbestos on the other, figure 11-7.

Some engines come from the factory with no exhaust manifold gaskets. The exhaust manifold is a heavy casting, and when it is fitted against the heavy head casting, there is little chance of distortion that may cause a leak. If there is a leak, the rusting of the metal will soon fill in any small gaps. If exhaust manifolds are removed and replaced without gaskets, it is important to clean the joint to prevent leaks. You can get a replacement gasket for an exhaust manifold, even if the original engine did not include one.

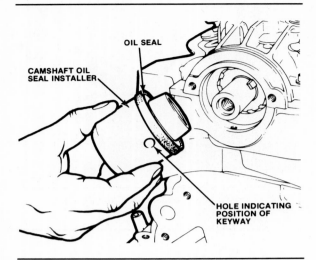

Figure 11-9. A seal is necessary when a rotating shaft protrudes from an engine or machine.

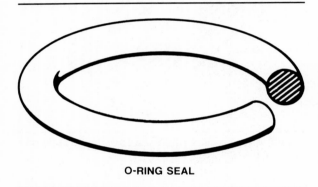

Figue 11-11. The O-ring seal is round with a circular cross section.

When the flange is held with capscrews that thread into a casting, the metal of the casting may be pulled up. A little ridge is raised around the tapped hole in the casting. When the capscrews are tightened, the cover tightens against the raised metal, leaving the gasket loose between screw holes. To prevent this, file the raised metal flush with the casting surface.

Capscrews may also bottom in their holes. This is caused by using capscrews that are too long, or by dirt and rust at the bottom of the hole. If the screw bottoms in the hole, there will be no tension on the flange and the fluid will leak past the gasket. Use the proper capscrews and clean out the holes with a tap before assembly.

SEALS

A seal is used to prevent leakage past a moving part. When a rotating shaft protrudes from a piece of machinery, a seal is usually necessary,

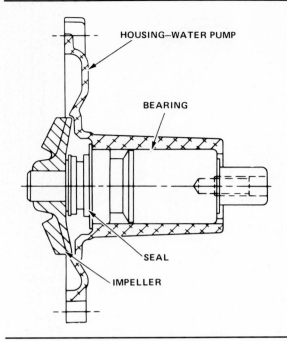

Figure 11-10. A seal is also used on a water pump.

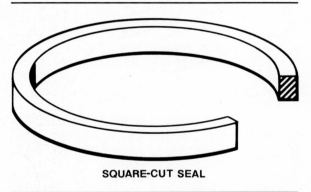

Figure 11-12. The square-cut seal has a square cross section.

figure 11-9. A crankshaft extends from an engine at both the front and the rear; seals are used at both ends to keep oil in and dirt out.

Seals are also used on other parts of the engine and accessories such as the water pump, figure 11-10, and power steering pump.

Synthetic Rubber Seals

The synthetic material used for most rubber seals is neoprene. This material is not as brittle as pure rubber, and it withstands more heat and chemical contact than pure rubber.

O-ring seal

The neoprene O-ring seal is round with a circular cross section, figure 11-11. An O-ring creates a seal when it is compressed between two

Gaskets, Fasteners, Seals, and Sealants

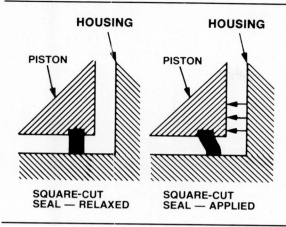

Figure 11-13. A square-cut seal compressed between a piston and a cylinder drags in the direction of the moving piston.

pieces of metal. The O-ring is fitted into a groove which is not quite as deep as the ring, causing the ring to be squeezed between its groove and the metal piece pushing against the inside of the O-ring. This forms a tight seal.

The O-ring is not used where rotational force would be placed on it, because the seal would not seat properly. The O-ring also tends to roll if it is subjected to much **axial movement**, that is, movement along the length of the shaft, as opposed to movement at right angles to the shaft. For example, if an O-ring was used between the land of a piston and the inside wall of a cylinder, the piston could not be allowed to move very far inside the cylinder. If the piston moved too far, it would damage the O-ring seal.

Square-cut (lathe-cut) seal

The square-cut neoprene seal is circular with a square cross section, figure 11-12. It is used between metal pieces affected by axial movement, but it is not used where rotational force might act on it. Rotational force would tend to pull the seal away from its seat much like the O-ring seal subjected to rotational force.

Hydraulic pressure applied to a piston, such as in a disc brake caliper, moves it through the cylinder fast enough to drag and bend the inside of a square-cut seal pressed against the cylinder wall.

The portion of the seal pressed against the cylinder wall does not move as far or as fast as the piston. When hydraulic pressure against the piston is released, the bent edge of the square-cut seal moves back to its original position, figure 11-13. This helps draw the piston back in the cylinder.

A square-cut seal responds to axial movement better than an O-ring. It will not roll in its groove and is not as subject to damage.

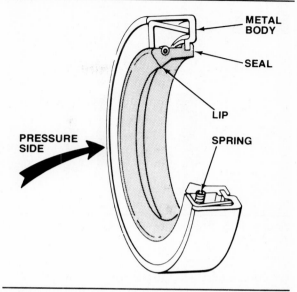

Figure 11-14. Cross section of a lip seal with a metal body and a garter spring.

Lip seal

The lip seal, figure 11-14, is circular like the O-ring and the square-cut seals. A lip seal is always installed with the lip facing the source of hydraulic pressure. Hydraulic pressure against

---

Axial Movement: Movement parallel to the axis of rotation or parallel to the centerline (axis) of a shaft. Endplay.

■ When An O-Ring Is Not An O-Ring

O-rings are commonly used for sealing between moving or non-moving shafts and housings. They are usually made of rubber, and designed so they are squeezed when installed, thus preventing leakage.

In racing engines, cylinder heads are sometimes O-ringed around the combustion chamber. But this is an entirely different kind of O-ring. A small groove is precisely machined into the cylinder head around the combustion chamber. Then a length of wire is laid into the groove so that approximately half of its diameter sticks up out of the groove.

When the head is installed on the engine with a head gasket, the wire exerts tremendous pressure against the gasket and cylinder block. This extra sealing pressure prevents the gasket from leaking or blowing out. Such tactics are thought necessary because of the high pressures developed in the combustion chambers of racing engines.

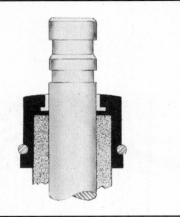

Figure 11-15. The positive valve seal clamps on the valve stem.

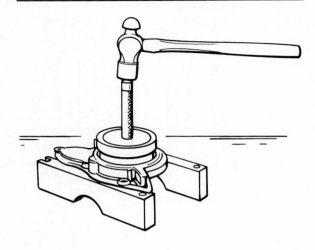

Figure 11-16. A bolt-on crankshaft seal.

the lip causes the lip to flare. As pressure on the lip increases, it presses harder against the shaft or cylinder it seals. When pressure against the lip is relieved, it slides easily against its shaft in either direction. In addition, many lip seals have a small circular coil spring placed behind the lip. This garter spring, as it is called, maintains pressure on the lip when the shaft is stopped and no oil pressure is present.

Lip seals are used where both rotational and axial forces are present. Because the lip that takes the rotational force is small, the seal will not distort with rotation as easily as an O-ring or square-cut seal.

The front and sometimes the rear crankshaft seals are the lip type.

Other Types Of Seals

The ends of many valve stems have an extra groove for a small square-section ring. This ring seals the valve stem and the keeper, so oil splashed on the keeper cannot run down the valve stem and into the guide. The positive type of valve stem seal clamps on the end of the valve guide, figure 11-15. The part that rubs against the stem is sometimes made of Teflon.

Some engines have bolt-on crankshaft lip seals, figure 11-16. The seals bolt to the rear face of the block and the main cap. A gasket between the seal and the block prevents leakage from behind the seal.

Other engines have a wick-type seal, figure 11-17, made of rope-like material that fits inside the rear main cap and web. Before installation it is soaked in oil so it will swell after it is in place.

SEALANTS AND CEMENTS

Sealants and cements are both packaged in tubes, spray cans, or small cans with a brush attached to the cap. But the two have very different purposes.

Cements

A **cement** acts as an adhesive to hold a gasket in place. Cements are usually classified as hardening and non-hardening. The hardening type sets up until it becomes brittle. It has to be chipped away when the parts are disassembled.

The non-hardening type remains pliable. It is used if there might be some movement of the parts due to heat expansion. This movement could break a seal of the hardening type.

To make engine disassembly easier, cement is applied to one side of a gasket and the mating side of a part, figure 11-18. The cement is allowed to dry and the parts are bolted together with the gasket between them.

Cement is often used on rubber oil pan gaskets and rubber rocker cover gaskets. Rubber gaskets must be cemented in place or they may shift when the flanges are tightened.

Cement can also be used on cork and paper gaskets.

Sealant

A **sealant** is often described as a gasket in a tube. A sealant is used to fill gaps and holes between mated parts.

When using a gasket, it may not conform to all the channels and ridges in a flange, but coating a gasket with sealant or using a sealant alone will fill the irregularities.

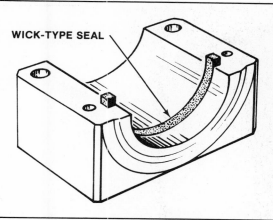

Figure 11-17. The wick-type seal is made of a material like rope that fits inside the rear main cap and web.

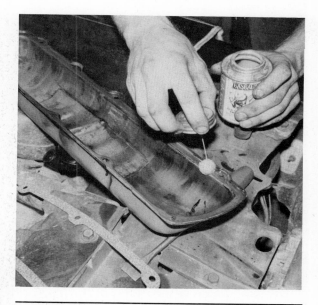

Figure 11-18. Using cement on only one side of a gasket makes engine disassembly easier.

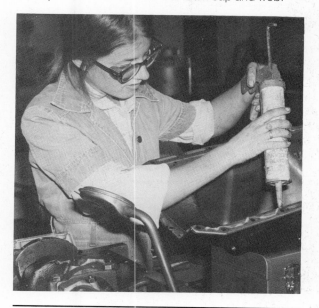

Figure 11-19. Silicone sealant can seal large gaps.

Formed-in-place gasket sealers are being used more and more by manufacturers and repair shops. They come in two basic types.

Silicone sealants

The silicone sealants can be used for metal and plastic covers such as the rocker and cam covers and engine and transmission oil pans. This material is thick and can seal large gaps, figure 11-19.

The temperature range of a silicone sealant usually reaches as high as 350°F (177°C), but will not withstand the 600°F (315°C) temperatures of the exhaust system.

One disadvantage of the silicone sealant is that you must fit the parts together within ten minutes of applying the sealant because it is an **aerobic** sealant. Another name for this compound is **room temperature vulcanizing (RTV) compound**.

Anaerobic sealants

Anaerobic means in the absence of air. This type of sealant cures, or sets up, only after the mating parts are bolted together, excluding air from the joint. This is ideal for overhaul work because you can make a gasket or seal on a number of parts and leave them on your workbench until you are ready to assemble them.

The anaerobic sealer is a good choice for cast timing covers, intake manifolds, and rear main bearing caps. This sealer is thinner than the silicone and is not practical for flexible covers. The use of anaerobic sealants is recommended for machined surfaces only.

RTV Compound: Room temperature vulcanizing silicone, available in a tube.

Sealant: A liquid or gel used with or without a gasket for preventing leaks between mated surfaces.

Cement: A liquid or substance which acts as an adhesive.

Aerobic: A sealant that sets up in the presence of oxygen (air).

Anaerobic: A sealer that cures in the absence of air.

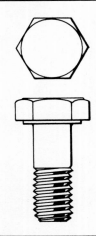

Figure 11-20. A fastener with a hex head and a diameter larger than one-quarter inch (6mm) is called a bolt.

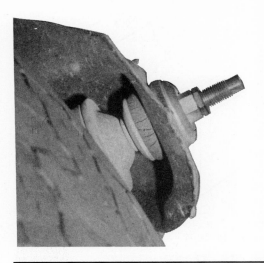

Figure 11-21. The Pal Nut is one example of a special nut developed to prevent bolts from loosening.

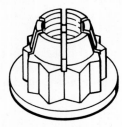

Figure 11-22. Self-locking nuts have teeth bent inwards to grip the threads.

Figure 11-23. Some lock nuts are oval at one end.

Where You Can And Cannot Use Sealants And Cements

You should never use a sealant or cement on any part of a seal that touches a moving shaft. However, seals that are pressed into place or clamped into position can use sealant on the back side of the seal to prevent leakage. When a sealant or cement is used, it must be used sparingly to prevent fouling the shaft or the seal lip.

It is normal to use sealant or cement on fittings with pipe threads, but do not use either of them on electrical sending units, even though they may be mounted with pipe threads.

An electrical sending unit must be installed dry so that the unit will have a good ground. Other types of fittings, such as flare-nut or ferrule-and-nut, are self-sealing. If they leak, the fittings should be changed, instead of trying to stop the leak with sealant or cement.

FASTENERS

Is it a bolt or a screw? Much confusion exists as to the correct terminology. Sometimes bolts are called screws and screws are called bolts. About 150 years ago most fasteners were called screw-bolts. Today fastener technology has advanced, but the names are still confusing.

Definitions

Technically, a fastener with a six-pointed or hexagonal head and a diameter of one-quarter inch (6 mm) or larger is called a bolt, figure 11-20. But there are exceptions.

In common practice, just about any threaded fastener with a hexagonal head is called a bolt.

A hex-head bolt is also called a capscrew. Capscrews are used extensively in repairing and rebuilding engines. The capscrew threads into a tapped hole. They are used on main bearing caps, cylinder heads, intake manifolds, water pumps, fuel pumps, bellhousings, and many other parts.

Machine screws are small versions of the bolt. They usually have a slotted head, a Phillips head, or a socket head; something other than a hex head. But they do exist in hex head, too. They are called machine screws, not because of their application, but because of how they are

Gaskets, Fasteners, Seals, and Sealants

Figure 11-24. Fiber lock nuts have a fiber insert.

Figure 11-26. Spring-type lockwashers look like a loop from a coil spring.

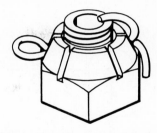

Figure 11-25. The castellated nut is used on a bolt that has a hole for the cotter key.

Figure 11-27. The teeth on a Shakeproof Washer can be internal or external.

made. Machine screws are made by a machining process, whereby each screw is cut from a rod.

Bolt Retention

Nuts are made to go on screws or bolts, securing them in place. Bolts and screws have external threads; nuts have internal threads.

To prevent bolts and screws from loosening, special nuts have been developed and patented. The Pal Nut is one example, figure 11-21. It is a thin nut stamped from sheet metal that screws on top of a regular nut. It jams against the regular nut to prevent loosening.

There are also self-locking nuts of various types. Some have threads that are bent inward to grip the threads of the bolt, figure 11-22. Some are oval shaped at one end to fit tightly on a bolt, figure 11-23. Fiber lock nuts have a fiber insert near the top of the nut or inside it; this type of nut is also made with a plastic insert, figure 11-24. When the bolt turns through the nut, it cuts threads in the fiber or plastic and this puts a drag on the threads which prevents the bolt from loosening.

One of the oldest type of retaining nuts is the **castellated nut**. It looks like a small castle, with slots for a cotter pin, figure 11-25. A castellated nut is used on a bolt that has a hole for the cotter pin.

Flat washers are placed underneath the nut to spread the load over a wide area and prevent gouging of the material. Flat washers do not prevent a nut from loosening.

Lockwashers are designed to prevent a nut from loosening. Spring-type lockwashers resemble a loop out of a coil spring, figure 11-26. As the nut or bolt is tightened the washer is compressed. The tension of the compressed washer holds the fastener firmly against the threads to prevent it from loosening. Lockwashers should not be used on soft metal such as aluminum. The sharp ends of the steel washers would gouge the aluminum badly, especially if they are removed and replaced often.

Another type of locking washer is the toothed washer. It is called a **Shakeproof Washer**, a registered term of Illinois Tool Works. The teeth on a Shakeproof Washer can be external or internal, figure 11-27. The teeth

Castellated Nut: A threaded fastener for a nut or bolt with slots for a cotter pin.

Shakeproof Washer: A type of lockwasher with external or internal multiple teeth.

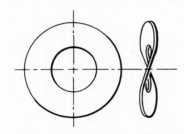

Figure 11-28. The spring steel lock washer.

bite into the metal because they are twisted to expose their edges. Shakeproof Washers are used often on sheet metal or body parts. They are seldom used on engines.

The spring steel lock washer, figure 11-28, also uses the tension of the compressed washer to prevent the fastener from loosening. This washer looks like a distorted flat washer because of its wavy appearance.

Thread And Bolt Sizes

Nuts, bolts, and machine screws are measured in several ways. Two basic measurements are the diameter or thread size and the pitch.

The bolt diameter is measured from the outside of the threads, figure 11-29.

The pitch is given in threads per inch in the U.S. measure. In metric measure, it is the center to center distance between two threads. As bolt diameters get larger, there are, generally, fewer threads per inch. Since all threads are cut at a 60-degree angle, figure 11-29, fewer threads per inch result in a deeper thread. A deeper thread is, of course, stronger.

Recognition and measuring, American sizes

In the U.S. system, bolt diameters of one-quarter of an inch and larger are given in fractions of an inch. The standard sizes are 1/4 inch, 5/16 inch, 3/8 inch, and up; increasing 1/16 inch at each step. Above 5/8 inch, the sizes increase in 1/8-inch steps.

Sizes below 1/4 inch are usually designated by the American National numbering system. The numbers start at zero and go through 12. A zero size screw is .060" in diameter. A number 12 is .216" in diameter. In the progression from size zero to size 6, each size is .013" larger then the preceding one. However, there are no screws in sizes 7, 9, or 11. This means that sizes 8, 10, and 12 are each .026" larger than the next smaller size.

In the American National system, most bolts or machine screws are available in two different thread pitches. The finer of the two is called National Fine; the coarser is called National

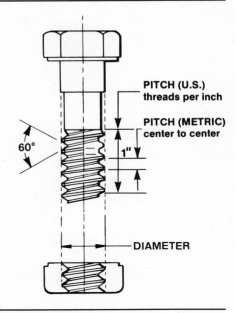

Figure 11-29. Bolt specifications and nomenclature.

Coarse. Thus, a number 6 screw is available in National Coarse with 32 threads per inch; in National Fine, it has 40 threads per inch. You would write is as 6-32 and 6-40. Figure 11-30 is a table of all National Fine and National Coarse diameters and pitches.

In addition to National Fine and National Coarse, there is National Special. National Special thread pitch may be more or less than the National Fine for a given bolt diameter. These special thread pitches are not universally available and will seldom be found on an automobile engine.

The third dimension to consider is length. Round-head and hex-head bolt length measurement does not include the thickness of the head. The bolt length is determined by putting a scale against the underside of the head and measuring to the end of the bolt.

Flathead screw length includes the head. To measure a flathead screw, stand it on end on a flat surface. Measure from the surface to the end of the screw. A flathead screw fits into a countersunk hole. After installation, the screw is level with or slightly below the surface.

The length measurements of flathead screws include the head because it is flush with the surface of the work. This means that the working length of a hex-head screw and a flathead screw are the same. If a designer finds that a hex-head gets in the way, he can specify a flathead screw of the same length, as long as he also specifies a countersink for the flathead.

Another standard size measurement is American Standard Pipe. It applies only to pipe

Gaskets, Fasteners, Seals, and Sealants

American National and Unified Coarse and Fine Thread Dimensions and Tap Drill Sizes

Size	Threads per inch NC UNC	Threads per inch NF UNF	Outside Diameter Inches	Size	Threads per inch NC UNC	Threads per inch NF UNF	Outside Diameter Inches
0	..	80	.0600	1	8	..	1.0000
1	64	..	.0730	1	..	12	1.0000
1	..	72	.0730	1 1/8	7	..	1.1250
2	56	..	.0860	1 1/8	..	12	1.1250
2	..	64	.0860	1 1/4	7	..	1.2500
3	48	..	.0990	1 1/4	..	12	1.2500
3	..	56	.0990	1 3/8	6	..	1.3750
4	40	..	.1120	1 3/8	..	12	1.3750
4	..	48	.1120	1 1/2	6	..	1.5000
5	40	..	.1250	1 1/2	..	12	1.5000
5	..	44	.1250	1 3/4	5	..	1.7500
6	32	..	.1380				
6	..	40	.1380	2	4 1/2	..	2.0000
8	32	..	.1640	2 1/4	4 1/2	..	2.2500
8	..	36	.1640	2 1/2	4	..	2.5000
10	24	..	.1900	2 3/4	4	..	2.7500
10	..	32	.1900	3	4	..	3.0000
12	24	..	.2160	3 1/4	4	..	3.2500
12	..	28	.2160	3 1/2	4	..	3.5000
1/4	20	..	.2500	3 3/4	4	..	3.7500
1/4	..	28	.2500	4	4	..	4.0000
5/16	18	..	.3125				
5/16	..	24	.3125				
3/8	16	..	.3750				
3/8	..	24	.3750				
7/16	14	..	.4375				
7/16	..	20	.4375				
1/2	13	..	.5000				
1/2	..	20	.5000				
9/16	12	..	.5625				
9/16	..	18	.5625				
5/8	11	..	.6250				
5/8	..	18	.6250				
3/4	10	..	.7500				
3/4	..	16	.7500				
7/8	9	..	.8750				
7/8	..	14	.8750				

Figure 11-30. All National Fine and National Coarse diameters and pitches. (Starrett)

threads. There is only one thread pitch for each diameter of pipe. As far as pipe sizes are concerned, there are no metric measurements. American Standard is also the metric standard.

Pipe threads may be tapered or straight. The tapered threads result in a wedging action when screwed into a fitting. This helps prevent leaks. Tapered threads are standard on all sizes of pipe or fittings. Straight pipe threads are seldom seen, and are used only for special purposes.

The hex-head bolts and capscrews have a customary head size relative to the bolt diameter. Quarter-inch bolts usually have a head that measures 7/16 inch across the flats. 5/16 inch bolts usually have a 1/2 inch head. The size of the head determines the size wrench to use.

■ British Whitworth

In 1841 there was no standard for making bolts and nuts. Each fastener factory manufactured bolts and nuts their own way, so a nut would fit a bolt only if both came from the same factory. In England, Sir Joseph Whitworth felt that bolts and nuts should be interchangeable. That same year, he presented an engineering paper describing "A Uniform System of Screw-Threads."

It took many years for England to adopt the new system, but by the time the automobile appeared, almost every screw thread in England was being made by the Whitworth Standard. It became known as the British Standard Whitworth or just Whitworth.

Later, another standard known as British Standard Fine was developed with finer threads. Both British Standards covered bolt sizes down to 1/8 or 3/16 of an inch. Below that, there was another standard known as the British Association. It used numbers ranging from zero to 7, but with the larger numbers for the smaller sizes. This is the opposite of bolts numbered in the American system. To avoid confusion, bolts in the British Association standard were identified by 0BA, 1BA, 2BA, etc.

The British felt that head sizes were not really important. Bolts were specified by thread diameter, and wrenches were stamped this size, not the size of the head. To make this system work, a bolt had to be made with a specific size of hex head for every size of the threaded portion.

Whitworth wrenches were made this way, and identified with the size of the bolt and a "W", such as 5/16W, or 1/2W. The opening in the wrench was much larger than the size stamped on it. It was made to fit the head of a bolt, the threaded diameter of which was stamped on the wrench.

When the British Standard Fine bolts were made, they came out with smaller size hex heads than those on the Whitworth bolts. Now the mechanic had to have another set of wrenches, because a wrench marked 1/2W would not fit the head of a 1/2SF bolt. However, the BSF bolt heads were exactly one size smaller than the Whitworth bolt heads. So to turn a 1/2BSF, you would use a 7/16W wrench.

To show that the wrenches would fit both sizes, they were marked with both sizes for distribution in England. Most British cars imported into the United States used Whitworth bolts. So the American tool makers marked their wrenches with the Whitworth size only.

This situation continued until 1951, when all bolt hex heads were made with the smaller BSF size. But wrenches continued to be marked with both "W" and "F" sizes.

In the middle 1950's all English cars started using American sized bolts. This was the Unified system, which was formed as a result of a conference between England, Canada, and the U.S. With American bolts on all the cars, they had to do something about wrench marking. The decision was made to use the standard U.S. marking, which gives the size of the hex head on the bolt. To show that it meant the diameter of the head, they put "AF" after the size, which means "across flats." Today, all British-made wrenches for use on American nuts or bolts have the standard fractional inch sizes with "AF" after them.

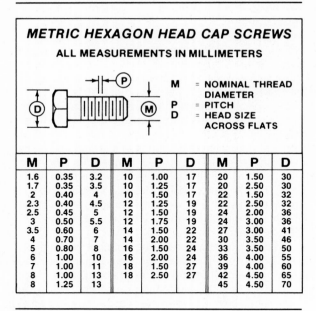

Figure 11-31. Metric bolt table.

Wrenches are made slightly larger than the actual size of the head so they will fit easily.

Although the customary size for bolt heads is well established, there is some variation. When ordering a hex-head bolt, the diameter, thread, and length are always specified, but the size of the head is usually not mentioned. If the head is a different size, you will have to use a different size wrench. It is important to remember that bolt sizes are always the diameter sizes. A hex-head bolt that takes a half-inch wrench is not a half-inch bolt. If you order a half-inch bolt, you will get a bolt almost twice the diameter you want.

Metric sizes

Measurements of metric fasteners are made in millimeters (mm). Metric bolts and screws are usually made with diameters in exact millimeter measurements. It is rare to find a metric bolt that measures in fractions of a millimeter, except in the very small sizes, figure 11-31.

Metric bolt designations list the diameter first, followed by the length and thread pitch. The pitch is listed as the center to center distance in millimeters between threads. For example, an 8 mm bolt that is 2 mm long with a thread pitch of 1 mm, may be listed an an M8 × 20-1.00. It would probably have a 13-mm head. A capital M is often used in front of the size to indicate that it is metric. In some cases the designation "mm" might follow the size, as in 10-1.25 mm.

The metric pitch measurement is similar in use to threads per inch. The larger the number, the coarser the thread. Metric bolts are available with more than one thread pitch. However, the difference between the pitches is not as great as it is in the American Standard system. The terms coarse and fine are not used in the metric system. Thread pitches are referred to simply by the number of the pitch.

Although diameters of metric bolts are almost always in even millimeter measurements, metric pitches come in fractions of a millimeter. The usual steps are .25 of a millimeter, such as 1.25, 1.50, 1.75, etc. Careful measurements with a pitch gauge must be made to avoid mismatching bolts and nuts.

The size of a metric bolt is the size of the threaded part of the bolt, the same as in Amer-

TABLE A—U.S. CUSTOMARY BOLTS, SCREWS & STUDS

SAE Grade Designation	Material & Treatment	Grade Identification Marking
1, 2	Low or medium carbon steel	
4	Medium carbon cold drawn steel	
5	Medium carbon steel, quenched and tempered	
5.1	Low or medium carbon steel, quenched and tempered	
5.2	Low carbon martensite steel, quenched and tempered	
7	Medium carbon alloy steel, quenched and tempered	
8	Medium carbon alloy steel, quenched and tempered	
8.1	Elevated temperature drawn steel, medium carbon alloy	
8.2	Low carbon martensite steel, quenched and tempered	

TABLE B—U.S. CUSTOMARY NUTS

SAE Grade Designation	Grade Identification Marking		
	Style A	Style B	Style C
2	None	None	None
5	Dot and radial or circumferential line 120° CCW from dot	Dot at one corner of nut and radial line at the corner 120° CCW from dot	One notch at each hexagon corner
8	Dot and radial or circumferential line 60° CCW from dot	Dot at one corner of nut and radial line at the corner 60° CCW from dot	Two notches at each hexagon corner

TABLE C—METRIC NUTS & BOLTS

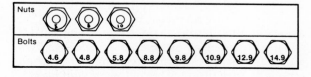

Figure 11-32.

Gaskets, Fasteners, Seals, and Sealants

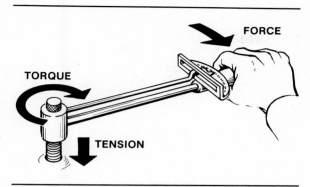

Figure 11-33. The torque wrench measures the tightness of a bolt.

ican Standard. Hex-head sizes in metric bolts are measured in millimeters.

Wrench length relationship to bolt head size

Look at any set of open end, box end, or combination wrenches. With wrenches of the same design, the smaller the wrench opening, the shorter the wrench, figure 11-32. This is to give the right feel to the wrench when the nut or bolt is tightened and to prevent overtightening.

Bolt size and strength are determined by the load the bolt must carry. The strength of the bolt must be higher than the force imposed on it. When the bolt is tightened, it is turned with more force than it will bear in service. For example, if a bolt is going to support a load of 100 pounds, it might be tightened to 150 foot-pounds. This ensures that the bolt will work without failing.

A torque wrench, figure 11-33, is used to measure how tight a nut or bolt is. Torque is described in Chapter 2. The wrench is used to tighten the nut or bolt to a specific amount of torque. Engine manuals give the torque specifications for every important bolt or nut on the engine. The specifications have been carefully calculated so that the bolts will carry their loads and not become loose.

Most torque specifications are written for fasteners that are clean and dry. If fasteners are lubricated, the pressure applied to the nut or bolt will be greater because there is less friction to overcome. If the torque specification chart does not mention any lubricant, then the threads should be clean and dry.

Fastener Strength

Nuts, bolts and screws must be strong enough to hold parts together as designed, and to withstand the maximum tension put upon them. Fasteners are made in different grades of strength, and are marked so you can easily determine this. Figure 11-32, Table A, shows U.S. customary bolt grade designations, materials and treatment, and grade identification markings. Notice that in the U.S. system grades are indicated by numbers. The larger the number, the greater the tensile strength of the bolt. Decimals are used to represent variations of the same strength level.

U.S. customary nuts are marked according to figure 11-32, Table B. Grade 2 nuts and some kinds of grade 5 and grade 8 nuts are not required to be marked. There are three possible styles of grade identification. Style A is applicable to all nuts. Style B is optional depending on the supplier, the purchaser, and the size of the nut. Style C is for nuts which are cut from a hexagon-shaped steel bar.

Metric fasteners generally have the strength classification embossed on the head of each bolt, or the face of each nut. The number indicates the strength in kilograms of force per square millimeter. This means the larger the number, the stronger the bolt or nut, figure 11-32, Table C.

When replacing a fastener, *always* use one of equal or greater strength. In some cases, manufacturers recommend that specific nuts and bolts be replaced instead of being reused, even though they appear to be in perfect condition. Always comply with these requirements, because a bolt that has been weakened by being torqued to a high value could fail if reused.

■ Floating Power

The idea of a damper attached between the engine and frame is not new. Some 50 years ago, Plymouth has a system called, "floating power." The engine had only two mounts. One was high in front, under the water pump; and the other was low, beneath the transmission. With this mounting, the engine could rotate slightly from side to side around its center of gravity. The first installation had an arm going from the engine to the frame. This arm would limit the movement of the engine and dampen vibrations, similar to the hydraulic damper GM is using.

"Floating power" was an improvement over the customary 4-point engine mounting of early cars. A 4-point mounting was strong and rigid, but transmitted a lot of vibration to the car body. "Floating power" was an attempt to make the passengers unaware of the engine vibrations. In later models the limiting arm was not used because the resistance of the rubber in the mounts was enough to keep the engine in place.

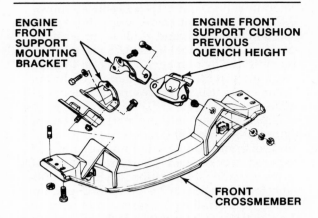

Figure 11-34. The engine mount bolts between the engine and the frame.

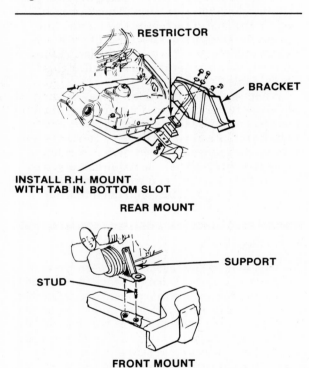

Figure 11-36. The Oldsmobile Toronado engine mounts.

ENGINE MOUNTS AND SHOCK ABSORBERS

Engine mounts and shock absorbers have one purpose in common; they both insulate the car body from engine vibrations. Engine mounts also hold the engine firmly in place.

Engine Mounts

An engine mount is a bracket that bolts between the engine and the frame or lower body of the

Figure 11-35. This mount consists of two metal brackets bonded to a piece of rubber.

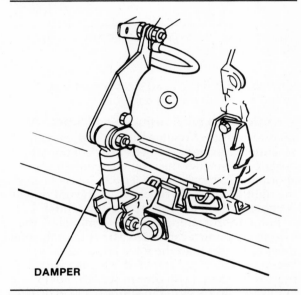

Figure 11-37. X-car hydraulic damper. (General Motors)

car, figure 11-34. As we have already learned, its main purpose is to hold the engine firmly in position. It should also insulate the car from engine vibrations.

Engine mounts are actually "tuned" to absorb vibrations from the engine. This is done by carefully determining the best shape of the mount, the right amount of rubber used to dampen vibrations, and the proper hardness of the rubber.

A typical engine mount consists of two metal brackets bonded to a large mass of rubber, figure 11-35. Holes in the brackets align with holes in the engine block and the frame or lower body. Some cars do not have a frame, so the engine mounts attach to a strengthened part of the lower body.

In time, the bonding between the rubber and the metal will weaken, and the engine may move around in the frame. A separated engine mount can allow the engine to move so much that it interferes with other moving parts, such as the steering mechanism. Some engine mounts have a bracket or tang that will prevent the engine from moving enough to cause dam-

Gaskets, Fasteners, Seals, and Sealants

age if the rubber fails. The extra bracket results in metal-to-metal contact when the rubber bonding fails.

Engine mounts are also used for locating the transmission. When talking about engine mounts, the engine and transmission can be treated as a single unit. The mount under the transmission may be referred to as the rear engine mount or the transmission mount.

In a conventional front and back engine-transmission mounting, three mounts are used. There is a mount on both sides of the engine which support most of its weight. These mounts are usually placed slightly forward of the center of the block.

The third mount under the rear of the transmission supports the weight of the transmission and part of the weight of the engine. This rear engine mount is usually supported by a crossmember that attaches to the frame rails or the lower part of the body.

Unusual designs, such as that of the Oldsmobile Toronado, require unique engine mounts, figure 11-36. The Toronado has front-wheel drive and the transmission is placed alongside the engine and is driven by a chain. The combination of V-8 engine and transmission makes a square-shaped package. It is supported by four mounts in the engine compartment, two at the front and two at the rear, attached to the sides of the engine and transmission.

Most transversely mounted engines in front-wheel-drive cars use a 3-point mounting. The Chrysler Omni and Horizon have a mount at the side of the engine which bolts to the front of the car. The other two mounts are at the front and rear of the engine, and bolt to the right and left sides of the car. The front-wheel-drive General Motors X-body cars have a 3-point mounting system. One mount is at the front of the engine on the right side of the car. The other two mounts are on each side of the transmission.

Rear-engine and mid-engine cars sometimes have different problems in isolating engine vibrations from the passengers. There have been rear-engine cars with engine mounts at the top of the engine. An upper mount does not really support the weight of the engine but acts as a steadying rest. Most rear- and mid-engine cars have a 3-point mounting system similar to front-engine cars.

Engine Shock Absorbers

Front-wheel-drive X-body cars from General Motors with a V-6 engine have an engine shock absorber attached horizontally between the lower part of the engine and the frame, figure 11-37. This hydraulic shock absorber is not to prevent the engine from moving, but to dampen vibrations. Buick calls it a hydraulic damper.

SUMMARY

A gasket is used to prevent leakage of oil, water, or pressure between two surfaces. Gaskets can be hard or soft.

Hard gaskets are a combination of metal and a softer material such as asbestos. Soft gaskets are made from cork or rubber. Embossed steel gaskets are all metal. Some gaskets are made from paper.

A seal prevents leakage past a moving part. Seals are commonly used on rotating or reciprocating shafts. Seals can be pressed or bolted in position. An O-ring is used on rotating shafts, while square section rings are used on reciprocating shafts such as valve stems. Lip seals are used where rotational and axial forces are present such as the front and rear crankshaft.

A sealant is used to fill minute spaces between mated parts and prevent leaks. A cement is used to hold a gasket in place prior to assembly. Sealants and cements come in tubes, spray cans and cans with a brush. Sealants can also be either silicone or anaerobic types.

A fastener with a hex head and a diameter of one-quarter inch or larger is usually called a bolt. Several special nuts have been developed to prevent loosening of nuts and bolts. Lockwashers prevent loosening by spring force or by digging into the work and bolt head.

Bolt diameters are given in fractions of an inch or in a number size, as in the American System. The metric system gives all diameters in millimeters. The size of the wrench is the size of the hex head, not the shank of the bolt.

A bolt has to be tightened to a certain torque specification to prevent it from coming loose or failing while in service.

Service manuals include torque specification tables for almost every bolt on the engine. These specifications are given for clean, dry threads.

Engine mounts hold and support an engine in the frame or unit body of the car. Engine mounts reduce vibration felt by the passengers. Common practice is to use three or even four mounts to support an engine or an engine-transmission combination. An engine shock absorber is used to dampen engine vibrations.

Review Questions

Choose the single most correct answer.
Compare your answers to the correct answers on page 268.

1. Another name for silicone sealant is abbreviated:
   a. RTV
   b. RVT
   c. VRT
   d. TVR

2. Gaskets are divided into two major types:
   a. Hard and soft
   b. Natural and synthetic
   c. Composition and metal
   d. Cork and rubber

3. An O-ring is used for:
   a. Axial movement
   b. Vertical movement
   c. Rotational movement
   d. Horizontal movement

4. A spring-type lock washer resembles:
   a. A part of a leaf spring
   b. A part of a coil spring
   c. Spring-type lock nuts
   d. None of the above

5. A square-cut seal is used for:
   a. Rotational movement
   b. Axial movement
   c. Horizontal movement
   d. Vertical movement

6. An anaerobic sealant:
   a. Seals when in contact with air
   b. Seals in the absence of air
   c. Seals when mixed with a catalyst
   d. None of the above

7. Most torque specifications are written for fasteners:
   a. That are clean and lubricated
   b. That are clean and dry
   c. Over 1/2" or 13 mm.
   d. In inch-pounds or Newton-meters

8. A seal is designed to prevent:
   a. Leakage past two stationary parts
   b. Leakage into the crankcase
   c. Leakage past a moving part
   d. All of the above

9. Which of the following is not a seal:
   a. O-ring
   b. Square-cut seal
   c. Do-nut seal
   d. Lip seal

10. Which of the following does not indicate a thread pitch:
    a. National Fine
    b. National Course
    c. National Standard
    d. National Special

11. In most front-engine designs with rear-wheel drive, _____ engine mounts are used.
    a. 2
    b. 3
    c. 4
    d. 5

12. Pipe threads are designed to:
    a. Be self-tapping
    b. Provide a leak-proof seal
    c. Only be used on pipe
    d. All of the above

13. The size of wrench to use on a bolt depends on:
    a. The shank of the bolt
    b. The torque of the bolt
    c. The head of the bolt
    d. Whether the bolt is English or Metric

14. Pitch for a metric bolt is the:
    a. Center to center distance between threads in millimeters
    b. Thread angle divided by the diameter in millimeters
    c. Number of threads per centimeter
    d. Number of threads per millimeter

15. A castellated nut uses:
    a. A castellated bolt
    b. A cotter pin that goes through the slots in the top of the nut
    c. A castellated washer
    d. All of the above

16. A shakeproof washer is:
    a. A lock washer
    b. A toothed washer
    c. Designed to bite into the surrounding material
    d. All of the above

17. A gasket is used to prevent leakage of:
    a. Oil
    b. Water
    c. Fuel
    d. All of the above

18. A bolt is usually larger than:
    a. 1/8 of an inch
    b. 1/4 of an inch
    c. 1 inch
    d. None of the above

PART THREE

Engine Service Tools and Equipment

Chapter Twelve
Tools and Precision Measuring Instruments

Chapter Thirteen
Engine Test Equipment

Chapter Fourteen
Cleaning and Inspection Equipment

Chapter Fifteen
Head and Valve Service Equipment

Chapter Sixteen
Block, Crankshaft, Piston, and Rod Service Equipment

Chapter 12
Tools And Precision Measuring Instruments

HAND TOOLS

Although engine rebuilding requires diagnostic machines and electronic analyzers, a mechanic's instincts, education and hand tools still play a major role in the procedure.

Open-end Wrenches

Open-end wrenches are used because they are the most convenient type of wrench. The open-end wrench has the head set at an angle to the wrench, figure 12-1. This allows you to move the wrench until it comes to an obstruction, flip it over, and then move it some more. Because of its angled head, an open-end wrench will work well in tight quarters.

An open-end wrench should never be used to break loose tight nuts or bolts. This can spread the end of the wrench or round the corners of the nut or bolt.

Box-end Wrenches

The box-end wrench has a full circle end, figure 12-2. For this reason it is much stronger than the open-end wrench.

The end of the box wrench is usually offset so it will clear obstructions. Care must be taken, however, to pull on it in a straight line, or it may lift off the fastener.

Box-end wrenches have 6- or 12-point openings, figure 12-3. The 6-point opening spreads the force over more of the fastener, so it is less likely to round the corners of the nut or bolt. But the 12-point wrench only has to be moved through 1/12 of a revolution (30 degrees) before it can be removed and repositioned, compared with twice that for the 6-point wrench.

Combination Wrenches

The combination wrench, figure 12-4, combines the open-end and the box-end wrench. The box end is usually angled at about 12 degrees, and will fit into places that a standard offset box wrench will not. A tight nut or bolt can be loosened with the box end and then run off quickly with the open end.

Flare Nut Wrenches

Flare nut wrenches are used on tubing connections are fittings, figure 12-5. For this reason they are often called tubing wrenches. Fittings can be easily damaged if an open-end wrench is used on them, but the flare nut wrench will loosen them without damage.

These wrenches are also available with 6- or 12-point openings. The 6-point wrenches are the least likely to round the corners of the fittings.

Tools and Precision Measuring Instruments

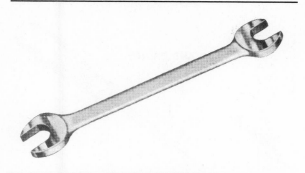

Figure 12-1. Open-end wrenches are the most convenient type to use.

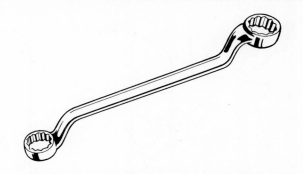

Figure 12-2. Box-end wrenches are stronger than the open-end wrench.

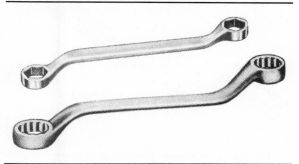

Figure 12-3. Box-end wrenches can have either 6-point openings or 12-point openings.

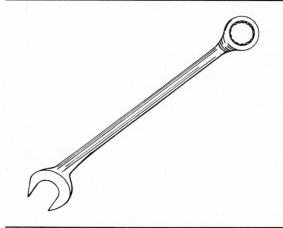

Figure 12-4. One open end and one box end are combined in the combination wrench.

Figure 12-5. The flare-nut wrench is used on tubing connections and fittings.

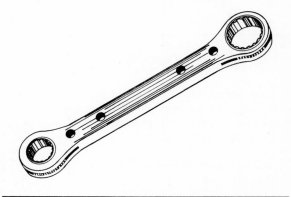

Figure 12-6. The ratcheting box-end wrench only has to be moved back and forth to turn the bolt.

Ratcheting Box-end Wrenches

Once a ratcheting box-end wrench is on the head of a nut or bolt, it only has to be moved back and forth to turn the bolt, figure 12-6. But like any ratchet, when the fastener is loose, the resistance of the ratchet will cause the nut or bolt to turn and the wrench will no longer ratchet.

These wrenches are also availabe in 6- and 12-point openings. They are very useful when space limitations preclude the use of a socket wrench.

Slot-type Screwdrivers

The size of the screwdriver tip should always be matched to the size of the screw slot. This will prevent damage to the screwdriver and the screw. A screwdriver blade that is too thin will contact the screw slot only at the front and back of the blade. The edges will cut into the screw and ride up out of the slot. A screwdriver blade of the correct thickness and width will be a tight fit in the slot and will not slip, figure 12-7.

If a screwdriver fits the slot tightly, but is too narrow, the blade may twist when it is turned.

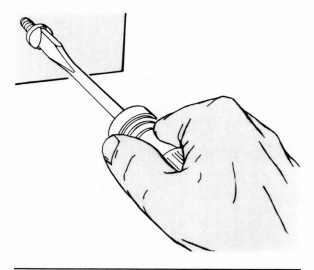

Figure 12-7. It is important to use a screwdriver with the correct blade width and thickness so it will be a tight fit in the screw slot and not slip.

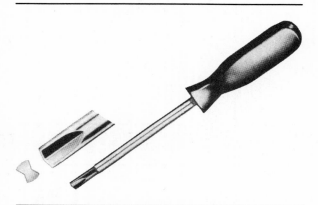

Figure 12-9. The clutch head screwdriver is also called a butterfly or a figure eight.

A blade too narrow for the slot may chew up the slot. The force on the screwdriver should always be as perpendicular to the screw head as possible to avoid slipping.

Phillips Head, Reed and Prince Screwdrivers

These screwdrivers, figure 12-8, are also called cross point or cross slot screwdrivers. The designs are very similar, yet are different enough that if the wrong screwdriver is used the screw head may be damaged. Cross slot screws used in the United States are generally Phillips.

Phillips screws and screwdrivers come in sizes numbered 0 through 4 from the smallest to the largest. When using a cross point screwdriver, look for identification as to type on the handle or shank.

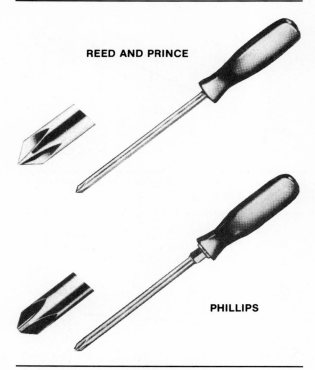

Figure 12-8. Phillips and Reed and Prince screwdrivers are also called cross-point or cross-slot screwdrivers.

Clutch Head Screwdrivers

Clutch head screwdrivers are also called figure eight or butterfly screwdrivers, figure 12-9. They are used mainly on body parts and accessories. The deep recess in the screw head is shaped like a butterfly or figure eight, and prevents the screwdriver from slipping.

Besides clutch head, Phillips, and Reed and Prince screwdrivers, there are a number of special recessed-head screws that you may encounter, figure 12-10. Each of these screws require the matching screwdriver made specifically for it. If an attempt is made to turn the screw with something else, the screw will probably be damaged and may be difficult to remove even with the right screwdriver.

Pliers

Pliers come in many shapes and sizes for almost any job imaginable. Diagonal pliers, figure 12-11, are frequently used in automotive work for cutting wire.

Ordinary gripping pliers are also very useful. They have a slip joint so they can be adjusted to narrow and wide positions, figure 12-12. Slip joint pliers, also called adjustable pliers, have a much greater range of adjustment, figure 12-13. They are available in many sizes.

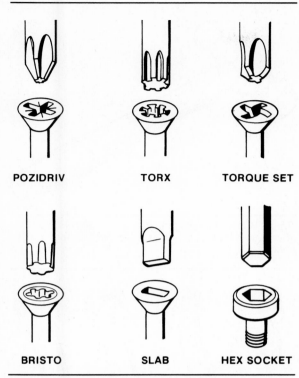

Figure 12-10. These are special recessed screws and drivers you may encounter. (Hand Tool Institute)

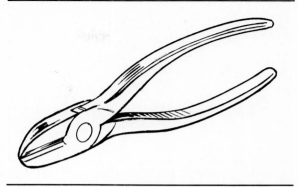

Figure 12-11. The diagonal pliers are used for cutting wire.

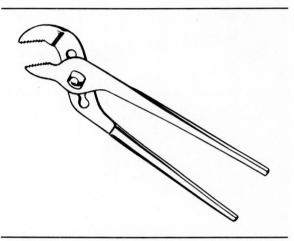

Figure 12-13. The slip joint pliers have a large range of adjustment.

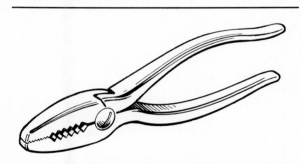

Figure 12-12. These gripping pliers can be adjusted to narrow and wide positions.

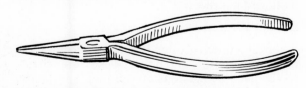

Figure 12-14. Needle nose pliers can reach into places inaccessible to ordinary pliers.

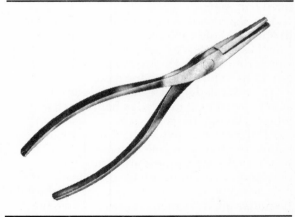

Figure 12-15. The duckbill pliers can get a good grip on wire or small parts.

Needle nose pliers, figure 12-14, are designed to reach into places where the wide jaws of ordinary pliers will not fit. Duckbill pliers, figure 12-15, have flat, thin jaws that can get a better grip on wire and other small parts than needle nose pliers.

Lockring, or snapring pliers, have pins at the ends of each jaw which are inserted into holes in the snapring. The ring can then be squeezed or expanded for removal or insertion. Some snapring pliers, figure 12-16, can be adjusted to expand or compress the ring.

Figure 12-16. Some types of snapring pliers can be adjusted to expand or compress the ring.

Figure 12-18. The shot-filled hammer was designed to replace the rubber hammer.

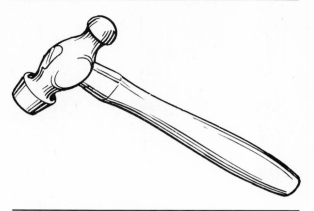

Figure 12-17. The ballpeen hammer has two uses: it drives chisels or punches and it is used to peen rivet heads.

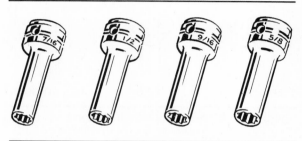

Figure 12-19. Sockets come in two lengths. The long ones are called deep sockets.

There are also pliers for pulling fuses, for removing hose clamps, for pulling spark plug wires, and many other special uses.

Hammers

Many types of hammers are used in auto repair today. The most common types are the ballpeen, mini sledge, brass, rawhide, and shot-filled hammers.

The ballpeen hammer is really two hammers in one: one side has a flat surface for driving punches, chisels, and similar objects; the other side is rounded and is used to peen rivet heads as well as to form and shape metal, figure 12-17.

The mini sledge, as its name implies, is a smaller version of the sledge hammer. It is used where a lot of force is needed.

The brass and rawhide hammers are designed to protect the object being struck. The heads on these hammers are softer than the objects on which they are used. Instead of the objects being marred or dented, the heads of the hammers are deformed, because of this, the heads on these hammers gradually wear out and have to be replaced periodically.

The shot-filled hammer was designed to replace the rubber hammer, and is becoming increasingly popular among mechanics, figure 12-18. The head of this hammer is filled with lead shot so that when the object is struck the hammer doesn't bounce, as a rubber hammer does. This allows more of the force to go into the object being struck.

Always observe the following safety precautions when using a hammer:
- Wear safety glasses
- Never strike hammers together that are made of hardened steel; they may shatter
- Never use a hammer that has a loose head; the head may fly off
- Ensure that the object to be struck is held securely and that your hands are away from the impact area
- Never use a hammer that has a cracked or damaged handle.

Socket Wrenches

A socket wrench is often the most efficient tool for removing nuts and bolts. Socket wrenches can be used with extensions and universal joints to reach between parts and into other inaccessible places.

The basic socket wrench consists of a socket attached to a handle. Sockets are made in two lengths. The long sockets, figure 12-19, called deep sockets, are used when a nut is on a bolt that is longer than the standard socket.

One end of the socket is made to fit over the head of the fastener. The other end is the drive end into which the handle or extension is in-

Tools and Precision Measuring Instruments

Figure 12-20. The basic socket handle is the reversible ratchet.

Figure 12-21. You use a breaker bar to break loose connections.

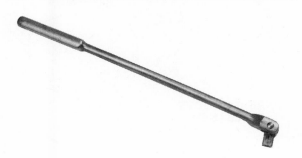

Figure 12-22. The socket flex handle can also be used to break loose connections.

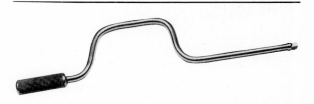

Figure 12-23. The speed handle uses crank action to remove or install a fastener.

Figure 12-24. Extension bars come in various lengths.

Figure 12-25. Sockets can be used with different size drives by employing a socket adapter.

serted. It has a square hole which may be 1/4 inch, 3/8 inch, 1/2 inch, or larger. The fractional inch drive sizes apply to all socket sets, even if the sockets are made to fit metric fasteners.

There is a range of socket sizes for each drive size. The 1/4-inch drive set usually covers 3/16-inch (4 mm) to 7/16-inch (12 mm) sockets. The 3/8-inch drive set covers approximately 3/8-inch (9 mm) to 3/4-inch (19 mm) sockets. The 1/2-inch drive set covers 7/16-inch (9 mm) to 1 1/4-inch (32 mm) sockets. For larger fasteners, you may have to use a 3/4-inch-drive set, but this is not usually necessary on automobiles.

Socket handles and extensions

Sockets can be used with a large assortment of handles. The basic handle is the reversible ratchet, figure 12-20. The lever is flipped to reverse direction. Ratchets should not be used to break loose stubborn nuts or bolts. Too much force on the ratchet mechanism will break it. A breaker bar, figure 12-21, used as a T-handle or L-handle should be used to break loose connections. A flex handle or hinge handle, figure 12-22, can also be used for this purpose. The ratchet can then be used to remove the fastener.

Another socket handle that is sometimes used by mechanics is the speed handle, figure 12-23. It uses crank action to remove or install a fastener very quickly.

All sockets and handles can be used with extension bars, figure 12-24. These come in different lengths, ranging from 1 1/2 inches (38 mm) to 20 inches (508 mm). Adapters are available to allow sockets to be used with different drives, figure 12-25. However, these should be used only in extreme circumstances, because it is easy to exert too much force and break the socket.

Special Or Unique Hand Tools

There are many tools available that make the job of engine rebuilding faster and easier. Some of these tools are made for one specific purpose and are nice to have, but are not necessary for the job. Others, such as torque wrenches, have no substitute and are absolutely necessary.

Figure 12-26. There are three types of torque wrenches: the beam type, the direct dial reading type and the micro-adjusting type.

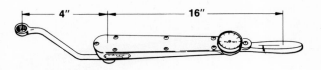

Figure 12-27. Using an adapter which lengthens the drive end of a torque wrench.

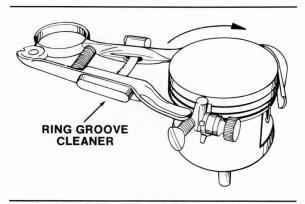

Figure 12-28. The ring groove cleaner is a hand tool.

Torque wrenches

In engine repair and rebuilding work, it is often essential that a nut or bolt be tightened to a specific amount, no more and no less. This is particularly true of fasteners in critical areas, such as head bolts, main bearing bolts, and connecting rod bolts or nuts. Adding a little more torque beyond the manufacturer's specifications "for good measure" can stretch the bolt beyond its limit, so that it does not clamp parts together as tightly as it should. The result can be leakage between the parts, or bolt failure.

A torque wrench indicates how much force is applied to the nut or bolt. There are three types of torque wrenches in common use, the beam type, the direct dial reading type, and the micro-adjusting type, figure 12-26. The torque is read from a scale on the beam type, and from a dial on the dial type, the torque is preset on the microadjusting type, and it clicks when that torque is reached.

Torque wrench scales are graduated in foot-pounds, inch-pounds, Newton-meters or kilogram-meters. They can be used with extensions or adapters to get around obstructions. However, adapters that increase the length of the wrench from the drive end, figure 12-27, and thereby increase the leverage, will apply more torque than the wrench indicates. Extensions of the handle will not affect the reading, but will make it easier to apply the torques because of the increased leverage. Extensions that simply increase the height of the wrench above the bolt or nut will not affect the torque reading.

When using an adapter that lengthens the drive end of the wrench, figure 12-27, use the following formula to determine what the reading should be for the torques required:

$$\text{Wrench Reading} = \frac{\text{Torque at Fastener} \times \text{Wrench Length}}{\text{Wrench Length} + \text{Adapter Length}}$$

If, for example, we wanted to tighten a bolt to 30 foot-pounds using a 16-inch wrench with a 4-inch adapter, we would substitute into the formula to get:

$$\frac{30 \times 16}{16 + 4} = \frac{480}{20} = 24$$

Therefore the torque on the bolt will be equal to 30 foot-pounds when the torque wrench shows 24 foot-pounds.

Engine Rebuilding Hand Tools

Pistons are often reused during an engine overhaul. Carbon collects in the ring grooves behind the rings and must be removed or the new rings will not fit properly into the grooves. The piston ring groove cleaner, figure 12-28, is a hand tool with scrapers to fit different sizes of grooves.

A piston ring expander, figure 12-29, is a tool that will spread the piston rings enough to allow them to slip over the head of the piston and into the grooves. Using this tool avoids deforming or breaking the rings.

Tools and Precision Measuring Instruments

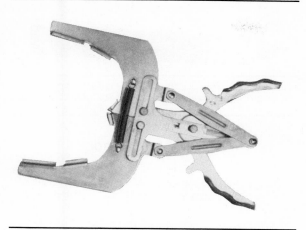

Figure 12-29. The piston ring expander spreads the rings enough to allow them to slip over the head of the piston.

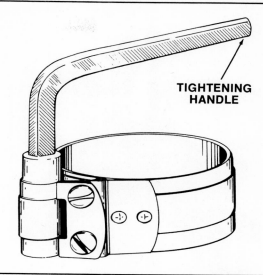

Figure 12-30. The ring compressor squeezes the rings into the grooves so the piston can be installed in the cylinder.

Figure 12-31. The piston pin drift is used to drive out piston pins.

Figure 12-32. Seal installers are used for installing oil pump, crankshaft, and similar type oil seals.

In order to install the pistons in the cylinders, the rings must be compressed. A ring compressor squeezes the rings into their grooves, figure 12-30. When the piston is pushed into the cylinder the rings pass into the cylinder in their compressed position.

A piston pin drift, figure 12-31, is a special tool for driving out piston pins. For pins that are a press fit in the rod, the drift may be part of the pressing fixture. Some pistons have snapring pin retainers that fit into grooves in the piston pin bore. These are removed with a snapring pliers, figure 12-16.

Seal installers are often necessary for installing oil pump, crankshaft, and similar type oil seals. The installer, figure 12-32, is slightly smaller than the seal. It fits over the seal and helps to drive it in straight.

Harmonic balancers fit tightly on the crankshaft and require a special puller for removal, figure 12-33. Because of the rubber ring between

■ Emergency Ring Compressor

Modern ring compressors do an excellent job. But what if a ring compressor is not available? In a pinch, a large hose clamp can be used. It may compress only one ring at a time, and then have to be shifted up to the next ring, but it will get the job done. If the available hose clamps are not large enough, two hose clamps can usually be joined end to end, making one large clamp.

Figure 12-33. This special puller is for use with the harmonic balancer.

Figure 12-34. This pry bar compresses the spring while compressed air holds the valve closed.

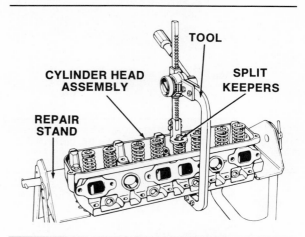

Figure 12-36. When the heads are off the engine, a C-type valve spring compressor can be used.

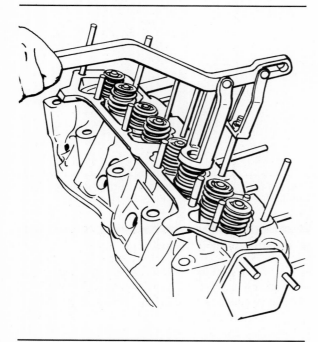

Figure 12-35. Overhead cam engines use a tool that bolts to the head and compresses the valve springs.

the hub and outside steel ring, the balancer will be pulled apart if a jaw type puller is used on the outer edge. To prevent this, tapped holes are provided in the hub of the balancer. Capscrews are threaded into the holes, and the special puller is used to pull on the capscrews while pushing on the end of the crankshaft. Some pullers are actually designed to remove or install the balancer or pulley.

There are many tools available for removing and replacing valves. The pry bar in figure 12-34 compresses the spring while compressed air holds the valve closed. Some overhead cam engines require a special tool that bolts to the head and compresses the valve spring so that the valve keepers can be removed, figure 12-35. It can be used with the head on or off the engine. A C-type valve spring compressor is used to compress the valve springs with the heads off the engine, figure 12-36. Other special cylinder head tools include rocker arm stud removers and installers.

Other special engine tools include drivers for removing and installing cam bearings, figure 12-37, and plastic sleeves which slip over connecting rod bolts, figure 12-38. These sleeves prevent scratching or marring the crankshaft when installing pistons in the block. New tools to assist mechanics are constantly being invented and manufactered as the need arises.

PRECISION MEASURING INSTRUMENTS

When working on crankshafts, cylinder heads, valves, and bearings, it is important to make

Tools and Precision Measuring Instruments

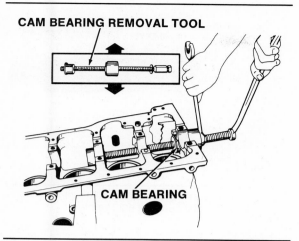

Figure 12-37. Another special engine tool is the cam bearing driver.

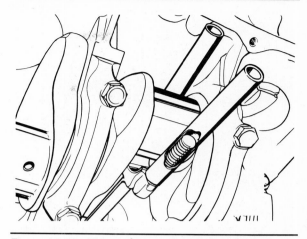

Figure 12-38. These plastic sleeves slip over the connecting rod bolts to protect the crankshaft.

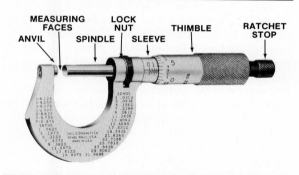

Figure 12-39. Micrometers indicate measure by revealing graduations as the thimble is turned.

sure of the exact dimensions of the parts. This is done with the use of precision measuring instruments either in U.S. or metric measure.

U.S. Customary And Metric Graduations, A Comparison

Most of the tools and measuring instruments used on automobiles today are made to be read in inch or millimeter dimensions. A measurement taken in inches can be converted to a measurement in millimeters by using a conversion factor or number. The dimension is multiplied by this number, and the result is the dimension in the other system.

For example, there are 25.4 millimeters in one inch. Therefore, the conversion factor from inches to millimeters is 25.4. To find the correct metric wrench to fit a 1/2-inch nut, multiply 1/2 by 25.4 to get 12.7 millimeters. Since metric wrenches are not available in fractional sizes, a 13-millimeter wrench would be used in this case, since it is only 0.3 millimeter too big.

When making precision measurements of crankshaft journals, valves, bearings, and similar items, the same conversion factors are used. If a journal measures 1.125 inches, we convert to millimeters by mutiplying 25.4, giving us 53.975 millimeters. This should be rounded to 53.98 millimeters, or one decimal place less than the inch measurement. This avoids a misleading appearance of precision by removing insignificant decimal places.

When converting from millimeters to inches, a similar rule applies: only retain one decimal place more than the number from which you are converting. For example, to convert 13 millimeters to inch measure, multiply by the conversion factor of .03937 to get .51181 and round to .5.

When measuring parts, either inch or metric measuring tools may be used. But if the specification or sizes are given in metric, and inch tools are used to do the measuring, then there is the added step of converting the measurement so that it can be compared to the specifications. For this reason, when working on metric-designed parts, use metric measuring instruments if possible. And when working on inch-designed parts, use inch measuring instruments.

Micrometers

Micrometers are designed to precisely measure the outside or inside of an object, and are called outside and inside micrometers respectively.

Although some micrometers are direct reading, most micrometers indicate measure by revealing graduations as the thimble is turned, figure 12-39. Additional graduations on the thimble itself are added to the exposed reading on the sleeve to get the total measure.

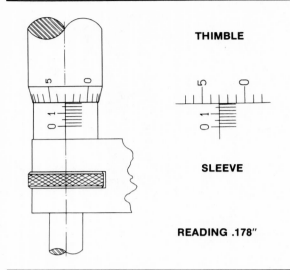

Figure 12-40. This figure shows how the total reading is obtained.

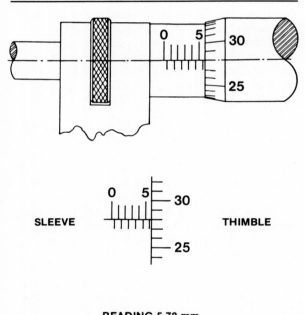

Figure 12-42. This shows how to obtain the total reading with a metric micrometer.

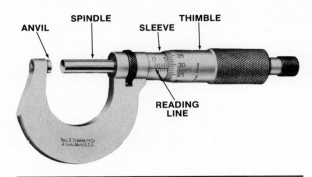

Figure 12-41. The metric micrometer operates the same way as an inch reading micrometer.

Micrometers usually measure in thousandths of an inch or hundredths of a millimeter, although more precise micrometers are available. The capacity of an outside micrometer ranges from one to six inches, and is determined by the opening in the frame. However, the range is only one inch less than its capacity. In other words, a 4-inch micrometer can only measure objects that are from 3 to 4 inches in size, while a 6-inch micrometer measures objects from 5 to 6 inches in size. Longer anvils can be substituted on some micrometers to increase their measuring range.

Refer again to figure 12-39 during the following description of how to use a micrometer. The object to be measured is placed between the measuring faces, the anvil and the spindle, of the micrometer. The thimble is turned to move the spindle until it contacts the object. The dimension of the object is then found from the micrometer by looking at the graduations on the sleeve and thimble. Each line exposed on the sleeve indicates 25 thousandths of an inch (.025″). Every fourth line then is 100 thousandths of an inch (.100″), and is marked by a number to indicate this. Since each revolution of the thimble moves the measuring face .025″, the edge of the thimble is graduated with 25 lines, each representing .001″. Every fifth line is marked with a number making it easy to read.

To obtain the total reading indicated in figure 12-40, multiply the number of graduations exposed on the sleeve by .025″, and add the number of the graduations aligned with the marker on the sleeve. In figure 12-40, there are 7 graduations multiplied by .025″ giving us .175″. The 3 graduation on the thimble is aligned with the mark on the sleeve giving us .003″. Adding the two gives us a total of .178″, which is the dimension of the object.

The metric micrometer, figure 12-41, operates exactly the same except for the difference in graduations. Each line on top of the sleeve scale indicates 1 millimeter (1 mm). Each line on the bottom of the sleeve scale indicates one half or .5 mm. The lines on the thimble each indicate one hundredth of a millimeter (.01 mm). To obtain the reading shown in figure 12-42, add the 5 mm indicated on the top sleeve scale and the .5 mm indicated on the bottom sleeve scale for a subtotal of 5.5 mm. To this add the 28

Tools and Precision Measuring Instruments

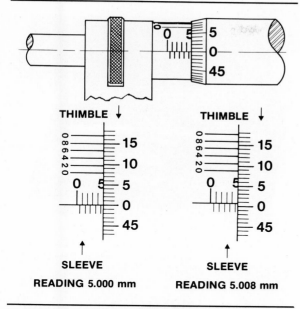

Figure 12-43. The vernier scale on this micrometer consists of a scale on the sleeve with five divisions.

Figure 12-45. The sled type of bore gauge is self-aligning.

hundredths or .28 mm indicated on the thimble scale, for a total of 5.78 mm. Micrometers which measure smaller than .001″ or .01 mm have a vernier scale in addition to the other scales. This consists of a scale on the sleeve with five divisions, each of which equal one-fifth of a thimble division, figure 12-43. For any position on the thimble, only one of the graduations on the scale will align with a graduation on the thimble. This additional reading as indicated on the scale on the sleeve is added to the main reading.

Inside micrometers use the same measuring thimble and sleeve system as the outside micrometers just described, figure 12-44.

Avoid overtightening a micrometer. Too much force will bend the instrument or the part it is measuring and give an incorrect reading. To prevent this, some micrometers are made with a ratchet stop or a friction thimble, figure 12-39. Overtightening the thimble will simply cause the ratchet or friction device to slip.

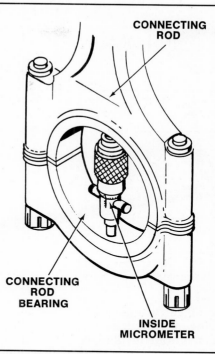

Figure 12-44. The inside micrometer uses the same measuring scale.

Cylinder Bore Gauges

Using a cylinder or dial bore gauge to measure cylinder bore diameter is much faster than using an inside micrometer.

There are two types of bore gauges. The sled type, figure 12-45, is self-aligning. It is inserted into the bore and the plunger spring pushes the sled against the cylinder wall. The dial is then set to zero. Sliding the sled up and down in the bore will indicate taper. Turning the sled around the cylinder will show if it is out-of-round.

■ The Good Old Days?

Well, yes, prices have increased in the automotive industry. Just take a look at this 1926 price list for doing repairs on the hugely popular Ford Model T.

These are the flat-rate labor charges recommended by Ford to all its dealers and mechanics:

Overhaul engine and transmission $25.00
Overhaul engine only (includes
   R&R, rebabbitting, reboring) $20.00
Overhaul transmission only $14.00
Grind valves, clean carbon . . . . $ 3.75
Replace head gasket . . . . $ 1.00
Overhaul carburetor . . . . $ 1.50
Adjust transmission bands . . . . $ 0.40

Figure 12-46. The rocking bore gauge measures more precisely than the sled type gauge.

Figure 12-48. Telescoping gauges are sometimes called snap gauges.

Figure 12-47. The setting micrometer for the rocking-type bore gauge gives it an advantage.

To get the instrument within the range of the bore being measured, extensions are added to or removed from the plunger. The sled-type bore gauge is not used to directly measure the diameter of the cylinder, but to measure variations in the diameter to determine taper and out-of-round. However, the actual diameter can be determined by zeroing the dial, removing the gauge, and measuring it with a micrometer.

The rocking-type bore gauge, figure 12-46, is more of a precision instrument than the sled type. It has two aligning slides that center a plunger in the bore. To get the diameter of the bore, the instrument is rocked up and down until the smallest reading is shown on the dial. The sled-type gauge usually reads to .0005'', while the rocking-type reads to .0001''.

A big advantage of the rocking-type bore gauge is the setting micrometer available for it, figure 12-47. The setting micrometer instantly aligns the plunger with the micrometer faces. The gauge dial is then zeroed at the standard bore size. Inserting the gauge shows if the cylinder is oversize, and moving the gauge around shows if it is tapered or out-of-round.

Telescoping Gauges

Telescoping gauges, figure 12-48, are sometimes called snap gauges because they are spring loaded and will snap out when released. They serve the same purpose as inside micrometers. A spring extends the gauge against the work, and the gauge is tightened by turning the handle. It is then rocked to make sure it is at the largest diameter. If not, the handle must be loosened and retightened. Then the gauge is removed in the locked position and measured with an outside micrometer.

Although telescoping gauges can be used to measure cylinders several inches in diameter, their primary use is in measuring small cylinders in which a hand holding an inside micrometer will not fit.

Feeler Gauges

Feeler gauges are flat pieces of material that are made to a precise thickness, figure 12-49. They range in thickness from .0015'' or .04 mm to .025'' or .70 mm. Blades of various sizes are inserted into the gap to be measured, until the correct size blade is found. Then the size of the blade is read to determine the size of the gap. Blades can be stacked to get additional measurements.

Dial Indicators

Dial indicators, as the name implies, have a plunger connected to a dial, figure 12-50. The amount of movement of the plunger is indicated on the dial in fractions of an inch or millimeter.

Many different kinds of holders, fixtures and accessories are available to enable dial indicators to measure depth, out-of-round conditions, run-out, and other dimensions.

Straightedges

Cylinder heads and blocks go through a heating and cooling cycle everytime an engine is started and then stopped. This cycle can cause warping of the head and block mating surfaces, especially if the engine is disassembled while warm.

Head and block gasket surfaces should always be checked for warpage with a straight-

Tools and Precision Measuring Instruments

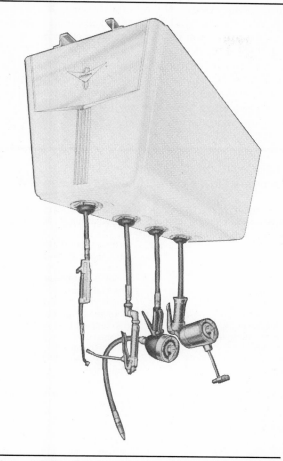

Figure 12-63. Overhead lubricant dispensers with retractable hoses are one of the many conveniences which may be powered by a compressed air system.

and the oil maintained at the proper level. Compressors usually have a dipstick or plug that is removed to check the oil level. A compressor that runs low on oil can destroy itself in a matter of minutes.

Compressors are usually belt driven. If one or more belts break, the remaining belts will be overloaded. Belts should be checked frequently and replaced if they show signs of slipping, cracking or wearing out.

Most compressor tanks have a valve on the bottom. This is a drain and should be opened regularly to drain the water that accumulates from condensation, and to minimize rusting of the tank.

SUMMARY

Tools make the job of engine rebuilding safer, faster and easier. They are designed for specific jobs, and the proper tool should be used for its designed job whenever possible.

Figure 12-64. The air compressor must be checked frequently and the oil maintained at the proper level.

Wrenches are made in open end, box end, combination open end and box end, and flare nut styles. The position of the fastener usually determines which kind of wrench to use.

Screwdrivers come in many lengths and tip sizes in addition to many tip configurations. Always use the screwdriver that is designed for the screw you are turning.

Pliers are used for both gripping and cutting. The jaw shape is designed for a specific use. Common hammers have steel heads, but in automobile engine rebuilding the brass, rawhide, plastic and shot-filled hammers are frequently used.

Socket wrenches are the most used tools in automotive work. They come in a variety of drive sizes, and have different handles available for different applications. A torque wrench is used to tighten a fastener to a specific degree of tightness.

There are many hand tools that are made specifically for use in rebuilding engines. These include piston ring tools, seal installers, pullers, and valve tools.

Engine parts must often be measured to determine tolerances and fit. Precision measuring tools are made to give accurate readings, but are only as accurate as the mechanic using them. Micrometers and dial indicators are especially fragile, and must be handled with care or they will lose their precision.

Safety is of prime importance when using power tools. Misuse of power tools can not only injure the operator, but also damage the work. Eye protection must always be worn when working with or near these tools.

Review Questions

Choose the single most correct answer.
Compare your answers to the correct answers on page 268.

1. Flare-nut wrenches are used:
   a. To break loose fasteners
   b. On the ignition system components
   c. When an open-end wrench won't fit
   d. On tubing fittings

2. Which of the following is not a type of torque wrench?
   a. Micro-adjusting type
   b. Indicator type
   c. Beam type
   d. Dial reading type

3. Air ratchets or speed ratchets have:
   a. An impact mechanism
   b. A 1/2" drive
   c. A reversing mechanism
   d. All of the above

4. Impact wrenches should be used with:
   a. 3/4" sockets
   b. Hardened chrome sockets
   c. Special impact sockets
   d. Only on bolts over 7/16" (11 mm)

5. A dial bore gauge is sometimes used to measure cylinder diameter because it is:
   a. Faster than an inside micrometer
   b. Easier to use than an inside micrometer
   c. More accurate than an inside micrometer
   d. Less expensive than an inside micrometer

6. Which of the following should never be used on an automobile?
   a. Clutch head screws
   b. Phillips head screws
   c. Straight head screws
   d. Butterfly head screws

7. Extensions that increase the height of the torque wrench above the fastener will:
   a. Not affect the torque reading
   b. Change the torque reading according to the wrench length/extension length
   c. Make the torque reading erratic
   d. Cause the fastener to be overtightened

8. To break loose a bolt, you can use a socket with a:
   a. "T" handle
   b. "L" handle
   c. Flex handle
   d. All of the above

9. An open-end wrench usually has its head:
   a. At an angle
   b. Curved
   c. Straight
   d. Flared

10. Box end wrenches:
    a. Have 8-point openings
    b. Are stronger than open-end wrenches
    c. Should not be used for breaking nuts loose
    d. All of the above

11. A Reed and Prince screwdriver can be used with a Phillips head screw if:
    a. The tip is ground down
    b. The screw is smaller than a No. 3
    c. It is not overtightened
    d. None of the above

12. The shot-filled hammer was designed to replace the:
    a. Ball peen hammer
    b. Rawhide hammer
    c. Brass hammer
    d. Rubber hammer

13. Which of the following is not a part of a micrometer?
    a. Anvil
    b. Stirrup
    c. Spindle
    d. Thimble

14. To measure a bearing journal you should use a:
    a. Micrometer
    b. Vernier calipers
    c. Telescoping gauge
    d. Dial bore gauge

15. Overtightening a micrometer with a friction thimble will:
    a. Break it
    b. Necessitate that it be recalibrated
    c. Bend the frame
    d. None of the above

Chapter 13
Engine Test Equipment

THE SCOPE FOR ENGINE CONDITION DIAGNOSIS

The **oscilloscope** is a sophisticated electronic test instrument that shows the changing voltage levels in an electrical system over a period of time. The oscilloscope or scope, as it is usually called, is used primarily for ignition and electrical system diagnosis.

The scope has a screen much like a television screen. On the screen is displayed a line of light called a trace. The trace indicates the voltage levels over a period of time. The automotive oscilloscope screen is marked with voltage and time scales, figure 13-1. The vertical scales at the sides of the screen indicate voltage, and are usually marked in kilovolts, meaning thousands of volts (abbreviated kV). Just as many voltmeter scales are marked with several voltage ranges, most oscilloscope screens have several voltage ranges marked on each side.

Some scopes also have a low-voltage scale on one or both sides of the screen, figure 13-1. The kV scales are used to measure voltage in the secondary circuit of the ignition system. The low-voltage scales can be used to measure voltage in the ignition primary circuit and the charging system. By operating a switch on the oscilloscope console, you can select the range that will give you the most legible voltage reading for the test being performed. On the screen shown in figure 13-1, the zero-voltage line is at the bottom. However, some scopes have the zero line above the bottom so that both positive and negative voltage can be shown.

The distance across the screen represents time. In figure 13-1, time measurements across the bottom of the screen are given in degrees of ignition distributor rotation. This is used to measure coil conduction time, or dwell.

Cathode Ray Tube

The oscilloscope screen is the front of a cathode ray tube, or CRT. As we said, the picture tube in your television is also a CRT. Although a cathode ray tube is a complex electronic device, its operation is quite simple.

A cathode ray tube, figure 13-2, has an electron gun which emits a beam of electrons. The inside of the oscilloscope screen has a fluorescent coating so that it glows at the spot where the beam hits it. The beam is controlled by two pairs of deflection plates. Depending upon the type of charge present upon the plates (+ or −), they will either attract or repel the electron beam, causing it to bend or deflect.

One set of plates bends the beam up and down to change the height of the trace. These are called the vertical deflection plates. They are

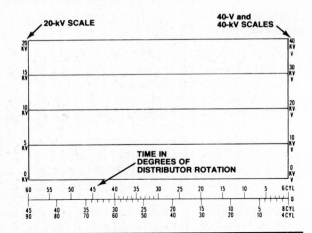

Figure 13-1. The oscilloscope screen has kilovolt scales on both sides and a low-voltage scale on the right side.

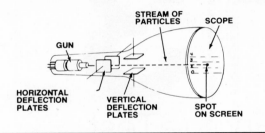

Figure 13-2. The electron gun traces a pattern of light on the inside of the cathode ray tube screen. (Marquette)

controlled by the voltage of the system being tested. The higher the voltage, the greater the beam is deflected, making the light appear farther up the voltage scale on the screen.

The second set of plates moves the beam from side to side so that the light travels across the screen from left to right. These are called the horizontal deflection plates. They are controlled by the speed of the engine being tested. The faster the engine runs, the faster the distributor turns, and the faster the light will travel across the screen.

Our eyes do not see the single dot of light on the screen, but instead see a trace of light across the entire screen. This is because the beam moves very fast, and the phosphors on the inside of the screen continue to glow for a short time after the beam hits them. We see only the left-to-right movement of the trace, because the electron beam is turned off during its return movement.

Oscilloscope Use

Most oscilloscopes are mounted in consoles that contain other pieces of test equipment, including voltmeters, ammeters, **dwellmeters**, **tachometers**, **vacuum gauges**, and other specialized scales. Many different tests can be made on a car while it is connected to the console. For this reason, the consoles are sometimes called engine analyzers.

Although connections differ from model to model, most scope consoles require some or all of the connections shown in figure 13-3:
1. Scope negative lead to the engine ground
2. Voltmeter positive and negative leads to battery terminals
3. Inductive ammeter clamp around specified battery cable
4. Positive lead either to battery positive terminal or to alternator output (battery) terminal
5. Scope positive lead to coil primary terminal (either + or − terminal, depending on equipment design)
6. Inductive clamp around coil secondary lead
7. Inductive clamp around number 1 spark plug cable
8. Vacuum probe to manifold vacuum source
9. Exhaust probe in tailpipe (if console has infrared exhaust gas analyzer).

POWER BALANCE TEST

If an engine is in good condition, all its cylinders should contribute equally to keep the engine running. If one cylinder is removed or cancelled, the engine should slow down unless that cylinder was not contributing.

The **power balance test** is usually made with the engine running at about 1500 rpm, but it can also be made with the engine under load on a dynamometer. Testing an engine under load is better because actual driving conditions are being duplicated.

To cancel a cylinder, the spark plug wire is grounded. This can be done by inserting grounding connectors between the wire and the plug at each spark plug. Then a screwdriver or wire is attached from the connector to the engine block. This grounds the high voltage to the spark plug and prevents that cylinder from firing. An electronic power balance tester does this at the touch of a button, figure 13-4.

Engine speed is measured with a tachometer during the balance test. Sometimes more accurate readings result if the engine is idling slightly higher than normal, or at about 1500 rpm. With the engine idling, each cylinder is grounded. The cylinders should all reduce the rpm by the same amount. If one or more cylinders do not reduce rpm as much as the others, those cylinders are weak. The amount of rpm

Engine Test Equipment

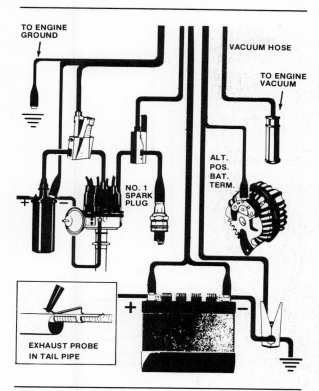

Figure 13-3. Typical oscilloscope console test lead connections. (Sun)

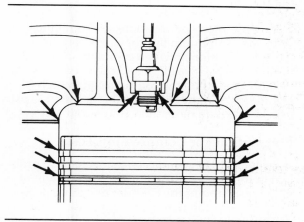

Figure 13-5. Good compression depends on proper cylinder sealing at the points shown.

drop has no significance if all cylinders are equal. A cylinder that is completely dead will not affect engine speed at all when grounded.

Power Balance Test Precautions

If the car you are testing has a valve-controlled EGR system, it must be disconnected during the power balance tests. Otherwise, the cycling of the EGR system would affect the engine rpm and interfere with your test results. A 1972-73

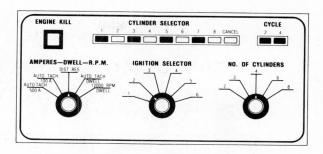

Figure 13-4. On this console control panel, the pushbuttons marked "Cylinder Selector" allow the operator to prevent certain cylinders from firing. The numbers refer to the firing order, not the cylinder number. (Sun)

Chrysler Corporation engine with the floor-jet EGR system cannot be reliably tested by the power balance procedure.

If the car you are testing has a catalytic converter, you must try to limit the amount of unburned fuel reaching the converter. Various test equipment makers have different methods of testing a converter-equipped car. Check the instructions for the equipment you are using.

COMPRESSION TEST

The power output of any car engine depends on the compression in its cylinders. This compression depends on how well each cylinder is sealed by the piston rings, the valves, the cylinder head gasket, and around the spark plug, figure 13-5. If any of these points is not sealed well, compression is lost and power output drops.

The compression pressure in each cylinder and variations in compression between cylin-

---

Oscilloscope: An electronic device used to visually observe and measure the instantaneous voltage in an electric circuit.

Dwellmeter: An instrument to measure the angle through which the distributor shaft turns while the distributor breaker points are closed.

Tachometer: An instrument that measures the speed of a rotating part in revolutions per minute.

Power Balance Test: A test to determine if all cylinders are contributing equally to the power output of the engine.

Vacuum Gauge: A pressure gauge which measures the amount of suction or vacuum created in the manifold.

Figure 13-6. Compression gauge.

ders are measured by a **compression test**. By measuring how well each cylinder will hold its compression, any pressure loss can be located. Compression pressure is measured in pounds per square inch (psi) or kilopascals (kPa) using a special compression gauge.

Compression Gauge

A compression gauge, figure 13-6, is able to measure very high pressures. One of the better gauge designs has a threaded adapter that screws into the spark plug hole. A hose connects the adapter to the gauge. Another gauge design has a tapered rubber tip that will fit into any size of spark plug hole. The gauge must be held firmly in the hole during testing. This design is harder to use and may not be as accurate.

The gauge needle will stay at its reading until a vent valve in the gauge is opened. This makes it easier to note the compression reading. Some testers have a built-in remote starter switch, so that you do not have to connect a separate switch to crank the engine.

Compression Specifications

Every carmaker gives compression specifications for their engines. They are usually stated in one of two ways:
1. Compression specifications may require the lowest-reading cylinder to be within a certain percentage (generally 75 percent) of the highest-reading cylinder. For example, if the highest cylinder reading is 160 psi (1103 kPa), the lowest reading must be at least 120 psi (830 kPa), since 75 percent of 160 (1105) is 120 (827).

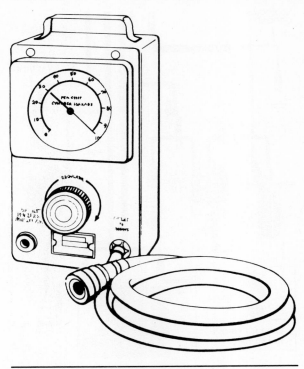

Figure 13-7. The cylinder leakage tester measures the percentage of compressed air that leaks from the combustion chamber.

2. Compression specifications are sometimes stated as a minimum figure, with a certain allowable variation between cylinders. If the minimum is 130 psi (896 kPa), and the variation is 30 psi (207 kPa), then the compression in each cylinder is all right as long as it is at least 130 psi (896 kPa) but no more than 160 psi (1103 kPa).

Preparation For Compression Tests

If earlier tests have indicated a mechanical problem in the engine, the compression test will help you pinpoint the cylinder or cylinders affected. The combination of compression test results and cylinder leakage test results can often tell you exactly what is wrong. The cylinder leakage test is described later in this chapter. The following paragraphs describe both wet and dry engine compression test procedures.

The dry test is done first, with the engine in its normal operating condition. The wet test is done if the engine fails the dry test.

Dry Compression Test

A dry compression test should be done on every cylinder. It should be done with the engine warm, the throttle fully open, and the spark plugs removed. An engine in perfect condition will have compression readings equal to those

Engine Test Equipment

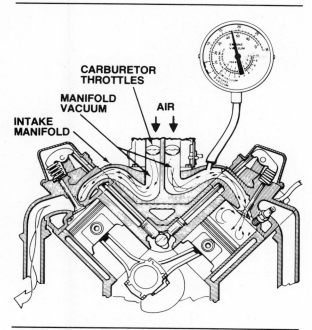

Figure 13-8. The vacuum gauge is connected to a tap on the intake manifold to measure manifold vacuum.

ALTITUDE	INCHES OF VACUUM
Sea Level to 1,000 Ft.	18 to 22
1,000 Ft. to 2,000 Ft.	17 to 21
2,000 Ft. to 3,000 Ft.	16 to 20
3,000 Ft. to 4,000 Ft.	15 to 19
4,000 Ft. to 5,000 Ft.	14 to 18
5,000 Ft. to 6,000 Ft.	13 to 17

Figure 13-9. A normal engine produces approximately these vacuum readings at the altitudes shown.

of a new engine. They will also be virtually the same for all cylinders.

If any engine cylinder reads low during the dry test, go to the wet compression test.

Wet Compression Test

A wet compression test should be made on all low-reading cylinders. Oil is added through the spark plug hole to temporarily improve the sealing of the piston rings. This test is not accurate on flat or pancake engines.

If the compression is higher on the wet test, the rings are at fault. If the compression is still low, the valves are at fault. This test is not valid if both valves and rings are bad, or if there is a head gasket leak. This test may not work if too much oil is squirted into the cylinder because the compression will increase when the oil fills up some of the space in the combustion chamber.

CYLINDER LEAKAGE TEST

The **cylinder leakage** test gives even more detailed results than the compression test. The leakage test can tell you:
• The exact location of a compression leak
• How serious the leak is in terms of percentage of the cylinder's total compression.

A special cylinder leakage gauge is used during this test, figure 13-7. The cylinder to be tested must have the piston at top dead center on the compression stroke, so that both valves are closed. Compressed air is applied to the sealed combustion chamber with the gauge installed in the line. The gauge needle indicates how much of the compression leaks out of the combustion chamber.

The gauge scale is marked from 0 to 100 percent. A reading of 0 percent indicates a perfectly sealed chamber with no leaks. A reading of 100 percent indicates that no pressure at all is being held within the chamber. The gauge must be calibrated before each test, following the equipment maker's instructions.

VACUUM GAUGE

A vacuum gauge connected to the intake manifold will measure the vacuum that the engine creates in the manifold, figure 13-8. Vacuum is measured in inches of mercury (in. Hg) in the American system. The metric system measures vacuum in millimeters of mercury (mm Hg) or kilopascals (kPA). The kilopascal is also used to measure pressure. When the kilopascal is written, a minus sign is used to indicate vacuum, and a plus sign or no sign at all to indicate pressure.

Normal engine vacuum at idle is from 13 to 22 in. Hg (330 to 560 mm Hg), depending on the altitude, figure 13-9. The reading varies with different engines because of the differences in cam timing and valve lift. Older engines will idle with several inches more vacuum than modern emission-controlled engines. Because of valve overlap, as explained in Chapter 7, the reading will also be steadier on early engines.

Compression Test: A test using a pressure gauge to measure the pumping pressure in a cylinder during cranking.

Cylinder Leakage Test: A test in which compressed air is forced into a cylinder and the percent of leakage measured.

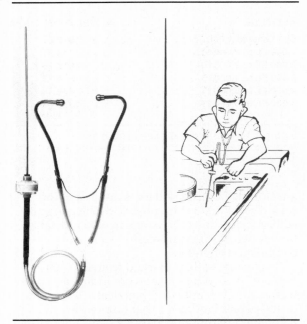

Figure 13-10. The stethoscope blocks out all noise except the noise coming through the tube.

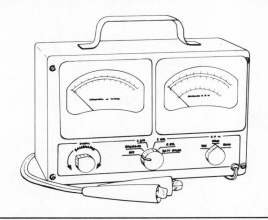

Figure 13-11. A typical tach-dwellmeter.

Generally, the higher and steadier the vacuum reading, the better the condition of the engine. Because of the great difference in readings between different designs of engines, comparisons between makes and types cannot be made. For example, readings of two new engines could differ: one might idle at 14 in. Hg (355 mm Hg) and the other at 20 in. Hg (510 mm Hg) because of differences in design.

However, when comparing engines of the same year, make, and design a vacuum reading is valuable. A low vacuum reading could be caused by something as simple as retarded timing, but it could be a much more serious problem and should be investigated and corrected.

In addition to the idle vacuum test, there are several other tests which can be performed. Cranking vacuum is checked by cranking the engine with the ignition disabled. It is valuable for diagnosing an engine that will not start.

The exhaust restriction test is done by checking vacuum at idle and at 2500 rpm. Normally, the reading at 2500 rpm should be higher than at idle. If it is lower, it indicates an exhaust restriction.

STETHOSCOPE

Engine noises are difficult to locate and pinpoint. A clicking, knocking, or any other noise might be from the bearings, fuel pump, alternator, or another source. A **stethoscope** conducts only those noises which are near the end of its tube, figure 13-10. When the end of the tube is near the source of the noise, the noise will be louder. Automotive stethoscopes come with a probe and a diaphragm that can be attached to the end of the tube. The probe will pick up noises from inside the engine if it is placed against the block or other metal surface. The diaphragm amplifies the noise so it can be heard more easily. A stethoscope eliminates guesswork as to the source of engine noise.

TACHOMETER AND DWELLMETER

A tachometer measures engine speed in revolutions per minute, figure 13-11. The instrument may have a dial and pointer or a digital readout. Power to run the tachometer may come from the engine, the car battery, or from shop current, depending on the tachometer design.

Many tachometers include or incorporate a dwellmeter, figure 13-11. A dwellmeter indicates the number of degrees of rotation that the distributor points are closed for each cylinder during each revolution of the distributor shaft. Dwell is a useful measurement because it is directly related to point gap. If the dwell is incorrect, so is the point gap and it should be corrected. Dwellmeters are of little value on most electronic ignition systems which have no points. Some dwellmeters will indicate if the ignition is performing properly or not.

Tachometers are made in three basic types: **primary**, **secondary**, and **magnetic**. The primary tachometer connects to the primary terminal of the distributor and to ground.

A secondary tachometer connects to any one spark plug wire. Some types have an electrical connection to the wire terminal; others are induction types that merely clamp around one plug wire.

The latest type of tachometer is magnetic. It will work only on engines that have a holder for a magnetic probe. The holder positions the mag-

Engine Test Equipment

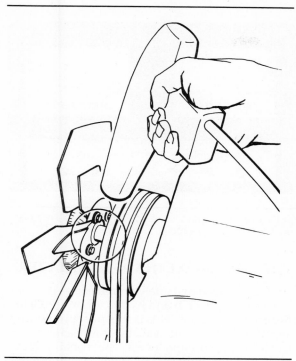

Figure 13-12. The strobe light of the fan tachometer is aimed at the hub and adjusted until the fan bolts appear to stand still.

netic probe of the tachometer next to a notch in a rotating part. Every time the notch passes the probe, the change in magnetic field is sensed and one revolution is counted. Most engines have the notch in the vibration damper, but some have it in the crankshaft or flywheel, and a plug must be removed to insert the probe.

Fan Strobe Tachometer

To check the operation of a cooling fan clutch, operate the engine with the air flow through the radiator blocked. A piece of cardboard in front of the radiator works well. When the radiator coolant gets warm enough, the fan thermostat will engage the fan and the increase in noise will indicate that the fan is working.

But this will not tell you exactly how much the clutch is slipping. The silicone in the fan drive allows some slippage even when fully engaged.

A strobe tachometer can be used to measure the exact fan speed. This is a light that can be adjusted to flash at any interval. When a mark on a rotating object appears to stand still, the strobe is flashing at the speed of the object.

To measure the fan speed, a mark is put on the water pump pulley, and the engine speed is adjusted until the pulley is turning at 2000 rpm, as indicated by the strobe light. Then the strobe

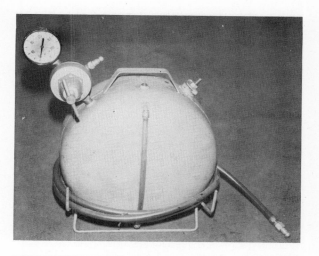

Figure 13-13. This small tank is used to lubricate the engine before final assembly or before testing the bearings for leakage.

light is aimed at the fan hub, and adjusted until the fan bolts appear to stand still, figure 13-12. The strobe will then indicate the speed of the fan. When the engine is overheated, fan speed should be about 80 percent of pulley speed. When the engine is cold or at normal operating temperature, the fan speed should be only about 25 percent of pulley speed.

BEARING LEAKAGE TESTER OR ENGINE PRELUBRICATOR

Removing crankshaft bearings and using a micrometer to inspect them is time-consuming and requires removing every bearing cap. A much faster method of checking for excessive bearing clearance is to use a bearing leakage tester.

The tester consists of a small tank, partially filled with engine oil, that can be pressurized

---

Stethoscope: A listening device used to pinpoint the source of engine noise.

Primary Tachometer: A tachometer which takes its signal from the ignition primary.

Secondary Tachometer: A tachometer which takes its signal from the ignition secondary.

Magnetic Tachometer: A tachometer that measures engine speed with a probe that magnetically senses the revolutions of the crankshaft.

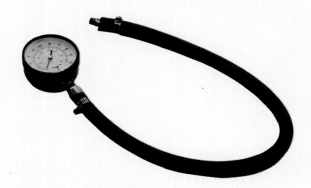

Figure 13-14. The oil pressure gauge is a direct reading gauge. It is used for testing engine oil pressure.

Figure 13-15. Typical dual-scale HC and CO meters on an infrared analyzer.

with compressed air, figure 13-13. It is sometimes called an **engine prelubricator** because it can also be used to lubricate the engine's pressure-fed bearings before starting.

The oil hose from the tank is connected to an oil passage on the engine block. The tapped hole for the oil pressure sending unit is usually the most accessible place. When the tank is connected, the pressurized oil will flow from the crankshaft bearings and out of the oil pump screen. Any source that drips at least 20 drops of oil per minute is considered to have correct clearance. If there is no dripping at all, the bearing is too tight, or the oil hole is blocked. If the oil comes out in a stream, the clearance is excessive.

The tester is generally used to doublecheck bearings in rebuilt engines before they are run.

OIL PRESSURE GAUGE

Some cars have an oil pressure gauge on the instrument panel, but it may not be accurate. It is usually an electric gauge with a sending unit screwed into an oil passage on the engine block. An oil pressure gauge for testing engine oil pressure is a direct reading gauge, figure 13-14. It is connected to the engine by removing an oil passage plug or the oil pressure sending unit. A hose or tube carries engine oil to the gauge head which measures the actual pressure of the oil in pounds per square inch. Even if there is air in the gauge, it will compress and have a negligible affect on the oil pressure reading.

If the engine knocks, the oil warning light comes on, or the dash-mounted pressure gauge reads low, a pressure gauge should be connected to the engine. It will tell you how much oil pressure there actually is, if any. If the problem only appears during driving, the gauge can be hooked up with a long hose so it can be read while driving.

THE INFRARED EXHAUST ANALYZER

An infrared exhaust analyzer measures the carbon monoxide (CO) and hydrocarbons (HC) in the engine exhaust. This is the most precise and sensitive instrument available for measuring CO and HC emission levels. The machine uses infrared light, light that our eyes cannot see, to measure these emissions.

CO is measured by a percentage, so a specifications table might show a range of 1 to 6%. HC is measured in parts per million (ppm), since it is found in smaller amounts in an exhaust sample. A specification list might show the HC range as 0 to 900 ppm. (By comparison, one ppm equals .0001 percent.)

The many models of infrared analyzers all work about the same. The analyzer measures CO and HC in engine exhaust by comparing the air from the tailpipe with clean air. It does this by sucking the two air samples into separate glass tubes. An infrared light beam shines through both tubes, but the beam is refracted (bent) by the impurities in the exhaust sample. The amount of refraction is measured by the analyzer and is converted to electrical signals at the meters.

These meters show the CO and HC concentrations. The meters may be single-scale or dual-scale meters, figure 13-15. They usually have an overall measurement range of 0 to 7.5 percent, or 0 to 10 percent, for CO, and 0 to 2,000 ppm for HC.

The analyzer picks up the exhaust sample from a probe inserted into the car's tailpipe. However, on cars that have catalytic converters, the probe should be inserted in the exhaust pipe *in front of* the converter whenever possible. Some Chrysler products have a probe access hole at the front of the converter, figure 13-16. If the car has an air pump, the pump outlet line must be disconnected and plugged during the test, figure 13-17.

Engine Test Equipment

Figure 13-16. Some vehicles have an access plug for the infrared analyzer probe in front of the catalytic converter.

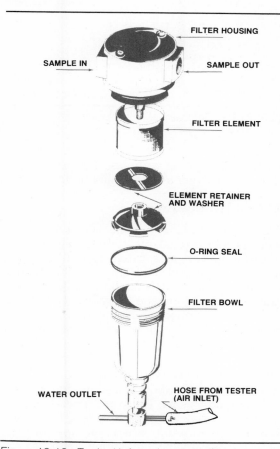

Figure 13-18. Typical infrared analyzer filter.

Filter Changes

Infrared exhaust analyzers draw the exhaust test sample through a filter system, figure 13-18. This removes moisture and particles from the sample. Routine filter service is required to ensure that the exhaust sample has no leftover particles from previous tests. Different ana-

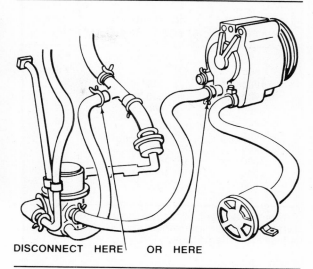

Figure 13-17. Because air-injection dilutes the exhaust with fresh air, the air pump must be disconnected before taking an exhaust sample.

■ Black Boxes Are Nothing New

Each year the new cars seem to have more and more driving operations controlled by electronic "black boxes." Many ignition, fuel, and air conditioning systems are governed by complex solid-state computers. And some carmakers are using electronic controls for transmissions.

Well, electrically controlled transmissions are not really new. Hudson did it back in the 1930's. The Hudson Electric Hand transmission didn't use a solid-state electronic computer, but it had an array of switches, circuit breakers, solenoids, and vacuum chambers to shift a conventional sliding gear transmission.

A selector switch was mounted on the steering column like an ordinary shift lever. The driver moved a small fingertip lever through a standard H-pattern for three speeds forward and reverse. This lever was a switch that controlled several solenoids. The solenoids, in turn, directed engine manifold vacuum through a selector valve to the vacuum chambers. The vacuum chambers operated the shift levers in the transmission. As a safety measure, a circuit breaker was attached to the clutch linkage so that the transmission couldn't shift unless the clutch was depressed. An interlock switch kept the vacuum chambers from trying to engage two gears at once.

Engine Prelubricator: An oil tank, pressurized with air, that is used to pump oil through the oil lines and passages of an engine while it is not running.

ELEVATION IN FEET	HC SPAN SETTING PPM	CO SPAN SETTING %
0	880	2.4
600	900	2.5
1200	920	2.6
1900	940	2.7
2400	960	2.8
3200	980	2.9
4000	1000	3.0
4400	1030	3.1
4800	1040	3.2
5300	1060	3.3
6000	1090	3.4

Figure 13-19. Typical altitude correction chart for infrared analyzer span adjustment. Correction factors will vary for different analyzers.

lyzers have different filters. Check the maker's instructions for filter cleaning directions

Most analyzers have an exhaust flow indicator of some kind to signal when filter service is necessary. This signal may be a flashing warning light or a simple visual gauge that contains a small floating ball.

Span Adjustment

Infrared analyzers are affected by altitude. The higher the altitude, the lower the density of the exhaust and reference samples. The span-set lines on most analyzer meters are calibrated for use at an altitude of about 600 feet (183 meters). Figure 13-19 shows the differences in span settings required to calibrate typical meters. Each test equipment maker has a procedure and altitude correction for span adjustment that must be followed.

Periodic Calibration

Most infrared analyzers can be electromechanically calibrated. They are also designed for certified gas calibration. Some states, such as California, have laws that require that this procedure be used. Special calibration equipment is built into or must be attached to the analyzer unit, figure 13-20. The equipment allows a certain amount of propane or hexane gas to enter the tester's infrared system. The gas used is certified to contain known and stated amounts of CO and HC.

The analyzer compares the CO and HC content in the gas against its reference sample and a reading on the display meter. Then the operator compares this reading with the stated value of the calibration gas. If the analyzer reading is within the specified limits of the gas sample concentration, the analyzer is correctly cali-

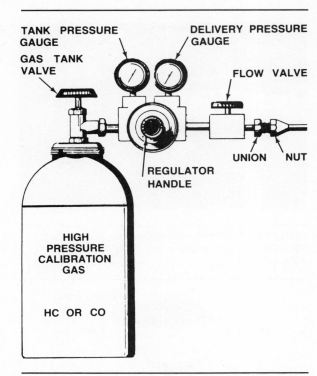

Figure 13-20. Gas calibration for an infrared exhaust analyzer. (Sun)

brated. If the reading is outside the specified limits, the analyzer must be calibrated.

CO And HC Readings

All exhaust samples should be taken when the engine is at its normal operating temperature. For carbureted engines, exhaust should be sampled both at idle and at 2,500 rpm. If the engine has fuel injection, only a reading at idle will be accurate. A dynamometer must be used to test fuel-injected engines at highway speeds. Of course, carbureted engines can also be tested on a dynamometer.

Normal CO and HC readings

The following table shows idle emission levels for older model years. These are general guidelines for *maximum* CO and HC levels. They are used in many tune-ups when exact specifications are not available. For engines with air injection, the readings should be taken with the air pump outlet line disconnected and plugged.

Model Year	CO%	HC ppm
Pre-1968 GM	1 to 6	0 to 900
Pre-1968 non-GM	1 to 6	0 to 600
1968-69	0 to 4	0 to 400
1970-74	0 to 3	0 to 300

Engine Test Equipment

Figure 13-21. Most temperature gauges have only a "C" for cold and an "H" for hot.

Abnormal CO and HC readings

The level of CO in the exhaust gas of a running engine is related to the air-fuel mixture. Too much CO emission is a direct result of an overly rich air-fuel mixture. There is either too much gasoline or not enough air in the mixture. The problem is in the fuel system and usually will be caused by one of the following conditions:
1. A dirty air cleaner filter
2. Clogged or dirty air mixture passages in the carburetor
3. Rich carburetor mixture adjustment
4. A stuck or improperly adjusted choke mechanism
5. Incorrect idle speed
6. A high float level in the carburetor
7. A leaking power valve in the carburetor.

High HC levels mean there is excessive unburned fuel in the exhaust. This is usually caused by a lack of ignition, or by incomplete combustion. The cause of high HC emissions usually can be traced to the ignition system. However, engine mechanical problems or carburetion problems can also increase HC emissions.

Some common causes of high HC levels are:
1. Incorrect ignition timing
2. Ignition system faults that cause a misfire (defective spark plug wires, breaker points, fouled spark plugs)
3. An excessively rich *or* lean air-fuel mixture
4. Leaking vacuum hoses, vacuum controls, or gaskets
5. Low engine compression
6. Defective valves, valve lash, valve springs, lifters, guides, or camshaft
7. Defective rings, pistons, or cylinders.

In addition to these, high readings of CO and HC at the same time may be caused by one of the following conditions:
1. Defective PCV system or catalytic converter
2. Defective manifold heat control valve
3. Defective thermostatic air cleaner.

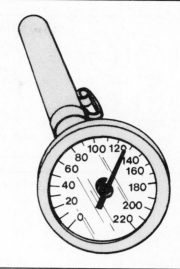

Figure 13-22. A gauge with markings in degrees on the dial can be calibrated to check the operating temperature of an engine.

ENGINE TEMPERATURE GAUGES

The temperature gauge on the instrument panel of the car is usually an indicator, not a precision instrument. Often, it is not even marked with numbers, merely "C" for cold and "H" for hot, figure 13-21. The ordinary instrument panel gauge should never be trusted in diagnosing engine temperature problems.

If it is necessary to check the operating temperature of an engine, a calibrated gauge should be used. This can be any gauge with markings in degrees on the dial, figure 13-22. To calibrate the gauge, the sending unit should be heated in a solution of antifreeze and water on the stove. As the temperature rises, the reading on the gauge should be checked with an accurate thermometer. If the gauge is not accurate, it should be marked to reflect the actual temperature. A 50-percent solution of antifreeze and water will boil at approximately 225°F (107°C) and an 80-percent solution will boil at 250°F (121°C). These boiling points are at sea level. High altitude will reduce the boiling point 2°F (1.1°C) per thousand feet (305 meters) above sea level.

Top Tank Thermometer

A quick check on engine thermostat performance is made with a top tank thermometer. In most cooling systems, it is possible to see coolant moving through the open filler neck when the thermostat opens.

If the test is started with a cold engine, the coolant should not move when the engine is

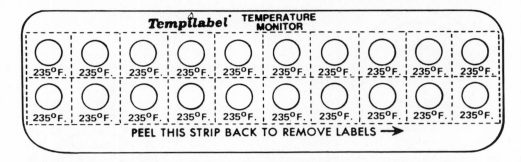

Figure 13-23. Temperature sensing labels turn black when the part to which they are applied reaches a specific temperature.

idling. The top tank thermometer is inserted into the coolant. As soon as coolant movement is noticed the reading is taken. This is the approximate opening temperature of the thermostat.

The advantage of this test is that it will detect thermostats that open either too high or too low. A thermostat that opens at a low temperature will reduce engine efficiency, decrease gas mileage, and may increase engine wear; one that opens at a high temperature may cause overheating on hot days.

Temperature Sensing Patches And Crayons

A temperature sensing patch is a self-stick label that is marked with an exact temperature, such as 235°F (113°C), figure 13-23. The patch will stick to any clean, dry surface. As soon as the surface reaches the marked temperature, the center of the patch will turn black.

The crayons are made to melt at a certain temperature. To check the temperature of a part, a crayon mark is made on that part. When the mark melts, it indicates that the part has reached that temperature.

Using patches or crayons of two different temperatures will give both the opening and the maximum temperature of the thermostat. For example, if a thermostat is supposed to begin opening at 195°F (91°C) and be fully open at 212°F (100°C), 190°F (88°C) and 220°F (104°C) patches might be used. If the coolant starts to flow when the 190°F (88°C) patch melts, it indicates that the thermostat has begun to open properly. If the 220°F (104°C) patch does not melt, it shows that the thermostat has fully opened and the system is not overheating.

SUMMARY

There are many marvels of the electronic age to help the mechanic find out exactly what is wrong with an engine so he can repair it quickly and efficiently.

The most important of the engine test machines have been described here. They include the oscilloscope which indicates the voltage levels in the electrical system of the automobile. Variations in the voltage levels can be used to diagnose engine conditions.

The power balance tester examines how much power is being put out by each cylinder. In this way, you can isolate the problem cylinder.

Compression testing measures the compression within the cylinder while the engine is cranking. The compression inside the cylinders is directly related to the amount of power put out by the engine. To get even more accurate information about cylinder leakage, you can do a cylinder leakage test. Leaking cylinders can be diagnosed by listening for leaks at the carburetor, the exhaust or the oil filler tube. A tool to help pinpoint engine noises and diagnose engine ills is the stethoscope.

A vacuum gauge measures the amount of vacuum that the engine creates in the intake manifold. Vacuum is measured in inches of **Mercury or kilopascals.**

The tachometer, which measures engine speed, and the dwellmeter, which measures distributor point dwell, are usually combined in one piece of equipment.

You can determine the amount the fan clutch is slipping, if any, with a fan strobe tachometer. This instrument measures the speed of the fan.

The bearing leakage tester is a pressure tank that forces oil through the engine oiling system to help locate excessive leakage. An oil pressure gauge is used to take an accurate measurement of oil pressure while the engine is running.

The infrared exhaust tester measures the amount of carbon monoxide and hydrocarbons in the exhaust. The measurements can be used to diagnose engine condition. Engine temperature gauges are used to measure engine operating temperatures.

Review Questions

Choose the single most correct answer.
Compare your answers to the correct answers on page 268.

1. When performing a wet compression test _____ should be added to the cylinder.
   a. Water
   b. Gasoline
   c. Oil
   d. Compressed air

2. An infrared test should always be done _____ the catalytic converter.
   a. In front of
   b. Behind
   c. Without
   d. None of the above

3. An oscilloscope is used to diagnose:
   a. Engine noises
   b. Compression problems
   c. Ignition problems
   d. All of the above

4. A power balance tests:
   a. Engine PRM
   b. Engine power
   c. Each individual cylinder for power output
   d. None of the above

5. An infrared exhaust analyzer measures:
   a. HC & CO_2
   b. O_2 & CO_2
   c. HC & CO
   d. CO & CO_2

6. A dwell meter indicates the degrees of cam rotation that the points are:
   a. Open
   b. Closed
   c. Energized
   d. Not conducting

7. Most oscilloscopes are mounted in consoles called:
   a. Oscilloscope consoles
   b. Engine Analyzers
   c. Motor Analyzers
   d. None of the above

8. A cylinder leakage test is performed with the piston:
   a. At top dead center
   b. At bottom dead center
   c. On the intake stroke
   d. On the exhaust stroke

9. An oscilloscope measures:
   a. Voltage and current
   b. Amperage and voltage
   c. Current over a period of time
   d. Voltage over a period of time

10. During a wet compression test, a raise in cylinder pressure indicates:
    a. Worn valves
    b. A bad head gasket
    c. Worn rings
    d. A cracked head

11. When power balance testing an engine, a cylinder that is completely dead will:
    a. Cause engine speed to drop
    b. Cause engine to speed to stay the same
    c. Cause engine speed to increase
    d. Cause the engine to die

12. The _____ scales on an oscilloscope indicate voltage.
    a. Vertical
    b. Horizontal
    c. Bottom
    d. All of the above

13. A CO reading that is high indicates:
    a. Rich air fuel mixture
    b. Lean air fuel mixture
    c. A carburetor float level that is too low
    d. A vacuum leak

14. A tachometer is used to measure engine:
    a. Dwell
    b. Timing
    c. Speed
    d. All of the above

15. CO is measured in:
    a. Parts per million
    b. Percentages
    c. Grams
    d. Kilopascals

Chapter 14

Cleaning and Inspection Equipment

CLEANING EQUIPMENT

To clean the various parts of an engine being repaired or rebuilt, a variety of tanks and chambers are used. These range from a tank in which parts are washed gently to a chamber in which the engine is blasted with tiny glass beads.

Parts Washer Tank

The parts washer tank, figure 14-1, is filled with a gentle solvent that is safe when in contact with skin. Never fill it with gasoline because of the fire hazard, or with a caustic chemical that could damage your hands.

Three methods of cleaning are available in the larger washer tanks:
1. parts can be submerged and soaked
2. parts can be held under the nozzle as the pump sends a stream of solvent over the part
3. parts can be gently agitated while submerged.

Parts washer tanks have a safety cover. The cover is held open with a chain and a fusible link, figure 14-2. If a fire starts in the tank, it will melt the link and the cover will close. Covers should always be held open with the safety link, and not with a stick or some other substitute.

Chemical Cleaning Dip Tanks

Carburetors and other small parts are usually cleaned by submerging them in a small tank or large can of chemicals. These chemicals are very harmful to your skin. To avoid skin contact, the parts are placed into a perforated basket, figure 14-3, and the basket is submerged.

Parts are left in the basket from one half to three hours. If an agitator is used, the time can be reduced. However, many chemicals should not be agitated. It is important to use only those chemicals that are recommended for use in an agitator. Follow the directions on the package for the best results.

Parts made of zinc alloy will discolor if they are left in the chemical too long. More importantly, the surface treatment on some parts may be removed by harsh chemicals. Check the surface treatment of a part and how it corresponds with a particular cleansing chemical before cleaning.

After cleaning, the basket is held over the tank to drain the fluid. The basket is then removed to a wash area where it can be hosed off with water or a cleaning solvent. An air nozzle is used to blow dry the parts.

Hot Tanks And Spray Tanks

Large cast iron engine parts, such as the block and cylinder heads, are cleaned in a hot tank,

Cleaning and Inspection Equipment

Figure 14-1. The parts washer is filled with a gentle solvent to soak parts clean.

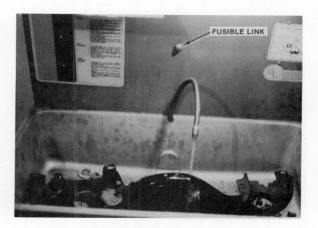

Figure 14-2. The cover is held open with a chain and a fusible link that melts and allows the lid to close if there is a fire.

Figure 14-3. Place the parts in a basket to lower them into the dip tank.

Figure 14-4. The larger engine parts are cleaned in a hot tank. (Graymills)

figure 14-4. The hot tank is filled with powerful chemicals that are heated with a gas burner. Most hot tank solutions are for iron and steel parts only. Aluminum, zinc, and other nonferrous metal parts require a different chemical solution.

Heating the chemicals makes them work faster. Small parts made of steel or iron may be cleaned in the hot tank.

To thoroughly clean engine parts, they are kept in the tank for several hours.

A variation of the hot tank is the spray tank, figure 14-5. It was developed to speed up the process of cleaning blocks and heads. The parts are put into the tank where dozens of nozzles spray hot chemicals under high pressure. In less

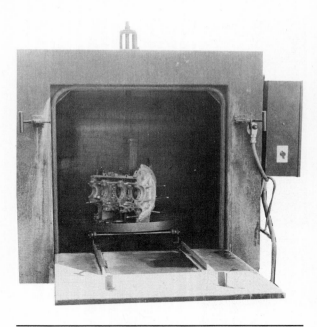

Figure 14-5. The spray tank speeds up the cleaning process by spraying hot chemicals onto the parts.

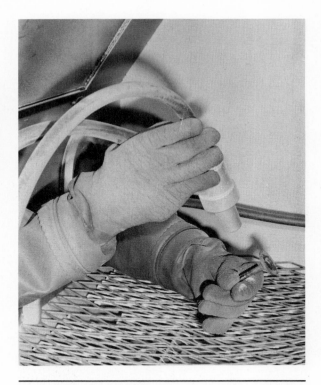

Figure 14-6. Large gloves are built into the chamber walls so the operator can work the blaster without injuring himself.

than an hour the parts are clean. However, there are some places inside an engine block that the spray cannot reach; those areas must be cleaned by hand.

The oil passages inside the engine should be cleaned with long brushes that will clean the full length of the passageway.

When using hot tanks of any type, it is important to remove all oil and water passage plugs from the block. This allows the chemicals to cleanse the whole engine and makes hosing out the engine much easier.

After the hot tank, the engine and parts are hosed off with hot water. The water must be run through every passage to remove all traces of chemicals and then the block is blown dry. To prevent rusting, the cylinder bores should be coated with a light oil immediately.

When working with hot tanks and spray tanks, always wear goggles or a face shield. When handling the parts, wear gloves, because the chemicals can cause skin injuries. Also, the parts are hot enough to burn you when they come out of the tank.

Blasting Chambers

One of the problems in rebuilding engines is removing hard carbon from valves, cylinder heads, and pistons. A hot tank can remove it from cast iron parts, but it is a lengthy process. To speed up the cleaning process, blasters are used.

Everything from walnut shells to sand has been used as a blasting medium, but most modern blasters use small glass beads. The beads are shot through a nozzle at high velocity.

Most blasting is done in enclosed chambers. The operator inserts his hands and arms into gloves that are built into the chamber wall, figure 14-6. The nozzle is picked up inside the chamber with a gloved hand and the machine is operated with a foot switch. A small window allows the operator to view the operation. The parts are removed and replaced in the chamber through a door in one end.

Blasters do a good job, but it is necessary to clean the parts thoroughly afterwards. All traces of the abrasive bead must be removed or the engine may be damaged. Washing is done with cleaning solvent or soap and water, followed by air drying with a blow gun.

Steam Cleaners And High Pressure Water Cleaners

These cleaners shoot steam or high pressure water out of a hand-held wand, figure 14-7. The wand is usually several feet long so it can reach inaccessible areas such as underneath a car or behind an engine.

Most of these cleaners have a soap or detergent container. As the steam or hot water runs through the machine it picks up a small amount of soap. The soap, added to the hot water or

Cleaning and Inspection Equipment

Figure 14-7. The steam cleaner has a hand-held wand with several feet of hose so the operator can reach around the engine without moving the cleaner. (Clayton)

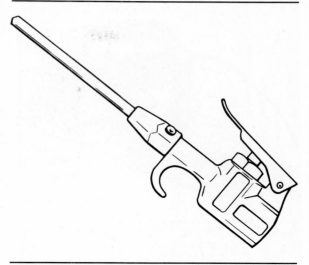

Figure 14-8. The air gun helps remove dirt from the smaller engine parts.

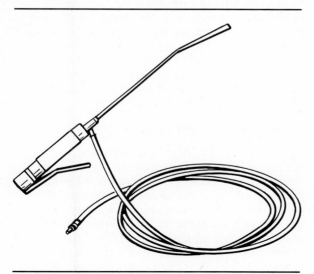

Figure 14-9. Some air guns have a hose that is put into solvent to allow the gun to spray solvent and air for additional cleaning.

steam, makes a much more effective cleaner. If soap is not available, the machine can be used without it.

These cleaners can be stationary or portable. Both types have a long hose so the operator can walk around the car without moving the cleaner.

A steam cleaner usually takes off more dirt and grease than a high pressure cleaner, but it may also take off the paint. The surface must be dried and repainted immediately or it will rust. Steam cleaners are ordinarily used on engines and greasy parts. High pressure water cleaners are used to wash away accumulations of mud, salt, road film, leaves, and other dirt. Never use a steam cleaner on a polished surface such as a painted car body.

Air Guns And Engine Cleaner Guns

The air gun, figure 14-8, is the only practical tool to remove dirt from small parts, either before or after washing with solvent. The air gun is also effective in drying parts after washing.

An engine cleaner gun has an attached hose that is submerged in a container of solvent, figure 14-9. When the gun trigger is pressed, the air pressure draws the solvent through the hose. The blast of air and solvent that comes out of the gun cleans the engine.

These guns work well if the grease and dirt are not too thick. The guns are more effective if something stronger than solvent is used. There are some commercial engine cleaners that can be used full strength. Engine cleaner guns are not as effective as steam or hot water cleaners.

Safety with air guns

Air guns are necessary tools for engine rebuilding, but they can be dangerous. An air gun should never be aimed at any part of the body. The dusting or cleaning of clothes with an air gun is not recommended. Goggles or safety glasses should be worn when using an air gun. A government-approved air gun has holes around the nozzle that prevent the pressure from rising above 30 psi (207 kPa), but even with this special gun, the possibility of injury still exists. Always handle an air gun with care.

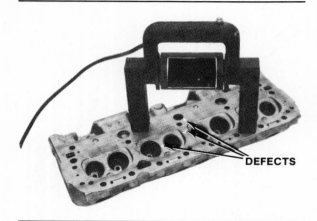

Figure 14-10. The iron particles cling to the crack in the magnetic particle inspection process. (Goodson)

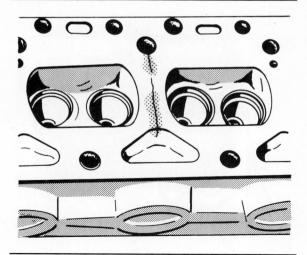

Figure 14-11. The nonmagnetic process features a dye that flows into the cracks and shows up under black light.

INSPECTION EQUIPMENT

Finding cracks, leaks, holes, or other defects in an engine being repaired is critical to the continued efficient operation of the engine.

In the first stages of development, a crack or defect in the block, crankshaft, or connecting rods is invisible to the naked eye; but if the part is installed in the engine it will eventually fail and perhaps damage other engine parts in the process.

It is also difficult to determine the location of a leak in an area of the engine that cannot be seen. Various pieces of test equipment aid the mechanic in finding those cracks and leaks.

Magnetic Particle Inspection Equipment

Magnetic particle inspection can be done only on metals that can be magnetized, such as iron and most steels. Some steels have such a high percentage of alloy that they are nonmagnetic.

To make a magnetic particle inspection of a part, the part must be magnetized. Then particles of magnetic iron are applied to the part. Any cracks will interfere with the magnetic field, causing it to start and stop at the crack. In effect, the crack creates a north pole and a south pole like the poles of a magnet. The iron particles will cling to the crack and outline it, figure 14-10.

There are two methods of applying iron particles. In the dry method, iron powder is dusted or blown on the part. The cracks can be seen with the naked eye. In the wet method, iron particles in liquid suspension are treated so they will glow under black light.

In either method, cracks will show up best if they are at right angles to the magnetic field. To find all cracks, both longitudinal and circular magnetization should be used. Longitudinal magnetization is done by placing a part inside an electric current carrying coil. The magnetic field is created so it runs lengthwise in the part. This will reveal cracks that run crosswise in the part.

Circular magnetization is done by running electric current through the part. This creates a circular magnetic field. It will reveal cracks that run along the length of the part.

Nonmagnetic Inspection Equipment

Metals that cannot be magnetized, such as aluminum, magnesium, and some steels, are inspected for cracks by the fluorescent penetrant method. A fluorescent penetrant, also called dye penetrant, is applied to the part; the special formula of the liquid makes it enter the cracks through capillary action. The liquid is washed off the surface, but remains in the cracks. Developer is then applied which draws the liquid out of the cracks. When the part is inspected under black light, the cracks will glow, figure 14-11.

Pressure Test Equipment

Visual inspection for cracks would be effective if all areas of the part were visible. But some parts, such as cylinder heads and blocks, are difficult to inspect completely. Also, a small hole or porous area that shows up in the inspection may be ignored because it does not appear to be a defect.

Because of these limitations, visual inspection methods cannot guarantee that a container under pressure will not leak. Cylinder heads and blocks operate with coolant under a

Cleaning and Inspection Equipment

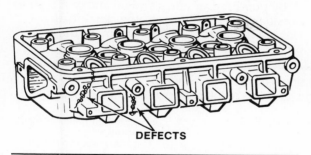

Figure 14-12. In a pressure testing rig, the openings in this cylinder head would be covered with plates. Here the head is shown after being removed from the rig. Notice the soap bubbled where air escaped from the head.

pressure of about 15 psi (103 kPa). The best way to test them is to apply air pressure to the coolant passages after applying a coating of liquid soap to the suspect areas. If there is a leak, the escaping air will cause bubbles to form, figure 14-12.

Pressure testing is not done routinely on every engine. Usually the purpose of this test is to find the location of a leak that is known to exist.

A pressure testing rig is used to test cylinder heads, figure 14-13. Place the cylinder head on a flat surface with the gasket surface facing up. Place adjustable arms with rubber-faced pads over the coolant openings and tighten them until they are sealed.

Seal the larger holes with plates and gaskets. Connect one of the rubber pads to a compressed air supply. After all passages are blocked off, pressurize the head. Apply liquid soap to the combustion chamber to check for leakage.

This pressure test can be used inside ports, behind valve seats, or in other areas that would be difficult or impossible to inspect for surface defects.

Another method of pressure testing the cooling system is to pump a liquid containing a dye penetrant into the water jacket and watch for a leak. The dye in the liquid will leave a permanent mark after the casting dries.

SUMMARY

Cleaning the engine and all its parts is an essential step in repairing or rebuilding the engine. It is also the first step in the inspection procedure. The engine must be inspected for cracks, leaks, or other defects that could cause damage if not corrected before the engine is reassembled.

As in engine diagnosis, there is equipment to help you clean and inspect the engine quickly and efficiently.

Figure 14-13. The pressure testing rig tests the cylinder head for leaks. (Van Norman)

The parts washer tank most often uses a gentle solvent to soak parts clean. The dip tank is used to clean carburetors and other small parts with caustic chemicals.

Larger, iron engine parts are cleaned in hot or spray tanks. Both tanks use powerful chemicals to speed up the cleaning process. In the hot tank the chemicals are heated; the spray tank takes the cleaning process one step farther by spraying the hot chemicals on the parts. Whenever working with dangerous chemicals, always wear protective shields and gloves to avoid injury.

The hard carbon that sticks to valves, cylinder heads, and pistons can be removed in a blasting chamber. Tiny beads of glass are shot through a nozzle at high velocity to knock off the carbon.

Steam and high pressure cleaners are good for removing the dirt and grease from the engine. But a steam cleaner will also remove some of the paint. It is important to dry and repaint the surface immediately after cleaning to avoid rust.

An air gun is used to clean the dirt from small engine parts. Never use an air gun to clean clothing and never point it at anyone. It should never be operated above the government-approved level of 30 pounds per square inch (207 kPa).

Cracks in the engine can be detected by magnetic or nonmagnetic inspection. The magnetic particle inspection can be done only on metals that can be magnetized. The part is magnetized and iron particles are applied to it, outlining the cracks in the surface.

In the nonmagnetic method, a dye is applied to the part and it flows into the cracks, making them visible to the mechanic.

By pumping air into a system and coating the outside with a liquid soap, you can find any leaks in the system. The air will make the soap bubble where it escapes.

Review Questions

Choose the single most correct answer.
Compare your answers to the correct answers on page 268.

1. The safety cover on a parts washer should be held open with a:
   a. Safety link
   b. Fusible link
   c. Chain link
   d. Missing link

2. Carburetors should be cleaned in a:
   a. Hot tank
   b. Spray tank
   c. Steam cleaner
   d. None of the above

3. To find cracks using magnetic particle inspection:
   a. Longitudinal magnetization should be used
   b. Circular magnetization should be used
   c. Both a and b
   d. Neither a nor b

4. Steam cleaners should not be used on:
   a. Chrome-plated surfaces
   b. Painted surfaces
   c. Zinc-plated surfaces
   d. All of the above

5. A parts washer tank is usually filled with:
   a. Gasoline
   b. Benzene
   c. A gentle solvent
   d. A caustic chemical

6. Modern blasters use:
   a. Glass beads
   b. Walnut shells
   c. Sand
   d. All of the above

7. A government-approved air gun limits pressure to:
   a. 15 kPa
   b. 30 kPa
   c. 103 kPa
   d. 207 kPa

8. In a large washer tank, parts can be:
   a. Gently agitated while submerged
   b. Submerged and soaked
   c. Held under a stream of solvent
   d. All of the above

9. Pressure testing should be done:
   a. On every engine
   b. Under a pressure of 107 kPa
   c. On cast iron cylinder heads only
   d. None of the above

10. Which of the following should not be cleaned in a hot tank
    a. Aluminum cylinder head
    b. Cast iron block
    c. Steel manifold
    d. None of the above

Chapter 15

Head And Valve Service Equipment

VALVE SEAT GRINDERS

Two types of hard-seat grinders are available. They are the **concentric** and **eccentric** types. The concentric valve seat grinder uses a large circular grinding stone which grinds the entire circumference of the seat in one motion, figure 15-1. The eccentric grinder employs a smaller circular grinding stone spinning on a different axis than that of the valve seat, figure 15-2. It cuts only a portion of the seat at a time. It is called eccentric because the center of rotation of the grinding stone is offset from the center of the valve seat.

Both types of seat grinders do a good job, but the eccentric grinder cuts faster and does not load up with metal dust as quickly as the concentric type.

The stones for hard seat grinders are mounted on a holder with a hollow center. The hollow center fits over a pilot that is inserted into the valve guide, figure 15-3. The pilot, located by the guide, positions the stone holder and the stone. If the guide is worn, the pilot may locate off center or may tip slightly. This will cause a poor fit between the valve and its seat. Before grinding the seats, the guides must be inspected and repaired or replaced.

The motor for a hard-seat grinder is made to run at high speed. It may look like an ordinary drill motor, but it is quite different. Seat grinding should never be attempted with an ordinary drill motor. It does not turn fast enough, nor does it have the power to do a good job.

VALVE SEAT CUTTERS

Within the last ten years, carbide cutters have been used in valve seat reconditioning, figure 15-4. The cutters are located by a pilot, similar to the pilots used on hard seat grinders. The big difference is that the cutters are turned by hand, or with a special motor that turns at one revolution per second, figure 15-5. This is extremely slow when compared with the hard seat grinders, which turn at several hundred revolutions per second.

Because the carbide cutters are very hard and turn slowly, they will last much longer than grinding stones. For example, the grinding stones on a hard seat grinder may last through only a few valve jobs, and must be dressed before beginning each job.

VALVE GRINDING MACHINES

A valve grinding machine has a chuck that accepts one valve at a time, figure 15-6. The

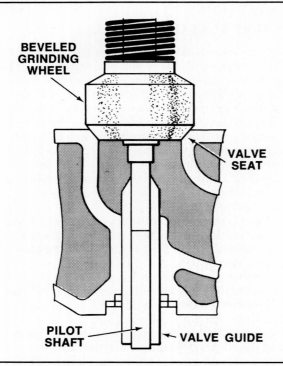

Figure 15-1. The concentric valve seat grinder grinds the surface of the seat in one motion. (Black and Decker)

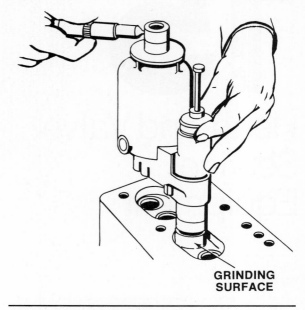

Figure 15-2. The eccentric valve seat grinder rotates on a different axis than that of the valve seat.

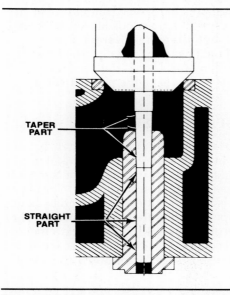

Figure 15-3. The stones for a hard-seat grinder are mounted on a hollow-centered holder. (Sioux)

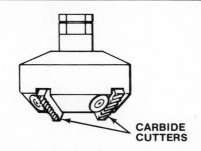

Figure 15-4. Carbide cutters have come into more frequent use within the last few years. (Neway)

chuck is mounted on a sliding table controlled manually with a lever. When the chuck is engaged with the motor, the valve rotates. The operator then uses the lever to move the table so the face of the valve moves across a high-speed grinding wheel. The operator controls the amount of metal removed from the face of the valve. Ordinarily, only enough metal is taken off to make the valve face smooth and concentric.

The valve face angle is adjusted where the chuck is mounted on the table, figure 15-6. Degree marks indicate whether the valve will be ground at 30 degrees, 45 degrees, or any other angle the operator selects.

Some valve machines have two grinding wheels. The second wheel is used for refinishing the bottoms of tappets, grinding the grooves out of rocker arms, or refinishing worn valve tips, figure 15-7. Special chucks or holders are used to ensure that the grinding is perpendicular to the part.

Some larger rebuilding shops have a machine that will refinish the bottom of a tappet, and put a slight convexity on it at the same time. The

Head and Valve Service Equipment

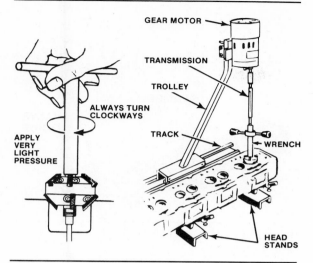

Figure 15-5. Carbide cutters are turned by hand or by a special motor. (Neway)

Figure 15-7. The second wheel on some grinding machines is used for grinding rocker arms or refinishing valve tips.

convexity helps to make the tappet spin in its bore while the engine is running, and spreads out the wear.

Figure 15-6. Valve grinding machines have a chuck mounted on a sliding table. (Winona-Van Norman)

VALVE GUIDE RENEWING

Valve stem metal is very hard, but valve guides are soft cast iron, bronze, or silicon bronze. Because of this, a valve guide will wear faster than the stem. When doing a valve job, check the guides for wear. The guides must be renewed or replaced if they are out of specification.

Knurling

There are many methods of renewing valve guides. The easiest method is **knurling**. A knurl is a special bit that is inserted into the valve guide and turned with a drill motor that has a drive reduction, figure 15-8. The knurl rolls a thread into the valve guide and raises the metal. The points of the thread will reduce the hole size so that it is smaller than the valve stem.

The next step in knurling is to run a reamer through the hole to take the tops off the threads

---

Concentric Valve Seat Grinders: A circular grinding stone that has the same center as the valve seat. It cuts the valve seat in a single motion.

Eccentric Valve Seat Grinder: A circular grinding stone with a center of rotation different from the center of the valve seat. It cuts only a portion of the valve seat at one time.

Knurling: Forming a series of ridges on the outer surface of a piston or the inner surface of a valve guide.

Figure 15-8. A knurl is a bit that is inserted into the valve guide and turned with a drill motor. (Winona-Van Norman)

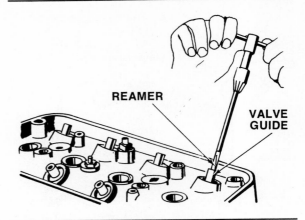

Figure 15-9. The reamer takes the tops off the threads.

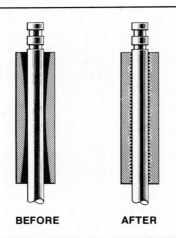

Figure 15-10. If the knurling is successful, the newly made threads will support the valve stem through its entire length. (Winona-Van Norman)

and make the hole fit the valve stem, figure 15-9. The remainder of the threads will support the guide, figure 15-10.

Sometimes the guide is worn to the extent that the knurl will not raise the metal far enough to support the valve. If this happens, some other valve guide renewal method will have to be used.

Guide Inserts

Another method of valve guide repair uses a bronze spiral insert. A special tap cuts an oversize thread in the guide, figure 15-11. Then a piece of spiral bronze is threaded into the guide. A **swaging tool** forces the bronze spiral into the guide. Finally, a reamer is used to bring the guide out to size.

Another guide renewal method employs a thin wall bronze sleeve, similar to a bushing, but thinner. The guide is first reamed to accept the sleeve. Then the sleeve is pressed or driven in. A swaging tool is run through the sleeve to force it into contact with the guide. Lastly, the sleeve is reamed to the correct size.

When the original guides are of the integral type, that is, a part of the cylinder head casting, a special insert can be used, figure 15-12. The original integral guide is bored out, and the new insert driven in. The final step is to ream the guide to size.

When the original guides are the insert type, they can be driven out and the new guides pressed in and reamed.

Although reaming has been used to size guides for many years, honing provides a closer fit, figure 15-13, and allows the machinist to finish a guide even if he doesn't have the proper size reamer. A reamer will usually produce a clearance between stem and guide of slightly less than .001'' (.025 mm). The hone will produce clearances as low as .0001'' (.0025 mm).

All of the methods discussed can be accomplished with handtools or portable electric tools. In addition, there are valve guide machines for production shops. The machine speeds up the work because the head is shifted on a special table to bring the guide under the drilling or reaming chuck. In some machines, the table floats on a cushion of air to make it easily movable. The head machine looks like an oversize drill press, figure 15-14.

Head and Valve Service Equipment

Figure 15-11. A special tap is used to cut a thread in the guide before inserting the bronze insert.

SEAT INSERTION TOOLS OR MACHINES

After several valve jobs, the valve seat may be ground so far that it should be replaced. If the seat is an insert type, it is knocked out, and a new, oversize insert is installed. New inserts usually come in a slightly larger oversize diameter than the original seats. This means that the **counterbore**, figure 15-15, or cut out area in which the seat inserts are placed must be enlarged for the oversize insert.

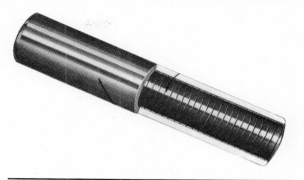

Figure 15-12. A special, thin wall insert can be used to replace integral valve guides. (K-line)

■ Ed Iskendarian

The name Iskendarian is synonymous with racing cams. Iskendarian is the world's largest manufacturer of racing camshafts.

The Isky reputation began with one man — Ed Iskendarian. He was born in 1921 in Tulare County, California, and if it hadn't been for a heavy frost a few years later, the name Iskendarian would probably occupy the label of a wine bottle instead of a camshaft.

Iskendarian became interested in cars after his family moved to Los Angeles. While he attended Polytechnic High School in L.A., one of his projects was building a Model T Ford roadster that he had converted from an old truck.

After graduating from high school, Iskendarian worked as an apprentice tool and die maker. He served a three year stint in the Army Air Corps during World War II and then moved back to California to prepare his Model T for the dry lake race meets. Isky started to rebuild a 1924 T, V-8, but he wanted a special camshaft. The hot rodding boom had hit California and he was faced with a five month waiting period for a special cam. Having already realized that the camshaft was the backbone of the racing engine, and that better power would come through better camshaft design and refinement, Isky went into the cam grinding business.

He progressed from a two-car garage in L.A. to his present four-building complex in Gardena, CA. He employs more than 100 engineers and technicians to constantly evaluate and improve the efficiency of Isky racing cams and valve train components.

Swaging Tool: A stamp or die used for shaping or marking metal. It is also called a swage.

Counterbore: The area that is cut or bored out for the valve seat insert.

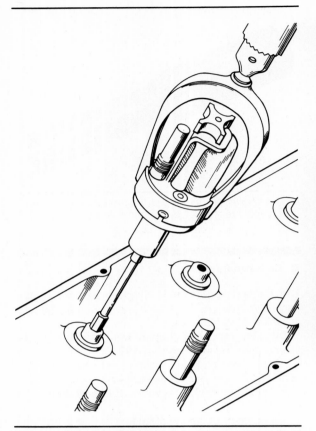

Figure 15-13. By using a guide hone, you can control clearance more precisely.

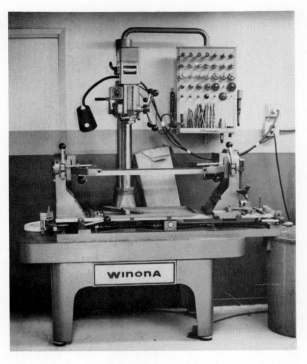

Figure 15-14. The head machine looks like an oversize drill press.

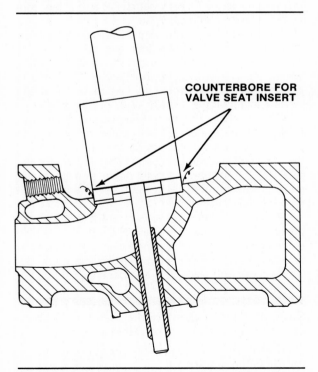

Figure 15-15. The counterbore must be enlarged for the oversize valve seat insert.

Inserts can also be installed in heads that originally had integral seats. The head material that formed the original seat is cut away until there is a counterbore to accept the replacement insert.

Machining of the counterbore is done with a special tool, figure 15-16. It is driven with an ordinary drill motor. Counterbores can also be cut on a valve guide machine.

When the counterbore is machined to fit the new insert, a driver is used to install the insert. Hammer blows to the driver will knock the insert into position and seat it in the counterbore. The driver must be guided by the valve guide machine or by the tool that clamps to the cylinder head. Using the driver without a guide is not recommended because it is easy to cock the seat and shave metal from the side of the counterbore. This metal then ends up under the seat and makes it sit at an angle. The result is poor heat transfer and eventual valve failure from overheating.

Since the insert is not made of the same material as the head, it will not expand at the same rate. To prevent the seat from working loose, the head metal around the seat is staked.

Head and Valve Service Equipment

Figure 15-16. The counterbore is machined with this special tool.

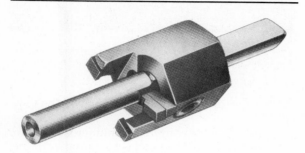

Figure 15-18. The machine that readies the valve guide for the seal consists of a pilot and cutters. (Sealed Power)

Figure 15-17. Positive valve stem seals prevent leakage. (TRW)

Figure 15-19. This tool is used to push the seal over the valve stem and onto the guide.

TOOLS FOR POSITIVE VALVE SEAT INSTALLATION

To install positive guide seals on an engine that did not have them originally, the guide must be machined. Machining the guide ensures that the seal will be concentric with the valve stem. It also means that the seal will be a tight fit on the guide and prevent leakage, figure 15-17.

The tool that machines the valve guide consists of a pilot and cutters, figure 15-18. The pilot is inserted into the guide and turned with a slow speed drill motor. As the tool is pushed down over the guide, it cuts the outside of the guide to the right diameter for the seal. If the head is mounted in a head machine, the tool can be operated by the machine.

The valve is inserted into the guide before installing the seal. A plastic cap goes over the end of the valve to prevent damaging the seal. A tool is then used to push the seal over the valve stem and onto the guide, figure 15-19.

Some engines come with positive valve seals as original equipment. If so, it is not necessary to do any machining as long as the replacement seals are the same size as the original equipment.

ROCKER STUD PULLERS

Rocker arm studs are subject to wear, especially if the geometry of the valve train changes after a valve job enough to allow the rocker to hit the stud.

If the rocker wears a groove in the stud, the old stud should be pulled out, and a new one driven in. Commercial stud drivers and pullers make the job much easier. A stack of washers on the stud will also act as a puller. Tighten a nut on top of the washers to pull the stud.

A new stud is driven in after tightening two nuts on its end to prevent damage to the stud

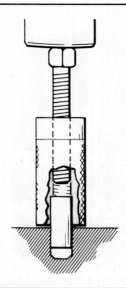

Figure 15-20. The stud driver presses in the new rocker stud.

Figure 15-21. A milling machine can use one or more cutting tools.

threads. The new stud should be driven in to the same depth as the old stud. A commercial stud driver is made for this job, figure 15-20. It is adjustable and bottoms out when the stud is at the correct depth. The valve train geometry should be checked and corrected to prevent stud wear. The valve train is discussed in Chapter 6.

If a stud is broken off in the head, the techniques just discussed will not work. The stud must be drilled out until there is nothing left but a thin shell. Then the shell is taken out carefully with pliers so the hole is not damaged.

HEAD SURFACING MACHINES

It is not uncommon for the machined surface of a cylinder head to become distorted after long periods of service. High and low spots may develop and cause leakage.

Whenever rebuilding an engine, check the head with a straightedge before assembly, as described in Chapter 12. If the head is distorted, it will have to be resurfaced using a precision machine.

Mills

A milling machine uses one or more large rotating cutting tools, figure 15-21. The cylinder head is mounted on a table. Either the table or the cutters are moved so that metal is taken off the gasket surface. This makes the head perfectly flat. The depth of the cut is regulated by the operator. Ordinarily, only enough metal is taken off to be sure the cutters reach all the low spots.

There are many different milling machine designs. In some, the head is mounted with the gasket surface up and horizontal. The cutters rotate in a horizontal plane, and move from one end of the head to the other.

Some mills mount the head on its side, with the gasket surface vertical. The cutter rotates, but does not travel. The table on which the head is mounted moves the head back and forth.

Grinders

The grinder uses a large diameter grinding wheel to resurface a head, figure 15-22. The head is mounted horizontally on a table with the gasket surface up. The grinding wheel is mounted so the flat side surfaces the head. The wheel is mounted on a sliding table that travels the full length of the head.

The depth of cut on the grinder is controlled by raising or lowering the table on which the head is mounted. When grinding, the wheel may not take off all the metal on its first pass. It is always allowed to make a return pass over the head before increasing the depth of cut. Automatic trippers will reverse the direction of the grinding table and send it on the return pass.

The maximum amount of metal removed with each round trip pass should be .002'' (.05 mm). The final pass should be done without changing the wheel height, to be sure that the entire surface has been cleaned up.

One of the big differences between the grinder and the mill is that the mill cuts dry. The grinder must have a constant stream of coolant splashing on the head to prevent overheating.

Head and Valve Service Equipment

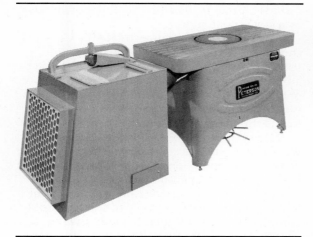

Figure 15-22. In resurfacing a head, the grinder uses a large diameter grinding wheel.

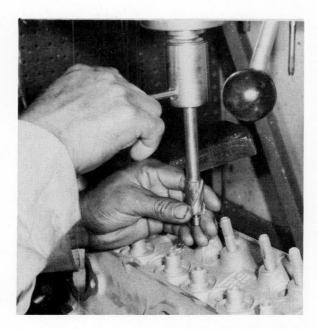

Figure 15-23. The spotfacer is a special valve spring seat cutting tool.

only one direction. When the table returns, the bit holder swivels up and rides over the work without cutting. Each time the table makes a cutting stroke, the bit holder moves over for the next cut. After many strokes, the bit has planed the entire surface of the head to the same level.

SPOTFACERS FOR ENLARGING SPRING SEAT

To keep valves from floating, stronger springs are sometimes installed which may be larger in diameter than the original equipment springs. The spring seat in the cylinder head has to be cut so the spring will fit.

Camshaft manufacturers have a special spring seat cutting tool called a spotfacer, figure 15-23. The tool not only cuts the seat wider in the head, it also cuts the valve guide to allow using another spring inside the main spring.

The tool is driven by a drill motor or head machine. A pilot centers the tool concentric with the valve guide. The tools are normally not adjustable, but are made for a particular spring diameter.

SPRING TENSION TESTERS

Two types of spring tension testers are on the market. One type is a scale with a dial reading, figure 15-24. It is similar to an arbor press. The handle moves the ram, which compresses the

Figure 15-24. This type of spring tension tester uses a scale with a dial reading. (Rimac)

The coolant is recirculated by a pump in the bottom of the machine.

Planers

A planer has a table that moves back and forth under a stationary cutting bit. It cuts in

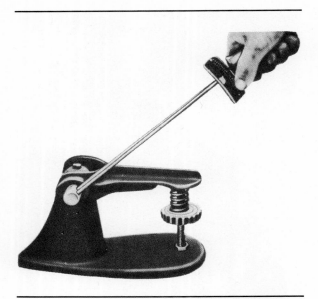

Figure 15-25. This spring tension tester uses a torque wrench with a scale reading.

spring against the top of the scale. The length of the spring is read from a pointer. The force is read directly from the dial.

The other type of spring tension tester uses a torque wrench which has a scale reading, figure 15-25. The micro-adjusting type of torque wrench, which clicks when a set torque has been reached, cannot be used with this tester.

A wheel on the tester sets spring height. Pressure is then applied with the torque wrench until the tester gives a loud click. At that point, the reading on the torque wrench should be noted and multiplied by two. This will give the spring tension at that length. Springs have both a closed and open specification. The important one is the closed spec. If the spring does not measure within ten percent of the closed specification, the valve will not be held tightly against its seat. The result will be leaking and burning valves.

SUMMARY

Hard seat grinders are used to restore worn or misshapen seats to the correct angle, flatness, and concentricity. The hard seat grinder makes the seat concentric with the valve guide by using a pilot inserted into the valve guide to support the grinding wheel. Seat cutters use carbide cutting blades, turned either by hand or by a slow speed motor.

Valve grinding machines rotate the valve in a chuck as the operator moves the face of the valve across a grinding wheel. The chuck can be adjusted to obtain the desired angle on the face of the valve. A second wheel on a valve grinding machine can be used to refinish valve stem tips, bottoms of tappets, or rocker arms.

Valve guides can be renewed by knurling, or with inserts. Inserts come in several types, but all require preparation of the old guide surface in the head with special tools.

Replaceable valve seats can be knocked out and replaced with new oversize seats. The old counterbore must be enlarged to fit the new, slightly larger seat insert. If the head does not have replaceable seats, the old seats can be removed by machining a counterbore into the head and driving in the new seats. Staking the metal around the seat prevents it from coming out during service.

Positive valve seals can be installed on engines that did not originally have them, but the guide must be machined first to fit the seal.

Rocker arm studs of the press-in type are sometimes subject to wear and must be replaced. They are removed by pulling with a nut and a stack of washers or with a special tool.

It is often necessary to resurface a head when rebuilding an engine. Basically, there are two types of machines used for resurfacing: a grinder and a milling machine. The milling machine uses a rotating cutter. Depending on the design of the machine, the head may travel under a rotating cutter or the cutter may travel across a stationary head. Head grinders use a large grinding wheel to resurface a head. On some machines, the head is held in one position while the grinding wheel rotates. Other grinding machines have the grinding wheel set in a table and the operator moves the head back and forth under a stationary single cutting bit.

When additional spring tension is needed, stronger springs are installed. If they are larger in diameter than the old springs the spring seat must be enlarged. This is done with a spotfacer.

Spring tension is an important specification because if the spring is not up to specs, it will not hold the valve tight against its seat and leaking will result. To make sure spring tension is correct, use a spring tension tester.

Head and Valve Service Equipment

Review Questions

Choose the single most correct answer.
Compare your answers to the correct answers on page 268.

1. Honing a valve guide gives a:
   a. Better finish
   b. Closer tolerance
   c. Rounder hole
   d. Savings in time

2. Valve seats are refinished by grinding or:
   a. Cutting
   b. Polishing
   c. Burnishing
   d. Lapping

3. Valve guides are made of:
   a. Bronze
   b. Cast iron
   c. Silicon bronze
   d. All of the above

4. The correct tension of the valve spring is important to:
   a. Ensure that the valve will close properly
   b. Ensure a tight valve seal
   c. Prevent valve float
   d. All of the above

5. Valve stems are always_____valve guides.
   a. Softer than
   b. The same hardness as
   c. Harder than
   d. Smoother than

6. Knurling a valve guide gives the guide:
   a. A smaller diameter
   b. A larger diameter
   c. Better wear characteristics
   d. More heat dissipation ability

7. To install positive valve guide seals, the valve guide must be:
   a. Cleaned
   b. Honed
   c. Machined
   d. None of the above

8. Valve seat cutters_____than grinders.
   a. Wear more quickly
   b. Last longer
   c. Leave a better finish
   d. Work much faster

9. Which of the following is a type of hard seat grinder?
   a. Eccentric
   b. Nonconcentric
   c. Concentric
   d. Centrifugal

10. The maximum amount of metal removed at each pass by a head grinder is:
    a. .002 mm
    b. .010 mm
    c. .05 mm
    d. .10 mm

Chapter 16

Block, Crankshaft, Piston, and Rod Service Equipment

BLOCK SERVICE EQUIPMENT

By now you know that the block is the basic framework of the engine. It is usually an iron or aluminum casting and is machined to accept the pistons, valve train parts, and bearings. When reworking an engine block, certain basic equipment is necessary.

Boring Bars

The automotive boring bar is used to bore cylinders to the next available piston size. The cylinder is bored to the required size minus about two thousandths of an inch for honing. Honing removes the tool marks left by the boring bar, brings the cylinder out to final size, and gives the cylinder walls a proper surface for ring seating.

The boring bar is a heavy piece of equipment that contains a long rigid bar, figure 16-1. It has a revolving cutting tool on the end, and centering shoes which expand to center the bar. The tool base is then bolted to the top of the cylinder block.

The bar has a feed mechanism that moves the cutting tool through the cylinder as it rotates. Light cuts may be taken in one pass. If a lot of metal must be removed, several passes are necessary.

Most boring bars have shutoff switches that turn off the motor when the bar stops cutting at the bottom of the cylinder. This prevents the bar from continuing to rotate and perhaps hitting the main bearing webs.

Boring bars all have limitations on the sizes of cylinders they are able to bore. Small cylinders are done with a small boring bar; larger cylinders with a large boring bar.

Air-cooled engines usually have individual cylinders. Since there is no block surface on which to locate and bolt the boring bar, a **jig** must be used, figure 16-2. The jig is a heavy plate with a hole in the middle. The cylinder bolts to the bottom of the plate, and the boring bar to the top. The bar is then used just as if it were bolted to the cylinder block. Air-cooled engines are seldom bored because the cost of the machine work is often more than the cost of a new cylinder

Hones

The hone uses abrasive stones to remove metal from cylinder walls. Hones have four "wings" that stick out from the body of the hone, figure 16-3. Two of the wings, opposite each other, are for the stones; the other two are for the centering shoes.

A nut at the top of the hone expands the shoes and stones until they press against the

Block, Crankshaft, Piston, and Rod Service Equipment

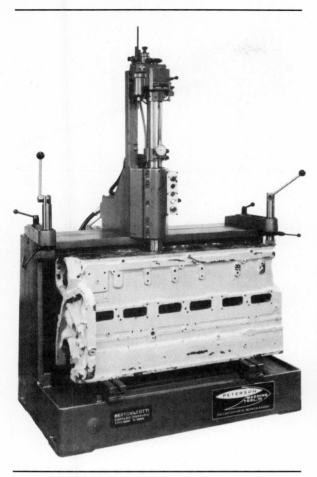

Figure 16-1. The boring bar is used to bore cylinders to the next available piston size. (Peterson)

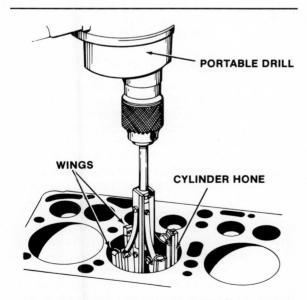

Figure 16-3. There are four wings on the rigid hone. Two are for the stones; two for the centering shoes.

Figure 16-2. When boring an air-cooled cylinder, a jig must be used. The jig is bolted to the cylinder with the boring bar on top.

cylinder wall. The hone is turned by a heavy-duty drill motor and moved slowly up and down in the cylinder.

Honing stones are available in different grits or degrees of coarseness. The coarse stones cut faster; the fine stones more slowly. The fine stones produce the type of finish required for good ring seating.

For the best finish, and to avoid loading up the stones with metal particles, all fine honing stones should be used with honing oil. Coarse grit stones may be used dry.

Glaze Breakers

The glaze breaker is a set of four stones mounted on a spring-loaded shaft, figure 16-4. A glaze breaker is used when re-ringing an engine. It removes the glaze from the cylinder walls so the new rings will be able to seat.

Jig: A device that holds the work and guides the tool during cutting operations.

Figure 16-4. The glaze-breaker is used when re-ringing an engine.

Figure 16-5. The Flex-hone® looks like a small brush with balls on the tips of the bristles.

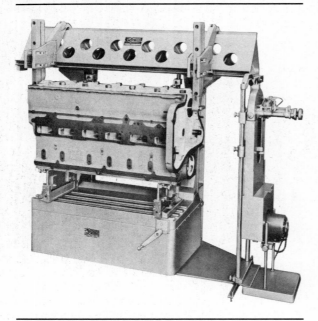

Figure 16-6. Main bearing bores, when out of alignment, must be straightened with a line boring machine. (Tobin-Arp)

A Flex-hone®, figure 16-5, looks like a brush with balls of abrasive particles at the end of each bristle. It can also be used for deglazing cylinder walls.

Line Boring Equipment

After long use, the expansions and contractions that the engine block goes through may cause the main bearing bores to become misaligned. When out of alignment, the main bearings put bending forces on the crankshaft. Main bearing bores that are out of alignment will cause the bearings to wear out much faster than those in alignment.

Line boring equipment consists of a boring bar that can be centered in the two end main bearing webs, figure 16-6. The cutting bit is adjusted to the size of the bearing, and all webs are bored to the same size. When using this method, an oversize bearing insert must be used to fit the oversize webs.

Another correction method is to grind the main bearing caps first. This reduces the size of the web hole slightly, so that the machine can bore back to the original standard size. The advantage to this is that standard bearings can be used since oversize bearings sometimes are not available.

The disadvantage of this method is that the location of the crankshaft is changed. The crankshaft then sits slightly higher in the block. However, if the cap grinding is kept to a minimum, the difference in crankshaft location may be only a few thousandths of an inch, which should have no effect on engine operation.

Excessive relocation of the crankshaft should be avoided because it changes compression ratio, causes slack in the timing chain, and moves the crankshaft out of line with the seal in the front cover. It can also cause misalignment between the crankshaft and the transmission.

Torque Plates for Boring and Honing

A torque plate is about the same thickness as a cylinder head. When bolted to the block, it sets up the same stresses and changes within the cylinders that will exist when the cylinder head is attached to the engine, figure 16-7. The torque plate has holes the same size as the engine cylinders so that when it is bolted to the block, all engine boring and honing operations can be done.

For many years, torque plates were used exclusively by high-performance and racing engine builders. Boring and honing with a torque plate was considered an excellent practice, but not really necessary for production engines. However, the Pontiac 301 V-8 engines require a torque plate (called a deck plate by Pontiac) to be used with all boring and honing operations. The 301 is a very light engine. It does not have the stiffness that heavier engines possess and so

Block, Crankshaft, Piston, and Rod Service Equipment

Figure 16-7. The torque plate is bolted to the block in place of the cylinder head to set up the same stress and distortion within the cylinders.

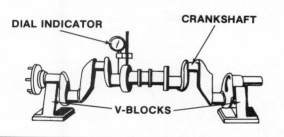

Figure 16-9. The V-blocks and dial indicator are used to check a crank for straightness.

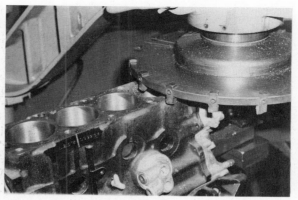

Figure 16-8. The same machine can be used to surface cylinder heads and blocks if the block fits into the machine.

requires the torque plate.

The torque plate is installed like a cylinder head. Every capscrew must be used and tightened to the same torque value as on the cylinder head. Without this precaution, the bores will distort when the head is installed, and the rings will not seat against the cylinder walls.

Block Surface Mills and Grinders

The same equipment used for surfacing cylinder heads can be used to surface blocks if the block fits into the machine, figure 16-8. A block, being heavier than a cylinder head, must be firmly supported. The supports or setup jigs used with cylinder heads are ordinarily not strong enough to support the block. If a mill or grinder is designed for block work, heavy duty supports are provided.

Block resurfacing is similar to head resurfacing. The block must be positioned so that the cut is taken parallel to the crankshaft, and square with the cylinders. If the cut slants in any direction, it may cause trouble in assembling the engine. If the slant is extreme, it can cause interference between valves and pistons, and varying compression ratios between cylinders. Proper setup ensures that the new surface will be aligned with the cylinders and crankshaft.

CRANKSHAFT SERVICE EQUIPMENT

The crankshaft, after the block and cylinder heads, is probably the largest, heaviest and most expensive part of the engine. The various tools and equipment used to straighten and restore the crankshaft are described in this section.

V-Blocks and Dial Indicator

The first step in restoring a crankshaft is to check it for straightness. To do this a dial indicator and two V-blocks are used, figure 16-9. The V-blocks must be mounted firmly and high enough so that the crankshaft can be rotated.

The crank is supported at each end main journal by the V-blocks. The dial indicator is positioned near the center on one of the other main journals, and the crank is rotated. If the dial indicator moves, it indicates an out-of-round main journal or a bent crankshaft. Measuring the main crankshaft journal with a micrometer will quickly show if it is out of round. Depending upon which side is out of round, the measurement should be added or subtracted to the dial indicator reading.

Cranks that are bent may be straightened two ways. The simplest method is to use a round nose chisel and hammer, figure 16-10. A blow in the fillet radius, next to the main journal will

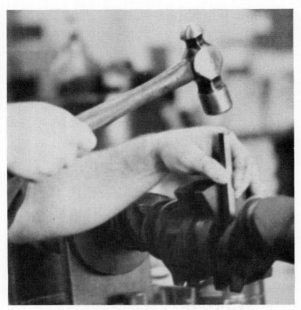

Figure 16-10. One method of straightening a crank is with a chisel and hammer.

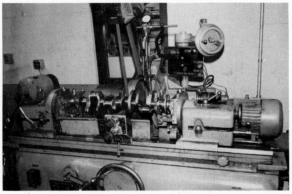

Figure 16-12. The crankshaft grinder restores journals by grinding away metal until the journal is round and smooth.

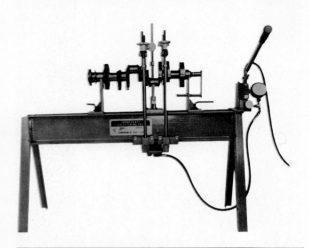

Figure 16-11. A crankshaft can be straightened in a press such as this one. (Storm-Vulcan)

straighten the crankshaft slightly. Succeeding blows will bring it into alignment.

The crankshaft can also be straightened in a press, figure 16-11. Either method can crack or break the crank if not done carefully. After straightening, the crankshaft is ground to the next available bearing undersize, if necessary.

Crankshaft Grinder

Rod and main journals will become worn after many miles of use or through neglect. The most common form of neglect is failure to change the oil. Particles of hard carbon and metal worn from other parts of the engine will damage the bearings. This is especially true if the filter is not changed also. It can become clogged and the oil will bypass it.

The crankshaft grinder restores worn journals by grinding away metal until the journal is perfectly round and smooth, figure 16-12. The grinder is actually a large lathe with special mountings for the crank. When grinding a main journal, the crank rotates around the shaft centers. A large grinding wheel spins at high speed. It is moved against the turning crankshaft until the shaft is ground to the next available undersize bearing.

When the rod journals are ground, the center of the crank is offset so the entire shaft rotates around the center of the rod journal. Because each rod journal is in a different position, the crank must be moved to a different mounting location for each one.

Hand grinders

A high speed grinder is sometimes used to put a slight radius on the oil hole in the journal, figure 16-13. This prevents the sharp edge of the journal from digging into the bearing.

Balancing Machine

A balancing machine, figure 16-14, is not only used for crankshafts. It can balance any rotating part. A bare crankshaft is purposely out of balance. Approximately one half of the weight of the piston, pin, rings, connecting rod, and bearing are considered to be rotating with the crankshaft, although this figure may vary.

The balancing machine spins the crankshaft with weights attached to the journal equal to one half the weight of the reciprocating parts. Unbalance of the crankshaft is indicated on a meter or by a light which tells where the shaft is heavy. The operator then drills a hole in the

Block, Crankshaft, Piston, and Rod Service Equipment

Figure 16-13. A hand grinder can be used to put a slight radius on the oil hole in the journal.

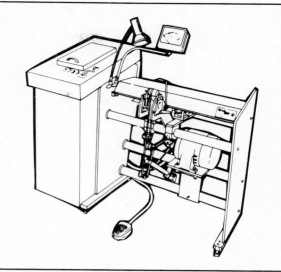

Figure 16-14. The balancing machine not only balances the crankshaft, but any rotating parts.

Figure 16-15. The crankshaft polisher turns the crank while the operator applies a belt sander to the journals.

counterweight to lighten the heavy side. If he has to add weight to the light side, he drills a hole and puts a plug of heavy metal in the hole. Lead is most commonly used, but tungsten is also used because it is slightly heavier than lead.

Balancing the crank in this way is done to a specific rod and piston weight. To be successful, all the pistons must weigh the same, and all the connecting rods must be of equal weight. To balance the pistons and rods metal is ground off in the areas where it will not affect strength.

Crankshaft Polisher

The surface left by the crankshaft grinder is too rough for long bearing life. After grinding, the crank must be polished. The polisher rotates the crank, while the operator applies a belt sander to the journals, figure 16-15. The polisher removes very little metal. Its only purpose is to create a smooth surface that will run on the bearings with the least amount of friction and wear. The polisher is mounted on the crankshaft grinder.

PISTON AND ROD SERVICE EQUIPMENT

Like the block, crank and cylinders, the pistons and rods must be machined to exact specifications so they can operate most efficiently. These machining procedures are done with a number of different machines.

Rod Honing Machine

A rod honing machine consists of a motor that rotates a hone or fine abrasive stone, figure 16-16. The operator moves the piston or connecting rod back and forth the length of the hone. Periodically, he checks the diameter of the piston pin hole and rod on a gauge mounted

Figure 16-16. This honing machine removes metal from pistons and rods with a hone or fine abrasive stones.

Figure 16-17. An arbor press can be used to push out old bushings and insert new ones.

Figure 16-18. The rod and cap parting faces are ground to make the opening in the connecting rod smaller. (Sunnen)

on the machine. The hone removes such small amounts of metal with each pass that the operator can make the hole fit the pin within a few ten thousandths of an inch or hundredths of a millimeter.

Honing is done with lubricant squirting on the part and hone at all times. A tray under the hone catches the honing oil. The oil drains into the machine and is recirculated by a pump. The hone is operated by a foot switch, so the operator can use both hands to hold the part being honed.

More elaborate honing machines have power stroking. Instead of the operator moving the part over the hone, the machine moves it.

Arbor Press

Engines that use full floating piston pins usually have a bushing in the end of the connecting rod. If a hydraulic press is not available to push the old bushings out and the new ones in, an arbor press can be used, figure 16-17. The arbor press must be large enough to do the job. Some small arbor presses do not have enough leverage because their handles are too short.

Rod Cap Grinder

The inside of the large end of the connecting rod can wear if a bearing insert comes loose and spins. If an engine is subject to heavy use, the big end of the rod can stretch. In either case, the rod must be brought back to the standard size, so it will fit a normal bearing insert.

This is done by removing the bolts from the rod, and grinding the rod and cap parting faces to make the hole smaller. This is done with a rod and cap grinder, figure 16-18. Then a hone is used to enlarge the hole to the standard size. The hone is centered, so that it just touches the sides of the rod at the parting line, but hones

Block, Crankshaft, Piston, and Rod Service Equipment

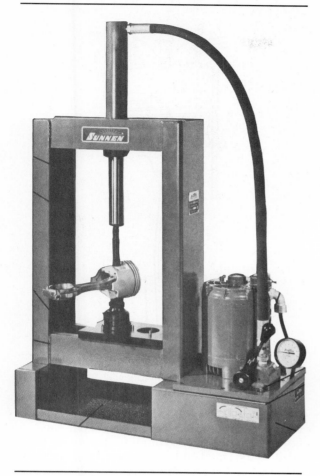

Figure 16-19. Pistons can be assembled to press-fit rods with a hydraulic press such as this one. (Sunnen)

Figure 16-20. The rod heater heats only enough of the rod to allow the pin to be pushed into the rod by hand.

material off the top and bottom of the hole to restore it.

If the bolts are in one piece with the rod, then only the cap can be ground to make the hole smaller. When the rod is honed, the hole moves up, and this shortens the rod. The amount that the rod is shortened is only a few thousandths of an inch or hundredths of a millimeter. It will not affect how the engine runs. But if a rod is rebuilt several times, the compression ratio of that cylinder will become progressively lower.

When both the rod and cap are ground to make the hole smaller, the rod is also shortened, but not as much. Since an equal amount of metal is removed from the cap and the rod, the hole moves up just half as much. Remember though, that each rebuild removes part of the bearing tang groove.

Hydraulic Press

A small hydraulic press, mounted on a table, is used to assemble pistons to press-fit connecting rods, figure 16-19. It can also be used to take them apart. The press usually has a capacity of about ten tons. A hydraulic ram does the actual pressing. The ram can be manually operated with a lever, or by an electric motor that drives a hydraulic pump. Pilots are used to guide the pin into position.

Rod Heater for Pin Assembly

Because of the difficulty in setting up a hydraulic press, and the possible damage to the pistons, rod heaters are sometimes used for assembling press-fit pins. The heater only heats the small end of the rod, figure 16-20. It heats it just enough to allow the pin to be pushed into the rod by hand. A small assembly jig ensures that the pin will not be pushed in too far.

After the pin is inserted into the rod and piston, it takes only a few seconds for the rod to cool and seize onto the pin. Then the rod and piston assembly can be handled without fear of dislodging the pin. Some rod heaters have two heat units. One rod can be heating while the other is being assembled. This makes the job go faster.

Balancing Scales

For an engine to be balanced, all the pistons must be the same weight. All the rods must also weigh the same. Different piston or rod weights would make it difficult, if not impossible, to balance the crankshaft.

The pistons are usually weighed to find the lightest one. Then metal is shaved off the others until they are all equal. The shaving is done

Figure 16-21. The scale weighs each end of the rod.

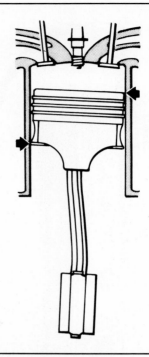

Figure 16-22. If a rod is twisted, the piston may be forced into the side of the cylinder wall.

Rods must be weighed on a double scale which weighs each end of the rod separately, figure 16-21. Metal is ground from both ends until all the small ends weigh the same and all the large ends weigh the same.

The scales that are used are very sensitive. They usually measure in tenths or hundredths of a gram, or to 1/32 of an ounce.

Rod Aligner

If a rod is bent or twisted, the piston will not operate correctly in the cylinder. A bend in the rod that cocks the piston pin will cock the piston in the cylinder. This will put stress on the pin and the connecting rod bearing that will cause early wear and failure.

If a rod is twisted, the rotation of the rod around the piston pin or in the piston will force the piston into the cylinder wall, figure 16-22. This will cause early ring failure or piston wear.

Bent and twisted rods can be straightened on a rod aligning fixture, figure 16-23. A mandrel expands in the big end of the rod and holds it firmly. The small end of the rod must have either the pin or the piston installed. If only the pin is installed, the aligner checks the alignment of the piston skirt.

Checking the alignment is done by sighting between the pin or piston and a flat plate or bar, figure 16-24.

Figure 16-23. The rod aligner checks the alignment of the piston skirt or the pin, depending on which part is installed in the machine. (Sunnen)

inside the piston, around the pin bosses or on a heavy part of the skirt.

Block, Crankshaft, Piston, and Rod Service Equipment

Figure 16-24. Alignment can be checked by sighting between the pin or piston and a flat plate.

The rod is straightened by inserting a bar into the hole in the piston pin and bending or twisting the rod. The fixture is strong enough so that the bending can be done without removing the rod.

Some rod aligners also check offset. An offset rod is one which is bent in two directions. The two bends cancel each other so that no bend or twist shows up. But the piston is offset from the bottom of the rod. This defect can result in the big end of the rod rubbing against the sides of the crank. Offset is difficult to correct. An offset rod must usually be replaced.

Piston Knurler

A high-mileage engine may have considerable wear on the pistons, resulting in too much piston clearance in the bore. The excessive clearance allows too much oil to get past the skirt. It also lets the piston rock in the bore, which wears out the rings.

To bring pistons up to size for the correct clearance, a knurler is used, figure 16-25. The knurler rolls a hardened wheel into the skirt of the piston, making little indentations. The edge of each indentation rises above the surface of the skirt, decreasing the clearance in the bore.

New pistons can also be knurled if they do not fit the cylinder bore. Knurling usually takes up a maximum of .006'' (.15 mm) on the diameter. The knurler will raise the metal more than that, but the fine points of the projections wear off quickly.

Knurling should be done with one pass of the roller. If the roller is passed over the skirt a second time, it may go on top of the previously raised metal and push it down again. After

Figure 16-25. The knurler is used to bring pistons up to size. (K-Line)

knurling, the pistons can be tailored to the proper clearance in each cylinder by lightly passing a file over the knurling. After each filing, a micrometer should be used to check piston diameter.

FLYWHEEL SERVICE EQUIPMENT

The rear surface of the flywheel is worn by the sliding action of the clutch disc. The flywheel face can be restored by mounting the flywheel in a special grinding machine. The flywheel is usually mounted flat with the face up. The grinding wheel can be lowered just enough to clean up the face of the flywheel without taking off excess metal.

SUMMARY

Block service equipment includes:
- boring bars
- hones
- line boring equipment
- torque plates
- block surfacers.

The boring bar is used to recondition cylinders to the next available size piston. A hone creates the surface finish on the cylinder that is necessary for fast ring seating.

Line boring equipment is used to align a series of holes, such as the main bearing bores, or the camshaft bearing bores. The torque plate simulates the clamping action of a cylinder head mounted to a block. The plate ensures that the cylinder bore will be round after the head is installed.

A block surfacer is similar to a head surfacer. It refinishes and restores the head gasket surface at the top of the block.

Crankshaft service equipment includes:
- V-blocks
- dial indicators
- crankshaft grinder
- hand grinder
- balancing machine
- crankshaft polisher.

V-blocks and a dial indicator are used to detect crankshaft out of round. The grinder resurfaces the main and connecting bearing journals on the crankshaft.

A hand grinder is used for touch up work on oil holes and other areas. The polisher provides a surface on the bearing journals that will be compatible with the bearings. The balancing machine shows where to remove weight from or add weight to the crankshaft to put it in balance.

Piston and rod service equipment consists of:
- honing machine
- arbor press
- rod cap grinder
- hydraulic press
- rod heater
- balancing scales
- rod aligner
- piston knurler.

The honing machine is used for fitting pins to rods or pistons, and for reconditioning the big end of the rod. An arbor press can be used for light pressing jobs, such as pressing bushings into full floating rods. The rod cap grinder grinds metal from the parting faces of the cap and rod so the assembled cap and rod can be honed back to its standard size. A hydraulic press is used for assembling press-fit rods to pistons.

The rod heater expands the small end of the rod so that the pin can be inserted by hand into a press-fit rod. Balancing scales measure the weight of rods and pistons for balancing. A rod aligner shows whether a rod is out of alignment. The piston knurler increases the diameter of a piston to reduce the clearance in the cylinder bore.

Flywheel grinders are used to resurface the face of the flywheel that is worn by the clutch friction disc.

Block, Crankshaft, Piston, and Rod Service Equipment

Review Questions

Choose the single most correct answer.
Compare your answers to the correct answers on page 268.

1. A boring bar is used to:
   a. Bring cylinders to their finished size
   b. Finish a cylinder after it is honed
   c. Increase cylinder diameter to the next piston size.
   d. None of the above

2. Air cooled engines are not usually bored to the next over size because:
   a. The cylinders would be too thin
   b. It's cheaper to buy new cylinders
   c. There is no way to mount the boring bar
   d. All of the above

3. Honing:
   a. Brings cylinder to finished size
   b. Provides the proper surface for ring seating
   c. Restores roundness in the cylinder
   d. All of the above

4. Line boring equipment:
   a. Restores main-bearing bores to their proper location
   b. Allows main bearings to maintain proper oil clearance
   c. Restores roundness to the main bearing bores
   d. All of the above

5. A torque plate is used to:
   a. Calibrate torque wrenches
   b. Mount the boring bar to the cylinder block
   c. Duplicate cylinder head on block while honing cylinders
   d. None of the above

6. Block resurfacing is similar to:
   a. Crank grinding
   b. Cylinder boring
   c. Head resurfacing
   d. None of the above

7. The first step in restoring a crankshaft is to:
   a. Check it for straightness
   b. Check it for cracks
   c. Regrind it
   d. None of the above

8. The most common reason for crankshaft failure is:
   a. Excess loads
   b. Improper machining
   c. Neglect
   d. None of the above

9. A crankshaft grinder is really a:
   a. Large lathe
   b. Small lathe
   c. Modified bench grinder
   d. Type of milling machine

10. A balancing machine balances:
    a. Reciproating weight
    b. Any engine part
    c. Rotating weight
    d. Centrifugal force

11. A rod honing machine:
    a. Balances the rod
    b. Straightens the rod
    c. Restores the proper size to the rod
    d. All of the above

12. A rod heater is used on rods with:
    a. Full floating pins
    b. Press fit pins
    c. Cold fit pins
    d. Hollow pins

NIASE Mechanic Certification Sample Test

This sample test is similar in format to the series of tests given by the National Institute for Automotive Service Excellence (NIASE). Each of the NIASE exams covers one of eight areas of automobile repair and service. The tests are given every fall and spring in about 250 cities throughout the United States.

For a mechanic to earn certification in a particular field, he or she must successfully complete one of these tests, and have at least two years of "hands-on" experience (or a combination of work experience and auto mechanic's training). Successfully finishing all eight tests earns the mechanic certification as a General Automobile Mechanic. More than 200,000 mechanics have taken these tests since the program began in 1972.

In the following sample test, some of the questions were provided by NIASE. Learning to take this kind of test will help you if you plan to apply for certification later in your career.

For more information write to:
National Institute for
Automotive Service Excellence
1825 K Street N.W.
Washington, D.C. 20006

1. Mechanic A says that with solid lifters, exhaust valves have more clearance than intake valves so that cylinder scavenging is complete.

   Mechanic B says that with solid lifters, exhaust valves have more clearance than intake valves to allow for expansion.

   Who is right?
   a. A only
   b. B only
   c. Both A and B
   d. Neither A nor B

2. When an engine is run at 2,000 rpm, manifold vacuum reads 20 inches, then falls to 14 inches. This could be caused by:
   a. A restricted exhaust system
   b. Worn valve guides
   c. An intake manifold leak
   d. A leaking carburetor power valve

3. Mechanic A says that solid valve lifters have a lash adjustment to compensate for wear of the valve guides.

   Mechanic B says that solid valve lifters have a lash adjustment to ensure correct valve seating.

   Who is right?
   a. A only
   b. B only
   c. Neither A nor B
   d. Both A and B

4. When reinstalling hydraulic valve lifters, some manufacturers recommend additional tightening after zero lash is achieved. What might be the reason for this?
   a. Proper positioning of the plunger in the lifter
   b. Quieter operation
   c. Better high-speed performance
   d. Better low-speed performance

5. When making a compression test, Mechanic A says the engine should be warm.

   Mechanic B says the throttle should be wide open.

   Who is right?
   a. A only
   b. B only
   c. Neither A nor B
   d. Both A and B

6. Which of these sequences is correct for a four-stroke cycle engine?
   a. Power, exhaust, compression, intake
   b. Intake, compression, power, exhaust
   c. Intake, power, exhaust, compression
   d. Exhaust, compression, intake, power

7. Piston rings should be checked for:
   a. End gap only
   b. End gap and ring groove side clearance
   c. Only ring and groove side clearance at one point in the groove
   d. Only end gap at the top of the cylinder bore
8. An engine with a cracked cylinder head may:
   a. Overheat
   b. Leak coolant into the cylinder
   c. Leak coolant into the oil
   d. Any of the above
9. A compression test will indicate:
   a. Worn bearings
   b. Low vacuum
   c. Defective oil pump
   d. Worn rings
10. Loose pistons may cause:
    a. Low oil pressure
    b. Oil burning
    c. Burned valves
    d. Noise
11. Crankshaft journals are measured and examined for:
    a. Roundness
    b. Taper
    c. Diameter
    d. All of the above
12. Broken ring lands may be caused by:
    a. Ring ridge not removed before pushing piston from cylinder
    b. Pistons in the wrong cylinder
    c. Excessive ring gap
    d. Leaving out one or more rings
13. To install a piston in a cylinder, use a:
    a. Piston pin vise
    b. Ring groove cleaner
    c. Ring expander
    d. Ring compressor
14. What is piston clearance?
    a. The clearance between the piston and the pin
    b. The clearance between the piston and the ring
    c. The clearance between the piston skirt and the cylinder bore
    d. The clearance between the piston and the cylinder head
15. Valve spring assembled height is measured from:
    a. Cylinder head to top of valve stem
    b. Valve head to top of valve stem
    c. Spring seat to underside of spring retainer
    d. Spring seat to topside of spring retainer
16. Which of the following is true about rocker arm shafts?
    a. They operate without lubrication
    b. They have coolant inside
    c. They are oil pressure lubricated
    d. They are used on all overhead valve engines
17. The best way to tighten head bolts is to:
    a. Tighten to twice the specified torque, then back off
    b. Judge by feel
    c. Approach the correct torque gradually following the recommended sequence
    d. Check the breakaway torque
18. Valve springs by themselves must be checked for:
    a. Tension
    b. Squareness
    c. Assembled height
    d. All of the above

Glossary of Technical Terms

API Service Classification: A system of letters signifying an oil's performance; assigned by the American Petroleum Institute.

After Top Dead Center: The position of a piston after it has passed top dead center. Abbreviated: atdc. Usually expressed in degrees, such as 5° atdc.

Air-Fuel Ratio: The ratio in pounds of air to gasoline in the air-fuel mixture drawn into an engine.

Alloy: A mixture of two or more metals.

Anaerobic: A sealer that cures in the absence of air.

Aneroid Bellows: An accordian-shaped bellows that responds to changes in coolant pressure or atmospheric pressure by expanding or contracting.

Anneal: To soften steel by heating to 1400° F (760° C) or above and allowing to cool slowly. Removes internal stresses.

Anodize: The process of oxidizing a metal such as aluminum, titanium, or magnesium by running electric current through a special solution containing the metal anode.

Antifriction Bearing: A bearing that rolls on the part it supports.

Antioxidants: Chemicals or compounds added to motor oil to reduce oil oxidation, which leaves carbon and varnish in the engine.

Atmospheric Pressure: The pressure on the earth's surface caused by the weight of air in the atmosphere. At sea level, this pressure is 14.7 psi at 68° F (760 mm Hg at 20° C).

Axial Movement: Movement parallel to the axis of rotation or parallel to the centerline (axis) of a shaft. Endplay.

Babbitt: Metal alloy of tin mixed with copper and antimony. Lead-based babbitt is lead with small amounts of tin and antimony. Used to make bearings.

Backpressure: A pressure that tends to slow the exit of exhaust gases from the combustion chamber; usually caused by restrictions in the exhaust system.

Balancing Shaft: A shaft that is used solely for the purpose of cancelling engine vibrations.

Bearing Bore: A full circle machined surface that supports the back of the bearing. It may consist of two saddles bolted together, as in rod or main bearings, or a bored hole, as in most camshafts.

Bearing Crush: The distance that a bearing is higher than its saddle or web.

Bearing Saddle or Web: The machined area which supports the connecting rod or main bearing insert.

Bearing Spread: The distance that the parting face of a bearing is wider than the diameter of its saddle or web.

Before Top Dead Center: The position of a piston as it approaches top dead center. Abbreviated: btdc. Usually expressed in degrees, such as 5° btdc.

Belt or Chain Tensioners: A wheel or pad, pressed against a belt or chain, to maintain its tension.

Billet: A forged piece of steel with a cross sectional area of 4 to 36 square inches (26 to 232 square centimeters).

Billet Crankshaft: A crankshaft cut on a lathe from a solid piece of steel.

Bimetal Temperature Sensor: A sensor or switch that reacts to changes in temperature. It is made of two strips of metal welded together that expand differently when heated or cooled causing the strip to bend.

Blowby: Combustion gases that get past the piston rings into the crankcase; this includes water vapor, acids, and unburned fuel.

Bore: The diameter of an engine cylinder; to enlarge or finish the surface of a drilled hole.

Bottom Dead Center: The exact bottom of a piston stroke. Abbreviated: bdc.

Brake Horsepower: The power available at the flywheel of an engine for doing useful work.

British Thermal Unit (Btu): The amount of heat required to raise the temperature of one pound of water one degree Fahrenheit at a pressure of one atmosphere.

Broach: To finish or change the shape of a roughly cut hole. A broach is the metal tool for finishing holes, usually used to form an irregular hole from a round hole.

Bushing: A small, full circle friction bearing.

Calorie: The amount of heat energy required to raise the temperature of one gram of water one degree Celsius at a pressure of one atmosphere.

Glossary of Technical Terms

Cam Thrust: The lengthwise movement of the camshaft to the front or rear of the engine.

Camshaft Bearing: The bearings that support the camshaft in the engine block or cylinder head.

Carbon Monoxide: An odorless, colorless, tasteless, poisonous gas. A major pollutant given off by an internal combustion engine.

Carburizing: A method of hardening camshaft lobes by heating the camshaft in a carbon atmosphere.

Castellated Nut: A threaded fastener for a nut or bolt with slots for a cotter pin.

Casting: The process of making metal parts by pouring the liquid metal into molds where it solidifies.

Cement: A liquid or substance which hardens to act as an adhesive.

Centrifugal Force: A force applied to a rotating object, tending to move the object toward the outer edge of the circle in rotation.

Circlips: A round spring-steel device that fits into a groove at the end of the piston pin hole to lock in the pin.

Clearance Volume: The volume of a combustion chamber when the piston is at top dead center.

Cog Belt: A flat rubber belt with teeth that mesh with cog wheels.

Coil Bind: The condition in which all the coils of a spring are touching each other so that no further compression is possible.

Compatibility: The ability of a bearing lining to allow friction without excessive wear.

Composite: A manmade, non-metallic material. The term is usually applied to advanced plastic or ceramic materials.

Compression Ratio: The total volume of an engine cylinder when its piston is at bottom dead center divided by its clearance volume.

Compression Test: A test using a pressure gauge to measure the pumping pressure in a cylinder during cranking.

Concentric Valve Seat Grinders: A circular grinding stone that has the same center as the valve seat. It cuts the valve seat in a single motion.

Conformability: The ability of a bearing lining to conform to irregularities in a bearing journal.

Counterweights: The weights opposite the rod journals on a crankshaft that balance the weight of the journal, bearing, connecting rod, and piston assembly.

Crankpin: Connecting rod crankshaft journal.

Crankshaft Counterweights: Weights cast into, or bolted onto, a crankshaft to balance the weight of the piston, piston pin, connecting rod, bearings, and crankpin.

Crankshaft Endplay: The end to end movement of the crankshaft in the bearings.

Crankshaft Offset: The distance that the centerline of the crankshaft is offset from the centerline of the cylinders.

Crossflow Cylinder Head: A head with the intake port and intake valve on the opposite side of the combustion chamber from the exhaust port and exhaust valve. The design provides an almost straight path for the mixture to flow across the top of the piston.

Cylinder Leakage Test: A test in which compressed air is forced into a cylinder and the percent of leakage measured.

Cylinder Sleeve: A liner for a cylinder which can be replaced when worn to provide a new cylinder surface.

Detergent-Dispersant: A chemical added to motor oil to break down and disperse sludge and other undesirable particles picked up by the oil.

Detonation: An unwanted explosion of an air-fuel mixture caused by high heat and compression. Also called knocking, pinging.

Displacement: A measurement of the volume occupied by a piston as it moves from the bottom to the top of its stroke. Engine displacement is the piston displacement multiplied by the number of pistons in an engine.

Drill: The cutting of round holes. A drill is an end-cutting tool.

Dry Sleeve: A sleeve that does not come in direct contact with the engine coolant.

Duration: The number of crankshaft degrees that an intake or exhaust valve is open.

Dwellmeter: An instrument to measure the angle through which the distributor shaft turns while the distributor breaker points are closed.

Dynamic Inertia: The tendency of a body in motion to remain in motion.

Dynamometer: A device used to measure the power of an engine or motor.

Eccentric: Off center. A shaft lobe which has a center different from that of the shaft.

Eccentric Valve Seat Grinder: A circular grinding stone with a center of rotation different from the center of the valve seat. It cuts only a portion of the valve seat at one time.

Embedability: The ability to absorb small particles into a bearing metal at a level with, or below, the surface.

Embossed steel gasket: A gasket made from thin sheet steel covered with a raised design.

Energy: The ability to do work by applying force.

Engine Prelubricator: An oil tank, pressurized with air, that is used to pump oil through the oil lines and passages of an engine while it is not running.

External Combustion Engine: An engine, such as a steam engine, in which fuel is burned outside the engine.

Firing Order: The sequence by cylinder number in which combustion occurs in the cylinders of an engine.

Force: A push or pull acting on an object; it may cause motion or produce a change in position. Force is measured in pounds in the U.S. system and in newtons in the metric system.

Forging: Pressing metal in a plastic state into molds under great pressure.

Four-Stroke Engine: The Otto-cycle engine. An engine in which a piston must complete four strokes to make up one operating cycle. The strokes are: Intake, compression, power and exhaust.

Friction: The force that resists motion between the surfaces of two bodies when they are in contact. The two basic types of friction are sliding friction and rolling friction.

Friction Bearing: A bearing that slides or rubs on the part it supports.

Friction Horsepower: The power consumed to overcome friction within an engine.

Full-Floating Pin: A piston pin that is free to move in both the piston and the rod.

Grind: Removing metal with an abrasive wheel.

Gross Brake Horsepower: The flywheel horsepower of a basic engine, without accessories.

Harmonic Balancer: A vibration damper.

Header: A free-flowing exhaust manifold, designed for a specific engine speed, with tubes of equal length which meet in a collector; it improves scavenging and increases engine efficiency at or close to its design rpm.

Helical Gears: Gears with teeth cut at an angle to the shaft instead of parallel to it.

Hone: To finish the interior surface of a cylinder with an abrasive device. A hone is the device used for finishing the internal surfaces.

Horsepower: 33,000 foot-pounds of work per minute equals one horsepower.

Hydrocarbon: A chemical compound made up of hydrogen and carbon. A major pollutant given off by an internal combustion engine. Gasoline, itself, is a hydrocarbon compound.

Ignition Interval: (Firing Interval): The number of degrees of crankshaft rotation between ignition sparks.

Indicated Horsepower: The theoretical maximum power produced in an engine's cylinders.

Induction Hardening: The use of a strong electromagnet to heat small areas on a large piece of metal. The area is then cooled rapidly to harden it. The process is used on crankshaft journals and valve seats.

Inertia: The tendency of an object at rest to remain at rest, and of an object in motion to remain in motion.

Insert Guide: A valve guide that is pressed or driven into the cylinder head.

Interference Angle: The angle or difference between the valve face angle and the valve seat angle.

Integral Guide: A valve guide that is formed as part of the cylinder head.

Internal Combustion Engine: An engine, such as a gasoline or diesel engine, in which fuel is burned inside the engine.

Jig: A device that holds the work and guides the tool during cutting operations.

Kilowatt (kW): The metric unit of power. One horsepower equals 0.746 kilowatt; one kilowatt equals 1.3405 horsepower.

Knurling: Forming a series of ridges on the outer surface of a piston or the inner surface of a valve guide.

Lap: Finishing two metal surfaces to each other by using a fine abrasive compound. Lapping is used to produce a gas-tight or liquid-tight seal.

Lifter Pump-up: The condition in which a hydraulic lifter adjusts or compensates for the additional play in the valve train created by valve float.

Linear: In a straight line.

Lobe Lift: The amount of lift at the cam lobe.

Magnetic Tachometer: A tachometer that measures engine speed with a probe that magnetically senses the revolutions of the crankshaft.

Major Thrust Face: The side of the piston that pushes against the cylinder wall during the power stroke.

Manifold Heat Control Valve: A valve that diverts part of the exhaust through a passageway under the carburetor to provide heat for the mixture that is in the intake manifold.

Manifold Vacuum: Low pressure, in an engine's intake manifold, below the carburetor throttle.

Mass: The measure of inertia of an object; sometimes defined as the "quantity of matter" in it.

Microinch: One millionth of an inch, abbreviated μ inch or μ in.

Mill: Cutting metals with a round, multiple tooth cutter; also used with a single cutting edge for flat surfaces.

Momentum: The force of motion. Momentum equals a moving body's mass times its speed.

Multigrade (Multiviscosity): An oil that has been tested at more than one temperature, and so has more than one viscosity number.

Glossary of Technical Terms

Mushroom Tappet: A tappet with a base larger in diameter than its body.

Net Brake Horsepower: The flywheel horsepower of a fully equipped engine, with all accessories in operation.

Newton-Meter (Nm): The metric unit of torque. One newton-meter equals 1.356 foot-pounds. One foot-pound equals 0.736 newton-meter.

Nitriding: Hardening of metal by dipping it into an acid bath.

Octane Rating: The measurement of the antiknock value of a gasoline.

Oil Clearance or Bearing Clearance: The gap or space between a bearing and its shaft that fills with oil when the engine is running.

Oscilloscope: An electronic device used to visually observe and measure the instantaneous voltage in an electric circuit.

Overlap: The period of crankshaft rotation in degrees during which both the intake and exhaust valves are open.

Oversize Bearing: A bearing made thicker with a larger outside diameter to fit an oversize web.

Oxidation Reaction: A chemical reaction in which oxygen is combined with another element or compound to form a new compound.

Oxides of Nitrogen: Chemical compounds of nitrogen given off by an internal combustion engine. They combine with hydrocarbons to produce smog.

Particulates: Liquid or solid particles such as lead and carbon that are given off by an internal combustion engine as pollution.

Pig Iron: Small solidified blocks of iron poured from the blast furnace into channels dug in sand. Used to make steel.

Pin Bosses: The areas surrounding the holes in the piston for the piston pin.

Piston Eyebrow: Small pockets cut into the top of pistons to provide clearance for the valve.

Piston Slap: A noise caused by the piston skirt hitting against the cylinder wall.

Piston Speed: The average speed of the piston, in feet or meters per minute, at a specified engine rpm.

Plateaued Finish: A finish in which the highest parts of a surface have been honed to flattened peaks.

Plating: Applying a thin coat of metal onto the surface of another substance.

Poppet Valve: A valve that plugs and unplugs its opening by linear movement.

Positive Crankcase Ventilation (PCV): Late-model crankcase ventilation systems that return blowby gases to the combustion chambers.

Power: The rate or speed of doing work (work × time = power).

Power Balance Test: A test to determine if all cylinders are contributing equally to the power output of the engine.

Pressure Differential: The difference in pressure between two points.

Primary Tachometer: A tachometer which takes its signal from the ignition primary.

Quench Area: An area in the combustion chamber that cools the mixture to extinguish combustion and reduce detonation.

RTV Compound: Room temperature vulcanizing silicone, available in a tube.

Ream: The process of smoothing and sizing round holes in metal. A reamer is the side-cutting tool used for finishing already-drilled holes.

Reciprocating Engine: Also called piston engine. An engine in which the pistons move up and down or back and forth as a result of combustion of an air-fuel mixture at one end of the piston cylinder.

Reduction Reaction: A chemical reaction in which oxygen is removed from a compound to form new compounds or to reduce a compound to its basic elements.

Reed Valve: A one-way check valve. A reed, or flap, opens to admit a fluid or gas under pressure from one direction, while closing to prevent movement from the opposite direction.

Ring Flutter: Rapid up and down movement of a piston ring that breaks the seal between the ring and the cylinder wall.

Road Draft Tube: The earliest type of crankcase ventilation; it vented blowby gases to the atmosphere.

Rocker Arm Ratio: The ratio of the valve side of the rocker arm to the cam side, measured from the pivot.

Roller Chain: A chain made of links, pins, and rollers. Each link is made of two plates, held together by pins, and separated by a roller around the pin. The width of the chain is determined by the length of the rollers.

Roller Tappet: A tappet with a roller that rolls on the camshaft lobe.

SAE Viscosity Grade: A system of numbers signifying an oil's viscosity at a specific temperature; assigned by the Society of Automotive Engineers.

Sandwich Type Gasket: A gasket with layers of different materials, such as asbestos and steel.

Scavenging: The ability of the exhaust system to create vacuum, which draws the exhaust from the cylinder when the exhaust valve opens.

Scuffing: The transfer of metal between two rubbing parts; caused by lack of lubrication.

Sealant: A liquid or gel used with or without a gasket for preventing leaks between mated surfaces.

Secondary Tachometer: A tachometer which takes its signal from the ignition secondary.

Shakeproof Washer: A type of lockwasher with external or internal multiple teeth.

Shot-Peening: The process of shooting small steel balls against metal to compress and strenghten it.

Single Grade (Straight Grade): An oil that has been tested at only one temperature, and so has only one SAE viscosity number.

Silent Chain: A chain made of links that pivot around pins. Each link is a series of plates held together by the pins. The number of plates determines the width of the chain.

Single-Plane Intake Manifold: A manifold design used on V-type engines, where each half of the carburetor feeds all the cylinders on one bank of the engine.

Slag: Impurities in iron ore. Removed before ore is processed into steel.

Sludge: A thick, black deposit caused by the mixing of blowby gases and oil.

Spring Height: The dimension from the spring seat to the bottom of the retainer. It is measured whenever springs are installed.

Squirt Hole: A hole in the side of a connecting rod that causes oil to squirt onto the cylinder wall.

Squish Area: A narrow space between the piston and the cylinder head. As the piston approaches top dead center, the mixture squishes or shoots out of the squish area across the combustion chamber.

Static Inertia: The tendency of a body at rest to remain at rest.

Stellite: The brand name for a very hard alloy of mostly chromium and cobalt applied to valve faces for longer wear.

Stethoscope: A listening device used to pinpoint the source of engine noise.

Stoichiometric Ratio: An ideal air-fuel mixture for combustion in which all oxygen and all fuel will be completely burned.

Stratified Charge: A nonuniform air-fuel mixture, arranged more or less in layers. The charge is rich in fuel in some areas, lean in others.

Stress Riser: A groove, scratch or other imperfection in the metal that could develop into a crack.

Stroke: One complete top-to-bottom or bottom-to-top movement of an engine piston.

Surface-to-Volume Ratio: The ratio of the surface area of a three dimensional space to its volume.

Swaging Tool: A stamp or die used for shaping or marking metal. It is also called a swage.

Synthetic Motor Oils: Lubricants formed by artificially combining molecules of petroleum and other materials.

Tachometer: An instrument that measures the speed of a rotating part in revolutions per minute.

Tap: The process of cutting threads in a hole. A tap is the tool used to form an inside thread.

Tetraethyl Lead: A gasoline additive used to help prevent detonation.

Throttle: A valve that regulates the flow of the air-fuel mixture entering the engine.

Thrust Bearing: The main bearing that controls crankshaft endplay.

Top Dead Center: The exact top of a piston stroke. Also a specification used when tuning an engine. Abbreviated: tdc.

Torque: The tendency of a force to produce rotation around an axis; it is equal to the distance to the axis times the amount of force applied expressed in foot-pounds or inch-pounds (customary U.S. system) or in newton-meters (metric system).

Torsional Vibration: The vibration set up by the piston thrust against the crankshaft causing it to twist slightly, and the crankshaft's efforts to straighten itself.

Two-Stroke Engine: An engine in which a piston makes two strokes to complete one operating cycle.

Two-Plane Intake Manifold: A manifold design used on V-type engines, where one half of the carburetor feeds half the cylinders on one bank and half the cylinders on the other bank.

Undersize Bearing: A bearing made thicker with a smaller inside diameter to fit an undersize crankshaft.

Vacuum: Air pressure lower than atmospheric pressure.

Vacuum Gauge: A pressure gauge which measures the amount of suction or vacuum created in the manifold.

Valve Float: The condition in which the valve continues to open or stays open after the cam lobe has moved from under the lifter. This happens when the inertia of the valve train at high speeds overcomes the valve spring tension.

Valve Lift: The amount of lift at the valve. It depends on lobe lift, rocker arm ratio, and the clearance in the valve train.

Valve Train: The assembly of parts that transmits force from the cam lobe to the valve.

Vaporize: To change from a solid or liquid into a gaseous state.

Variable Rate Spring: A spring that changes its rate of pressure increase as it is compressed. This is achieved by unequal spacing of the spring coils.

Glossary of Technical Terms

Varnish: An undesirable deposit, usually on the engine pistons, formed by oxidation of fuel and of motor oil.

Venturi: A constriction in an air passage, such as in a carburetor, that increases the airflow and creates a vacuum.

Venturi Vacuum: Low pressure in a venturi caused by fast airflow through the venturi.

Viscosity: The tendency of a liquid such as oil to resist flowing.

Volumetric Efficiency: The comparison of the *actual* volume of air-fuel mixture drawn into an engine to the *theoretical maximum* volume that could be drawn in. Written as a percentage.

Weight: The measure of the earth's gravitational pull on an object.

Wet Sleeve: A sleeve that comes in contact with the engine coolant.

Work: The application of energy through force to produce motion.

Wrist Pin: The hollow metal rod that holds the piston to the connecting rod; also called a piston pin.

Index

Anodizing, 58
Antioxidants, 168

Babbitt, 54
Balancing Shafts, Engine, 128-129
Balancing Machines, 250
Bead Blasting, 230
Bearings, 156-164
  antifriction, 157
  construction and installation, 161
  crush, 162
  friction, 156-157
  materials, 158-160
  oil clearance, 162-164
  spread, 162
  undersize and oversize, 164
  web, 161
Billets, 57, 135
Boring Bars, 246
Bottom Dead Center (bdc), 17, 24
Broaching, 61
Bypass Oil Filters, 175

Cam Thrust, 121
Camshafts, 114-122
  drive, 115-118
  lubrication, 122
  materials, 118
  overhead cams, 11-12
  thrust, 121-122
Carbide Cutters, 236
Castellated Nut, 187
Catalytic Converters, 76
Composites, 55
Compression Gauge, 218
Compression Ratio, 15-18
Compression Test, 217-219
  dry compression test, 218
  wet compression test, 217

Connecting Rods, 138-142
  lubrication, 142
  materials and construction, 141
Cooling System, 65-74
  components, 67-74
  engine coolant, 67
  operation, 65-67
Counterbore, 239-240
Crankpin, 133
Crankshaft, 131-136
  firing impulses, 133-135
  lubrication, 136
  materials and construction, 135-136
  offset, 131-133
Crossflow Head, 87
Cylinders
  arrangement, 7-9
  bore, 17
  bore gauge, 207
  combustion chamber design, 85-89
  heads, 85-91
  sleeves, 82-83

Datum Line, 83-84
Desmodromic Valves, 103
Detonation, 31
Dial Indicator, 249
Diesel Engine, 47
Displacement, 15-17
  metric specs, 16
Drilling, 60
Dwellmeter, 220
Dynamometer, 35

Emission Control, 42-45
Energy, 27
Engine Mounts, 191
Engine Shock Absorbers, 192
Engine Temperature Gauges, 225-226

Exhaust System
  components, 76-77
  types, 75-76

Fasteners, 186
Flexplates, 137
Fluorescent Penetrant Inspection, 232
Flywheels, 137
Four-Stroke Cycle, 4-6
Fuel System, 18-20
  air-fuel distribution, 19
  air-fuel ratio, 40-42
Full-Floating Pins, 140

Gaskets, 179-182
Glaze Breaker, 247
Grinders, 242, 249, 250, 252
Grinding, 62
Guides, 109-112
  lubrication, 110-112

Hammers, 200
Helical Gears, 116
Hemispherical Combustion Chamber, 86
Honing, 59-60, 246-247, 251-252
Horsepower
  brake, 34-35
  engine, 33-34
  torque-horsepower relationship, 34
Hydraulic Press, 253

Ignition System, 21-24
  synchronization, 23-24
Induction Hardening, 112
Infrared Exhaust Analyzer, 222-225
Interference Angle, 99

L-Head Valve Train, 101
Lapping, 62-63
Line Boring, 248

Index

Magnetic Particle Inspection, 232

Manifolds
exhaust, 12-14, 94-95
heat control valve, 94
intake, 12-14, 91-94
vacuum, 36

Mass, 28

Micrometers, 205-207

Milling, 62

Mills, 242, 249

Momentum, 30

Motor Oil
composition and additives, 167-168
designations, 168-171

Non-Rotating Crankshaft, 139

Oil Pressure Gauge, 222

Oil Pump, 172-174

Oscilloscope, 215-217

Overhead Valve Train, 102

Overhead Cam Valve Train, 102

Oxidation Reaction, 30

Pin Bosses, 142

Pistons
constuction and material, 142-143
crown shape, 146-147
expansion, 144
knurler, 255
offset, 146
pins, 147-148
rings, 148-152
speed, 145-146

Planers, 243

Plateaued Finish, 64

Pliers
diagonal, 199
duckbill, 199
gripping, 199
needle nose, 199

Poppet Valves, 9

Power, 28

Power Tools, 210-213

Prelubricator, Engine, 221

Pushrods, 125-126

Quench Area, 85

Ring Compressor, 203

Reaming, 60

Reciprocating Engine, 4

Rocker Arms, 126-128

Rocker Stud Puller, 241

Rod Aligner, 254

Rod Heater, 253

Roller Chain, 116

Rotary Engine (Wankel), 4, 47-48

SAE Viscosity Grades, 169

Screwdrivers
clutch head, 198
Phillips head, 198
Reed and Prince, 198
slot-type, 197

Sealants and Cements, 184-186
aerobic sealants, 185
anaerobic sealants, 185

Seals, 182-184
lip, 183-184
square-cut, 183

Shot-Peened, 141

Silent Chain, 116

Single-Plane Intake Manifold, 93

Sodium-Cooled Valves, 100

Spotfacer, 243

Springs, 106-109
design, 106-107
installation, 108-109
material, 107-108
tension tester, 243

Squish Area, 85

Stamping, 56

Stethoscope, 220

Stoichiometric Air-Fuel Ratio, 41

Stratified Charge Engine, 48-49

Swaging Tool, 238

Synthetic Motor Oils, 170

Tachometer, 220
fan strobe tach, 221

Tapping, 60-61

Temperature Sensing Patches, 226

Thermal Energy, 31-32

Thermostats, 72

Throttle, 20

Timing, 25-26

Top Dead Center (tdc), 17, 24

Torque
engine, 32-33
metric power and torque, 35-36
torque-horsepower relationship, 34

Turbocharging, 13

Two-Plane Intake Manifold, 93

Two-Stroke Engine, 45

V-Blocks, 249

Vacuum Gauge, 219-220

Valves
angles, 98-100
cooling, 100-101
guides, 237-238
lifters, 102, 120-121, 123-125
lubrication, 103-104
materials and construction, 104-105
rotators, 109
seats, 112, 235, 241
timing, 119-120
valve trains, 101-102

Venturi Principle, 37-38

Vernier Calipers, 209

Vibration Dampers, 137-138

Volumetric Efficiency, 139-140

Wankel Engine (See Rotary)

Water Jacket, 80

Water Pumps, 70

Weight, 28

Work, 27

Wrenches
box-end 196
combination, 196
flare-nut, 196
impact, 211
open-end, 196
ratcheting, 197

Answers to Review and NIASE Questions

Chapter 1: Engine Operation and Construction
1.(d) 2.(b) 3.(b) 4.(b) 5.(d) 6.(c) 7.(d) 8.(d) 9.(c) 10.(a) 11.(a) 12.(d) 13.(c) 14.(b) 15.(d) 16.(c) 17.(c)

Chapter 2: Engine Physics and Chemistry
1.(b) 2.(c) 3.(d) 4.(d) 5.(b) 6.(a) 7.(a) 8.(d) 9.(d) 10.(c) 11.(d) 12.(d) 13.(c) 14.(c) 15.(a) 16.(b)

Chapter 3: Basic Metallurgy and Machine Processes
1.(d) 2.(c) 3.(d) 4.(a) 5.(b) 6.(a) 7.(a) 8.(d) 9.(a) 10.(d) 11.(a) 12.(c) 13.(d) 14.(d) 15.(d)

Chapter 4: Cooling and Exhaust Systems
1.(a) 2.(d) 3.(b) 4.(b) 5.(c) 6.(d) 7.(a) 8.(c) 9.(d) 10.(c) 11.(d) 12.(b) 13.(b) 14.(c) 15.(c) 16.(c) 17.(d) 18.(d)

Chapter 5: Engine Construction and Operation
1.(c) 2.(a) 3.(d) 4.(d) 5.(b) 6.(a) 7.(d) 8.(a) 9.(b) 10.(d) 11.(b) 12.(c) 13.(b) 14.(c) 15.(d) 16.(d)

Chapter 6: Valves, Springs, Guides and Seats
1.(d) 2.(d) 3.(b) 4.(c) 5.(d) 6.(b) 7.(d) 8.(d) 9.(c) 10.(d) 11.(b) 12.(c) 13.(d) 14.(d) 15.(d) 16.(c) 17.(a) 18.(d)

Chapter 7: Camshafts, Lifters or Followers, Pushrods, and Rocker Arms
1.(a) 2.(d) 3.(d) 4.(b) 5.(a) 6.(d) 7.(a) 8.(b) 9.(a) 10.(b) 11.(d) 12.(b) 13.(d) 14.(c) 15.(d)

Chapter 8: Crankshafts, Flywheels, Vibration Dampers, Pistons, and Rods
1.(d) 2.(d) 3.(c) 4.(a) 5.(b) 6.(c) 7.(a) 8.(d) 9.(c) 10.(b) 11.(d) 12.(d) 13.(d) 14.(d) 15.(c) 16.(d) 17.(b) 18.(a) 19.(d) 20.(d) 21.(b) 22.(b) 23.(c) 24.(a) 25.(b)

Chapter 9: Engine Bearings
1.(c) 2.(a) 3.(b) 4.(a) 5.(d) 6.(d) 7.(a) 8.(c) 9.(b) 10.(a) 11.(b) 12.(c) 13.(d) 14.(d) 15.(a)

Chapter 10: Engine Lubrication and Ventilation
1.(c) 2.(d) 3.(b) 4.(d) 5.(a) 6.(b) 7.(c) 8.(a) 9.(d) 10.(b) 11.(d) 12.(a) 13.(d) 14.(c) 15.(a) 16.(c)

Chapter 11: Gaskets, Fasteners, Seals, and Sealants
1.(a) 2.(a) 3.(c) 4.(b) 5.(b) 6.(b) 7.(b) 8.(c) 9.(c) 10.(c) 11.(b) 12.(b) 13.(c) 14.(a) 15.(b) 16.(d) 17.(d) 18.(b)

Chapter 12: Tools and Precision Measuring Instruments
1.(d) 2.(b) 3.(c) 4.(c) 5.(a) 6.(c) 7.(a) 8.(d) 9.(a) 10.(b) 11.(d) 12.(d) 13.(b) 14.(a) 15.(d)

Chapter 13: Engine Test Equipment
1.(c) 2.(a) 3.(c) 4.(c) 5.(d) 6.(b) 7.(b) 8.(a) 9.(d) 10.(c) 11.(b) 12.(a) 13.(a) 14.(c) 15.(b)

Chapter 14: Cleaning and Inspection Equipment
1.(b) 2.(d) 3.(c) 4.(b) 5.(c) 6.(a) 7.(d) 8.(d) 9.(d) 10.(a)

Chapter 15: Head and Valve Service Equipment
1.(b) 2.(a) 3.(d) 4.(d) 5.(c) 6.(a) 7.(c) 8.(b) 9.(c) 10.(c)

Chapter 16: Block, Crankshaft, Piston, and Rod Service Equipment
1.(c) 2.(b) 3.(d) 4.(d) 5.(c) 6.(c) 7.(a) 8.(c) 9.(a) 10.(c) 11.(c) 12.(b)

NIASE Mechanic Certification Sample Text
1.(b) 2.(a) 3.(b) 4.(a) 5.(c) 6.(b) 7.(b) 8.(d) 9.(d) 10.(d) 11.(d) 12.(a) 13.(d) 14.(c) 15.(c) 16.(c) 17.(c) 18.(d)